Rocketing into the Future
The History and Technology of Rocket Planes

Michel van Pelt

Rocketing into the Future

The History and Technology of Rocket Planes

 Springer

Published in association with
Praxis Publishing
Chichester, UK

Michel van Pelt
Space Analyst
Leiden
The Netherlands

SPRINGER–PRAXIS BOOKS IN SPACE EXPLORATION

ISBN 978-1-4614-3199-2   ISBN 978-1-4614-3200-5 (eBook)
DOI 10.1007/978-1-4614-3200-5
Springer New York Heidelberg Dordrecht London

Library of Congress Control Number: 2012932254

Cover design: Jim Wilkie
Project copy editor: Christine Cressy
Typesetting: BookEns, Royston, Herts., UK

Printed on acid-free paper

Springer is part of Springer Science+Business Media (www.springer.com)

Contents

Illustrations

Chapter 8

Chapter 9

Appendix

To Giulia

Author's preface

As a Dutch kid growing up in the early 1980s I devoured the 'Euro 5' science fiction series by Bert Benson. They were typical boy's adventures: in each book a secret team of European policemen had some 200 pages to track down a menagerie of rampaging robots, mutant criminals and murderous scientists hell-bent on terrorizing Earth and the solar system. The bad guys were usually seeking world domination, which they inevitably intended to obtain via some overly complicated but fascinating scheme. As required by the genre, the good guys always managed to arrest the interplanetary villains before they could bring their devious schemes to fruition. Just in the nick of time of course. It was ideal literature for a certain somewhat nerdy would-be aerospace engineer. The books were all written in Dutch, and much later I found out that writer Bert Benson's real name was Adrianus Petrus Maria de Beer, which sounds about as futuristic in Dutch as is does in English. In spite of this, the stories breathed a kind of cosmopolitan atmosphere, with a diverse team of agents from various European countries (the leader was Dutch, naturally) flying to exotic countries, forgotten islands and hostile moons and planets. Their means of transportation was the Euro 5, a wedge-shaped rocket spaceplane with a set of large wings at the back and smaller 'canard' wings in front, four rotating ray-guns, and a small boat-shaped plane for short reconnaissance trips (which I now know looks a lot like NASA's M2-F3 'lifting body' experimental rocket plane of the early 1970s). In the final pages of each book, it was usually this marvelous machine that saved the day, if not the entire universe. For me this gigantic blue vehicle was really the centerpiece of the stories, rather than the colorful team of international heroes. I guess kids in other countries at the time were reading similar adventure books with rocket planes in a starring role.

To me, part of the appeal of the Euro 5 books was that the idea of a spacefaring rocket plane did not seem to be very far-fetched. After all, when I was reading them the Space Shuttle had just entered service and promised to make all those boringly tube-shaped, expendable rockets obsolete. In 1976, NASA predicted that it would be launching around seventy-five Space Shuttles per year; more than six per month. Because most of the system's hardware could be reused, launch prices would be lower than for the old-fashioned single-use rockets. The low cost and flexibility of the Shuttle would even make it economically feasible to repair satellites or bring them back to Earth for major refurbishments. Companies would be able to use the Shuttle to build microgravity factories in orbit, and scientists would be able to fly all manner of experiments and instruments. The Shuttle was going to be everything for everybody, a kind of real-life, American Euro 5. And that would be just the first step in reusable launch vehicle development. Many space experts, members of the public, and for sure a certain twelve-year-old, expected that real rocket planes would soon follow, taking off from normal airports, operating more or less like normal aircraft and flying ordinary people (and even Dutch teenagers) into space.

Like many childhood fantasies, that did not come true. The Space Shuttle proved to be a very complex, extremely expensive and dangerous launch vehicle. Instead of taking off like a plane, a Shuttle lift off is a major event with an enormous checklist of things that can easily postpone the launch if not working perfectly. If you headed to Kennedy Space Center to watch a Shuttle lift off, you could count yourself very lucky if it got off on time. You ran the risk of spending many days at nearby Cocoa Beach, watching a continuous parade of launch delay messages on NASA TV and eating lunches that would likely ground you for being overweight if you were an astronaut. Instead of witnessing two launches within a week, as had been predicted by NASA's optimistic forecasts of the early 1980s, you were more likely to return home for work and other commitments without having seen even one (ruining yet another childhood dream).

The huge numbers of flights per annum did not materialize mainly because the Shuttle system took much more time than anticipated to prepare for each flight, and because NASA had been extremely optimistic in the number of satellite launches and other missions the Shuttle would be required to perform. Although the Space Shuttle performed some spectacular missions, like repairing the Hubble Space Telescope in orbit, retrieving malfunctioning satellites and putting a large space station up module by module, it did not grow into the cheap, regularly flying 'space truck' envisioned in the early 1980s. In fact, during its thirty-year operational lifetime it was the most costly way to launch anything or anyone into space, even without taking into account the costs of new safety measures introduced in response to the loss of Challenger in 1986 and Columbia in 2003.

In his 1986 State of the Union address, only five years after the Shuttle's debut, President Ronald Reagan already called for a successor, "...a new Orient Express that could, by the end of the next decade, take off from Dulles Airport, accelerate up to 25 times the speed of sound, attaining low earth orbit or flying to Tokyo within two hours". The design for the X-30 experimental precursor of this National Aerospace Plane (NASP) initially looked very much like the famous Concorde supersonic airliner: delta wings on a long and slender fuselage with a pointy nose. However, while Concorde's maximum speed record was 2,330 km per hour (1,450 miles per hour, 2.2 times the speed of sound), the X-30 would have to accelerate to an orbital velocity almost 14 times faster. Unlike the Space Shuttle, it would not shed rocket stages and propellant tanks on the way up, but would be a true single-stage-to-orbit, fully reusable spaceplane. It was also to be much safer to fly than the Shuttle, which only a few days prior to Reagan's speech had experienced its first disaster, resulting in the loss of the Challenger Orbiter, killing seven astronauts and throwing the entire program into complete disarray.

Although the X-30 was initially envisioned to use only (exotic and very complex) airbreathing engines, meaning it wouldn't really be a rocket aircraft, its development would be founded on a long history of rocket plane and mixed-propulsion jet/rocket aircraft experiments. Prior to the advent of reliable, powerful jet engines, rockets enabled revolutionary Second World War fighters to climb quickly and intercept high-altitude bombers. Soon after the war, rocket planes were the first to break the dreaded 'sound barrier' and then the lesser-known 'heat barrier', establishing them not to be real barriers at all. Then they became the first (and only) aircraft to reach the edge of space, breaking record after record by flying faster and higher than any airbreathing type of aircraft. And large rocket boosters were used to shoot aircraft straight into the air without a runway. Step by step, often involving considerable danger and consequently some disasters, engineers and pilots thus learned how to design and fly vehicles that were part airplane, part missile and part spacecraft. In fact, up until the early 1960s many people expected the rocket plane rather than the expendable ballistic missile to represent the near future of manned spaceflight. The vehicle envisioned by president Reagan therefore appeared to be long overdue.

However, in spite of the impressive aeronautical and spaceflight heritage and optimistic expectations, only a couple of years after Reagan's speech it became clear that the X-30 as well as its contemporary European and Japanese competitors (the German Sänger-II, the British HOTOL and the Japanese JSSTO), were technically too ambitious and would be extremely expensive to develop. By the mid-1990s all these spaceplane projects had been killed off by alarmingly rising costs and ever mounting technical difficulties.

It now seems that the concept of the reusable spaceplane has reached a kind of dead end, due to a lack of technology breakthroughs, political will and economic rationale. Work on spaceplanes is still ongoing, but at a rather slow pace and at a relatively modest level, and generally with a focus on hypersonic military missiles rather than orbital crewed launch vehicles. The Space Shuttle was retired in 2011, and its currently planned successors are old-fashioned-looking capsules that will be launched on conventional expendable launchers. Also European and Russian plans for new launchers and crewed vehicles focus on next-generation expendable rockets and relatively simple re-entry capsules. This is a very sad situation, considering the long history of rocket plane and spaceplane development, in addition to the huge amounts of time and money already invested in their development.

There is however also some good news. A new industry of companies developing and marketing suborbital rocket planes has recently emerged. Scaled Composites' rocket plane SpaceShipOne reached an altitude of just over 100 km (330,000 feet) in June 2004, making it the first fully privately funded human spaceflight mission. Less then four months later it won the $10 million Ansari X Prize by flying beyond 100 km in altitude twice in a two-week period. As per the rules of the X Prize, no more than ten percent of the empty weight of the spacecraft was replaced between these flights, making SpaceShipOne a truly reusable spaceplane, albeit a suborbital one. It is also interesting to note that the flight on 4 October 2004 not only earned Scaled Composites the X Prize, but with a maximum altitude of 112 km (367,441 feet) also the unofficial record for the highest altitude reached by a manned aircraft (unofficial because the plane did not take off under its own power). The previous unofficial record of 108 km (354,199 feet) had been set on 22 August 1963 by the air-launched X-15 rocket plane, and thus had stood for over 41 years!

At the time of writing, commercial space tourism flights onboard the larger SpaceShipTwo plane, also developed by Scaled Composites but marketed by Virgin Galactic, are scheduled to start no earlier than 2012. Several other suborbital rocket planes are under development, and there are even plans to run rocket plane races to boost their development and commercial value.

Nevertheless, suborbital rocket planes are a far cry from the large orbital spaceliners we were promised. We can't yet book tickets for an orbital cruise around the planet on a luxury spaceplane, or fly from New York to Tokyo within two hours as president Reagan announced over a quarter of a century ago. Even the incredible Concorde supersonic jetliner, which during its 27 years of operations notched up more supersonic flying hours than all the world's air forces together, is no longer flying and there is no successor in sight! In the second decade of the twenty-first century, astronauts are still launched vertically on top of expendable missiles, and planes with rocket engines are only used for suborbital tourist trips. A true rocket spaceplane, taking off under its own power from a runway, using its wings to fly into orbit and then to glide back to Earth for an elegant landing at an airport seems to a remote vision, as far away from reality as it was in the 1950s.

Why is it so hard to develop a true rocket propelled spaceplane to fly us into orbit "the way it was meant to be", perhaps even doing a playful roll on the way up? And why has the continued evolution of rocket planes, which has been progressing for over 80 years, seemingly run into a brick wall? Is there any hope left for would-be rocket plane passengers and pilots, and if so, what might an operational spaceplane look like? Is Euro 5 ever going to fly? These are some of the questions I will address in this book. In search of answers, but also just out of curiosity, we will journey through history and discover all the wonderful, exciting and weird rocket planes that people have dreamed of, have designed, and in some cases have actually flown (and crashed). Rocket planes have always been at the forefront of technology, pushing the established boundaries of aviation and spaceflight. Because of this they were often highly secret projects, which just adds to their appeal. Whatever rocket aircraft in history you look at, you will always find that at the time it represented a daring leap in technology, and involved much adventure for their brave pilots. Rocket planes were showing a glimpse of the future, even if that future changed continuously and often did not materialize as expected. Hence the title of this book, *Rocketing into the Future*.

In this book many key technical issues for rocket planes are described, such as how rocket engines work or how the speed of sound varies with altitude. Books on spaceflight of the 1950s and 1960s tended to explain such things, as it was rightly considered that the general public did not know much about such novel technologies. Nowadays not many books and articles about high-speed aircraft and spaceflight bother to explain the technology and physics involved, assuming that readers are either already familiar with or would be bored by such 'details'. However, to truly understand the enormous challenges facing rocket plane designers, and the often brilliant solutions that have been implemented, some understanding of the basics of air- and spaceflight is necessary. It is certainly possible to explain these without complicated mathematics, as you will see (but you are allowed to skip the passages concerned of course, since there is no exam at the end).

To limit the scope of this book to a reasonable level, I defined 'rocket planes' as manned aircraft that use lift-producing wings and rocket power to fly. Spaceplanes that are launched vertically into orbit on top of conventional rockets, without the use of lift-producing wings, and only truly fly upon return to Earth (such as the Space Shuttle) are only described where appropriate; I consider these mere winged gliding payloads rather than true rocket planes. The 'brute force' launchers on which such shuttles are bolted ascend almost vertically out of the atmosphere as rapidly as possible in order to quickly escape aerodynamic drag, rather than exploiting the lift-producing capabilities of the air to fly up to high altitude prior to rocketing into orbit; what is called a 'lifting ascent'. Conventional aircraft that use added rocket boosters to help them to take off are also mostly considered beyond the scope of this book. Unmanned winged missiles and rockets that have small wings only for steering and stability purposes are also only described when there is an important link to the main topic.

There is of course a grey area associated with this definition of a 'true' rocket plane. For example, one could argue that the vertically launched Natter interceptor and the Ohka manned missile of the Second World War are not really rocket aircraft; that the 'Zero-Length Launch' jet fighters catapulted into the air with the help of huge rocket boosters represent merely an extreme form of rocket assisted take-off; and that the X-30 was not a rocket plane because it used only airbreathing engines. Nevertheless this book does include relatively detailed descriptions of them, in part because they play a role in the overall story of the rocket plane, but also due to the writer's fascination with these exotic aircraft. This isn't a very scholarly approach, but this book does not pretend to be an academic report. Similarly, vehicles like the X-20, the Space Shuttle and the Russian Buran also clearly fall outside the main scope, but it is necessary to discuss them in broad strokes because they represent important intermediate steps that link twentieth century rocket aircraft to possible future orbital rocket spaceplanes; in fact, the Space Shuttle has been mentioned several times in this introduction already. Moreover, several early Space Shuttle concepts were actually true rocket planes.

However, even focusing on rocket aircraft that fit my definition requires further selectivity: there is a vast amount of information 'out there', certainly enough for a hefty encyclopedia of rocket aircraft. In particular, the assortment of rocket planes designed during the Second World War in Germany, Japan and Russia is simply bewildering. Selections thus had to be made, and this book is not intended to be an all-encompassing database of rocket airplanes: if you are interested in the diameter of the tail wheel of the Me 163 or have an urgent desire to know the liquid oxygen boil-off rate for the Bell X-1, then you will need to look elsewhere. Whole volumes have been written about the individual rocket planes and pilots that are mentioned in the following pages. The aim of this book is to offer a concise account of the amazing history of rocket aircraft, the perseverance of their designers, the bravery of their pilots, the logic of their technical evolution and how they paved the way to future spaceplanes.

Another thing that I quickly found out is that for early rocket planes, say up to the end of the Second World War, it is possible to concisely describe their development, since it was usually short and involved only a few key individuals. But after the war, aircraft and rocket technology increased in complexity so rapidly that large numbers of people and long development times were required; even the initial design became a team effort rather than the work of a single, brilliant 'lone wolf' inventor. The Opel RAK-1 rocket glider for instance, involved three key people and several technicians, and flew a few months after the project was conceived; in contrast the Space Shuttle took over a decade to develop, involved six main contracting companies and over 10,000 engineers, technicians, managers and support personnel. It is impractical to completely describe the entire history of an extremely complex vehicle such as the Shuttle or even the earlier X-15 in only a few pages, and thus the further we move through time, the more I had to focus on a rocket aircraft project's most important issues and events.

Researching this story has led me to an astonishing wealth of books, magazine articles, websites and technical papers, many of which are listed in the bibliography. Sometimes different sources provide conflicting pieces of information. Faced with such conflicts, I either sought the original source or used what appeared to be most plausible. My search also often led to surprising finds, such as descriptions of truly weird and suicidal designs that thankfully for the pilots never left the drawing table; and websites selling such must-have items as an intricately detailed scale model of a rocket propelled Opel car, and a wooden made-in-the-Philippines model of a 1920s' spaceplane design (both of which now decorate my bookshelves). I hope the readers of this book will have as much fun and amazement as I had in writing it. Ready for take-off?

Thanks to Shamus Reddin for information on the Me 163A's RII-203 engine and also for the interesting information on Hellmuth Walter's rocket technology on his website (*www.walterwerke.co.u*k). Bruno Berger of the Swiss Propulsion Laboratory provided some clarifications on the Mirage SEPR 844 rocket boost pack. Alessandro Atzei and Rogier Schonenborg accompanied me on two fruitless efforts to witness a Shuttle launch, during which we nevertheless had a lot of fun and saw much of Cape Canaveral and the Kennedy Space Center. My father allowed himself be talked into making a long car ride to see the Soviet Buran shuttle in a museum in Germany. The European Space Agency team of the Socrates study, of which I was part, I thank for all the interesting discussions and information on the design of a rocket plane (even although it never left the Powerpoint stage). To Georg Reinbold, thanks for all the interesting discussions over many years about the costs of spaceplanes and reusable launchers. I thank Arno Wielders, Stella Tkatchova, Ron Noteborn, Dennis Gerrits and Peter Buist for their suggestions and many interesting ideas concerning space transportation, life, the universe and everything. And David M. Harland, himself an accomplished writer, did all the editing required to put this text into shape. And last but not least, a big thank-you to all those writers who managed to tell the stories of many almost forgotten rocket aircraft projects; without their books, reports, papers and websites the 'big picture' presented in this book would not have been possible.

1

Introduction

The history of powered flight started on 17 December 1903 at Kitty Hawk, North Carolina, when the Wright brothers were able to keep their feeble plane in the air for some 12 seconds. In that short time pilot Orville Wright managed to cover a distance of 36.5 meters (120 feet); less than the wingspan of a modern Boeing 747 airliner. A bit over 10 years later, airplanes had already become effective military machines and were used for reconnaissance, air defense and attack over war-torn Europe.

The history of rocketry goes back much further, maybe even to thirteenth century China. Gunpowder-filled tubes were initially used for fireworks, but were soon also applied as artillery to rain fiery arrows on the enemy and, perhaps more importantly, to scare them to death. In the early nineteenth century the English inventor William Congreve greatly improved the propellant and structure of powder-based rockets, and his designs were used on European battlefields during the Napoleonic Wars. In 1814 the ship HMS *Erebus* fired rockets on Fort McHenry during the Battle of Baltimore, and inspired Francis Scott Key to write about "the rockets' red glare" in the national anthem of the United States of America.

ROCKET PLANES, ROCKET TRAINS AND ROCKET AUTOMOBILES

At the beginning of the twentieth century rocketry based on solid propellants like gunpowder was well established, while the golden age of flight had just begun. It was only a matter of time before someone would think of using rockets to propel a plane. The first to suggest it may have been French aviation and rocket pioneer Robert Esnault-Pelterie, who in 1911 proposed a winged, rocket propelled aircraft. The idea was a bit ahead of its time though, as planes were then still rather flimsy contraptions ill-equipped to harness the brute power of rocket motors. But the need to observe hostile armies from the air and shoot down their reconnaissance planes during the First World War soon gave an enormous boost to aircraft development. By the end of that conflict the aeroplane had been transformed from an unreliable curiosity to a sturdy, fast and maneuverable machine exemplified by fighter planes like the Fokker DVII and the Sopwith Camel.

In the 1920s and early 1930s science fiction stories became very popular and, inspired by this, rocketry societies were formed in the US, Germany and Russia. These began experimenting with rockets and even speculated about their use for interplanetary travel. It is thus not too surprising that during those years several concepts for rocket propelled planes were conceived. Some we can now proclaim to be highly impractical, such as Fridrikh Tsander's 1921 self-consuming spaceplane design. According to the Russian's concept, metallic parts of the vehicle no longer needed during the flight would not be discarded (as in a conventional multi-stage rocket), but be fed into a furnace to be converted into rocket fuel. The structures of the empty tanks and disposable wings would thus actually help to propel the plane. Only an essential part of the structure and a small set of wings would be retained for the return to Earth. Powdered metals do actually burn at very high temperatures, and are often added to the solid propellants of rocket boosters in order to increase thrust. But the exhaust products of the all-metal propellants would certainly have clogged up the engines of Tsander's aircraft, and of course feeding them with scrap metal at a sufficiently high rate would anyway have been a major challenge. However, there were also some very practical concepts, such as rocket expert and spaceflight enthusiast Max Valier's idea to equip already existing airplanes with rocket engines to obtain cheap and relatively reliable rocket plane test vehicles.

Valier is the first to kick off a project that brings airplane and rocket technology together in a real prototype, so let's start our story with him. In 1927 he approaches Fritz von Opel, grandson of the famous German automobile pioneer Adam Opel. At that time Fritz was director of testing at his family's car factory and, more importantly for our story, in charge of the firm's publicity. Valier has long been planning to experiment with vehicles that are equipped with rocket motors, and urges von Opel to organize some spectacular demonstrations involving rocket propelled cars. Rockets were able to accelerate a vehicle up to high speed more rapidly than the car engines of the 1920s, and more importantly make more noise and smoke doing so. Von Opel, himself a racing car driver and pilot, quickly recognizes the advertising value for his family's company. Although several German engineers are developing rocket motors based on liquid propellants that are more potent than solid propellants, these are rather complicated and still experimental. Valier proposes to use the rockets manufactured by Friedrich Wilhelm Sander, based on compressed black powder. In contrast to liquid propellant designs, Sander's rockets are simple and have already been successfully applied in signal rockets and the line-throwing missiles used to assist ships in distress. The amount of thrust can easily be varied by using and igniting different numbers of rockets in parallel. However, using Sander's rockets in a manned vehicle has its risks: they are sensitive to storage conditions. In particular, cracks develop in the powder propellant charge as it ages and, following ignition, produce sudden increases in burn rate and therefore pressure, resulting in an explosion. And, of course, it is very difficult to inspect the propellant for defects right before use.

In April 1928, after a series of secret tests at the Opel factory in Rüsselsheim, von Opel organizes his first rocket car run for the press. The RAK-1 (Raketenwagen-1,

Fritz von Opel in the RAK-2 rocket car [Opel].

or Rocket car 1) is basically a standard racing car with the engine removed and a dozen solid propellant rockets fitted in the back. To force the vehicle firmly down onto the track and prevent it from taking off, a pair of stubby, downwards-pointing wings are fitted (this is the first use of such airfoils, which are now customary in Formula 1 and Indy racing cars). Although five of the rockets fail to ignite during the demonstration, the driver, Valier, nevertheless stuns the journalists and

Von Opel drives his RAK-2 rocket car [Opel].

photographers assembled at the Rüsselsheim track to witness the test by achieving a top speed of over 110 km per hour (70 miles per hour).

Just over a month later Fritz von Opel himself, wearing an aviator's jacket and goggles but no helmet, drives the RAK-2 at the high-speed AVUS track in Berlin, watched by some 3,000 guests from show business, sports, science and politics. The RAK-2 car is based on the chassis of the regular Opel 10/40 PS car, fitted with two down-force wings. In contrast to the fixed wings of the RAK-1, these airfoils are connected to a lever in the cockpit that enables the driver to control the amount of down-force by changing the angle of the wings. They are also much larger in order to handle the anticipated higher speed of the new rocket car, which is propelled by two dozen solid propellant rockets, each of which has a thrust of 25 kg (55 pounds). The amount of propellant is impressive: "120 kilos of explosives, enough to blow up a whole neighborhood" von Opel later recalls. The rockets are ignited by a pedal-activated electrical system, with each press on the pedal igniting another rocket. In the event, all of the rockets ignite and make the RAK-2 run a huge success. Von Opel later described the sensation of driving the powerful RAK-2: "I step on the ignition pedal and the rockets roar behind me, throwing me forward. It's liberating. I step on the pedal again, then again and – it grips me like a rage – a fourth time. To my sides, everything disappears. All I see now is the track stretched out before me like a big ribbon. I step down four more times, quickly – now I'm traveling on eight rockets. The acceleration gives me a rush." Trailing a thick column of smoke, the RAK-2 accelerates to 238 km per hour (148 miles per hour). Not enough to break the contemporary speed record for cars, which then stood at 334 km per hour (208 miles per hour), but still very fast in a world where normal cars had a top speed of around 110 km per hour (70 miles per hour). In spite of the negative lift of the two large side wings, the high speed of the RAK-2 makes the front end of the car almost leave the ground, but von Opel is an experienced racing car driver and manages to keep the machine on the road. In less than three minutes the spectacle is over. The adrenalin still pumping through his veins, von Opel announces that his next goal is to fly a rocket propelled airplane, and urges the spectators: "Dream with us of the day in which the first spaceship can fly around our earth faster than the Sun." Fritz von Opel is an overnight sensation. The magazine *Das Motorrad* (The Motorcycle) reported: "No one could escape the impression that we had entered a new era. The Opel car with a rocket engine could be the first practical step toward the conquest of space." Silver-screen darling Lilian Harvey confided to a reporter: "I'd like to ride in the rocket car with Fritz von Opel." Even now, some 85 years later, the RAK-2 run is remembered as one of the most spectacular events in car history. The Technik Museum Speyer in Germany has a beautiful shiny black replica of the bullet-shaped RAK-2, and even today the car looks fast and stunning.

Even faster is the unmanned RAK-3. When mounted on train rails and equipped with ten rockets it manages to accelerate up to 290 km per hour (180 miles per hour) on its first run. The vehicle is retrieved several miles down the track, towed back and prepared for the second sprint. Thirty rockets are fitted this time, but it is too many and almost immediately after ignition the car jumps off the rails and is destroyed. During the first test of the subsequent RAK-4 rail car, one of the motors blows up and sets off the other rockets in a massive explosion. The debacle only kills the cat that is carried as the sole passenger, but the railway authorities prohibit any further rocket vehicle runs because they don't want to ruin their railroad track. This seals the fate of the already planned RAK-5 rail vehicle. The rocket propelled motorcycle that von Opel has already begun to test is soon also deemed to be too dangerous by the government, which prohibits von Opel from using it to try to break the motorcycle world speed record. However, even without the RAK-5 and the crazy motorcycle von Opel has by then already earned all the publicity that he sought, as well as the nickname 'Rocket Fritz'.

Rocket cars have no practical use other than publicity stunts and record breaking, and most German rocket pioneers, worried that he is threatening the credibility of rocketry and its potential for spaceflight, frown on von Opel's stunts. However, for Valier the cars and railcar are an essential part of his plan to achieve spaceflight. As early as 1925 he devised a step-by-step 'roadmap' for achieving space travel using rocket propelled aircraft. In his view, it would start with early test stand experiments with rocket motors, then experiments with rocket ground vehicles such as cars, sleds and railcars would open the way for a rocket propelled airplane that would in turn lead to stratospheric flight and ultimately the construction of a rocket spaceship. With von Opel's assistance Valier has reached the third step of his master plan. So the team starts work on a project with potentially an important future: the announced rocket propelled plane.

An Ente replica in the German Sailplane Museum [Martin Bergner and Deutsches Segelflugmuseum].

In March 1928 von Opel, Valier and Sander visit the Wasserkuppe plateau, then a focal point for glider flying in Germany. During the 1920s and 1930s virtually every German aeronautical engineer and test pilot of note was building, testing, and flying aircraft at the Wasserkuppe. The light planes were launched from the plateau to fly down into the valleys below, gaining altitude by using updrafts caused by the wind rising up the slopes. The Opel team is seeking a suitable glider onto which it can fit rocket motors, and at Wasserkuppe they encounter some of Alexander Lippisch's revolutionary, tailless gliders. Instead of having a horizontal stabilizer at the back, they have them at the front, giving them a rather duck-like appearance when seen from the ground (hence these types of planes are called 'canard' designs, after the French word for duck). This unusual configuration offers sufficient space to mount rockets at the back without the risk of setting the plane on fire. In June von Opel's team strikes a deal with a local glider society, the Rhöhn-Rositten Geselschaft, in which von Opel finances the Sander rockets and the society furnishes a Lippisch-designed aircraft called the 'Ente' (Duck, in German). For the Opel company, the flight of a rocket aircraft will be simply another spectacular publicity stunt, but the society's goal is to develop rocket propelled take off into an alternative for launching glider planes. Normally gliders are either towed into the air by a rope attached to a car, or launched down a rail by a rubber catapult system with an eight-man crew. A rocket assisted take-off would enable a glider to get airborne without assistance.

The plan is to fit two black powder rockets to the Ente and link them electrically to a firing switch in the cockpit. The first is a powerful boost rocket that supplies a thrust of 360 kg (790 pounds) for 3 seconds. The second will fire immediately after the first burns out. It is less powerful but longer burning to keep the plane in the air: 20 kg (40 pounds) of thrust for 30 seconds. However, tests with model aircraft and scaled rockets show that the high-thrust motor would be too powerful for the plane, so it is decided to use a standard rubber-band rail launcher in combination with two of the less powerful sustainer rockets, which will fire in succession to provide one

The two rocket motors can be seen at the back of the Ente replica [Martin Bergner and Deutsches Segelflugmuseum]

minute of continuous thrust. To prevent an overexcited pilot from making a potentially deadly mistake, the electric ignition is rigged so that it is impossible to ignite both rocket motors at the same time. The team also devises an ingenious counterweight system that is placed under the cockpit floor, and which automatically adjusts the center of gravity of the aircraft as the fuel of the rockets is burned, for otherwise the center of gravity would continuously shift forward as the rocket propellant in the back is consumed, making the glider unstable and very difficult to fly.

Fritz Stamer, who has long been a test pilot for Lippisch's designs, is selected to fly the aircraft. On 11 June 1928 (shortly before the first test of the RAK-3 rail car) and after two false starts, Stamer takes off and in just over a minute flies a circuit of about 1.5 km (1 mile) around the Wasserkuppe's landing strip. His verdict was that the world's first rocket plane flight had been "extremely pleasant" and that he "had the impression of merely soaring, only the loud hissing sound reminded me of the rockets".

The plane appears to be very easy to keep under control, so for the second flight the team decides to increase the thrust by firing both of the rockets simultaneously. Unfortunately, a well-known problem of Sander's rockets pops up again: one of the rockets explodes, punching holes in both wings and setting the aircraft alight. Stamer reported: "The launching went alright and while the plane took to the air I ignited the first rocket. After one or two seconds it exploded with a loud noise. The nine pounds of powder where thrown out and ignited the plane instantly. I let it drop for some sixty feet to tear the flames off." He manages to bring the plane down from a height of around 20 meters (65 feet). Just after landing the second rocket catches fire, but it does not explode. Stamer is able to walk away unhurt. However, the Ente is severely damaged, and the fiery crash scares the sponsoring glider society into abandoning the project. In September of that same year the magazine *Scientific American* rightly tells its readers: "On the whole we are inclined to think that the rocket as applied to the airplane might be a means of securing stupendous speeds for a short interval of time, rather than a method of very speedy sustained flight."

In spite of its problems, the Ente project has inspired Max Valier to develop a plan for crossing the English Channel with a rocket plane based on a more sophisticated rocket motor that uses liquid propellants. He expects his harpoon-shaped design to reach a top speed of 650 km per hour (400 miles per hour) and cover the 30 km (20 mile) distance in only three or four minutes. As if this isn't sufficiently ambitious, he is already thinking about an even bolder plan. This is described in an article in *Die Umschau* in Germany in 1928 and later in an article by Hugo Gernsback entitled 'Berlin to New York in less than One Hour!' in the November 1931 issue of the American magazine *Everyday Science and Mechanics*. In the same year Harold A. Danne presents a very similar concept to the American Interplanetary Society, with reference to Valier. The plan is to fly a rocket plane across the Atlantic in record time. In the plan published in 1928 the aircraft leaves from Berlin Tempelhof airport, but in the 1931 concepts the airliner would take off from the water to preclude a long, expensive runway (at that time there were not many large airports, so many large aircraft were built as floatplanes). At an altitude of 50 km (30 miles) Valier's 'Type 10' (his tenth rocket plane design) would reach the amazing speed of 2 km/s (7,200 km per hour or 4,500 miles per hour) and cross the ocean in less than an hour. The speed and altitude for his design were absolutely spectacular, as the air speed record in 1929 stood at only 583 km per hour (362 miles per hour), and the altitude record was close to 12.8 km (42,000 feet). By comparison, it had taken Charles Lindbergh 33.5 hours to cross the Atlantic in 1927. Suspecting that the atmospheric drag as his rocket aircraft descended would be insufficient to decelerate to a reasonable landing speed, Valier foresaw the need for forward-firing braking rockets. In spite of the extreme speed and altitude, he didn't think the flight would be very interesting for the passengers: they were to be kept comfortable in a pressurized cabin, and he expected there would be little to see of the Earth below because of "the vapor and light cloud formations". During the unpowered coasting phase at high altitude the passengers would be weightless, but apparently Valier didn't consider that to be a unique selling point. According to him, the richest reward would be the sight of the black sky and the Sun that would appear "surrounded by glowing red protuberances and the silvery corona", as can be seen during a total solar eclipse. He was wrong about the view of the Sun, which in space appears brighter but otherwise just the same as from Earth. He was also wrong about the appeal of such a trip, as space tourists are currently willing to pay hefty sums of money for a suborbital rocket flight to the edge of the atmosphere; the marvelous view of the Earth and the few minutes of weightlessness being the key selling points. What Valier did foresee was the large amount of propellant an intercontinental rocket plane would have to carry. Of the 80 ton (180,000 pound) total weight of his Type 10 design, 58 tons (130,000 pounds) would be propellant: 73

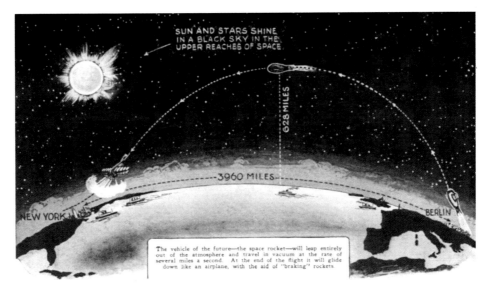

Trajectory of Valier's transatlantic rocketplane shown in the November 1931 issue of the magazine *Science and Mechanics* [*Science and Mechanics*]

percent! With today's fuel prices, such an appetite would be difficult to combine with commercially profitable intercontinental flights (let alone with today's carbon dioxide footprint minimization demands).

In the meantime von Opel, not having been able to fly a rocket plane himself in public owing to the crash of the Ente, orders a new rocket plane from Julius Hatry, a well-known German glider builder. As Hatry recalled in an interview in 2000: "From 1927 to 1929, I had been working on models for rocket-fuelled airplanes. I had flown models successfully and had decided to build a manned craft." Hatry initially refuses Opel's offer to team up, but relents when he finds that the car magnate is negotiating to purchase one of his airplanes for modification. Hatry delivers a more conventional glider design than the Ente, featuring a high upper wing and a twin tail arrangement which is attached by twin booms. This allows rockets to be fitted on the back of the fuselage, because the tail is far back and high up with respect to the main body of the aircraft, with the booms leaving sufficient clearance for the rocket's hot exhaust. To increase the thrust and endurance, instead of the two motors of the Ente this plane is fitted with 16 sustainer rockets that will be fired in pairs. Each motor is able to deliver about 24 kg (53 pounds) of thrust for about 24 seconds. Overall, they should get the plane airborne without the use of a catapult launcher. Confusingly, the plane is also called the RAK-1, same as the earlier rocket car.

Von Opel, Hatry and Sander conduct their first tests on 10 September 1929 in a hunting field just outside Rüsselsheim. A handful of onlookers, including a *New York Times* photographer, watch the plane go nowhere while burning up its boosters. The problem seems to be in the initial launching. A second attempt is made the same day, this time using a standard rubber-band launch catapult. About a meter or two in the air von Opel launches his rockets and flies about 1,400 meters (4,600

Fritz von Opel at the back of the RAK-1 rocketplane [Opel].

In 1999 Opel produced this replica of the rocket-propelled aircraft RAK-1 [Opel].

feet). The paper publishes a photograph of the flight in its Sunday edition on 6 October. But the team is not satisfied, because they want the plane to be able to take off without any catapult assistance. After some adjustments they make another test, secret this time, on 17 September. With Hatry in the cockpit and with the help of a rail-sled equipped with a pair of rockets together delivering 700 kg (1,500 pounds) of thrust, the plane manages to launch itself. It travels roughly 500 meters (1,600 feet), with a maximum altitude of about 25 meters (80 feet). Satisfied, von Opel calls another publicity event. On 30 September the team prepares the plane at Frankfurt's Rebstock airport. Sixteen rockets are packed into the back of the RAK-1, and two larger ones in the sled. In front of a large crowd and several cameras, von Opel takes the controls and lights the rockets. He knows he is taking quite a risk, considering the rockets have a tendency to explode. On the first and second attempts the rockets on the plane don't ignite correctly, causing the aircraft to meekly jump off the launch rail and skid over the ground for a short distance. Now they have a problem: Sander has not reckoned on a third attempt and only 11 rockets for the plane are left. Von Opel decides to try anyway, and this time the rocket glider launches successfully from the 20 meter (65 feet) long slide rail and takes to the sky. Igniting one rocket after the other he keeps the plane airborne for 80 seconds. Even although six of the 11 motors fail to ignite, he flies 25 meters (80 feet) above the ground and reaches a speed of about 150 km per hour (90 miles per hour). Due to the relatively high weight (270 kg, 600 pounds) with respect to the size of its wings, von Opel has to land the plane at high speed to enable the wings to generate sufficient lift for a controlled landing. The fast landing ends badly in a crash; he is able to walk away but the plane is a write off.

Fritz von Opel in front of the wrecked RAK-1 rocketplane [Opel].

Von Opel is nevertheless extremely exited about the flight and the future he sees for rocket planes and spaceflight, and euphorically comments: "It is magical, flying like that, powered by nothing more than the combustion gases streaming out of the engines at 800 km an hour. When will we be able to harness the full power of these gases? When will we be able to fly around the world in five hours? I know this time will come and I have a vision of future world travel which will bring together all the people of the Earth to live as one. So I race towards this vision like a dream with no sense of space and time. A machine flying almost by itself I hardly need to touch the controls. I feel only the borderless intoxicating joy of this first flight."

In December the *Denton Record-Chronicle* in Texas writes: "The day of 1,000 passenger airplanes propelled by rockets at 5,000 miles per hour was envisioned by Fritz Von Opel, German auto manufacturer and authority on rocket planes. Von Opel, who flew a rocket propelled plane in Germany last October a distance of a mile in 75 seconds, said he expected to see the 5,000 mile per hour airplane in operation within the next generation and possibly the next decade." Like Valier, von Opel sees a great future for rocket plane transportation, and expects to see his vision come true soon. Unfortunately the single RAK-1 flight proves to be Opel's last rocket vehicle experiment. One month after his fiery take off the world stock market crashes even worse than his last rocket plane, and the Opel company is prohibited by its majority owner, General Motors, from pursuing further expensive rocketry work. Von Opel quits the company and goes to live in Switzerland, where he dies in 1971, aged 71. Max Valier had been killed prior to the flight of the RAK-1 plane, when an advanced rocket engine that used liquid oxygen and alcohol as propellants blew up during a test run in his laboratory, curtailing his ambitious plans for flying across first the English Channel and later the North Atlantic.

Although ocean-crossing stratospheric rocket planes were still science fiction to many people in 1929, the use of rockets to assist planes take off either with a heavy cargo or from a shorter runway do find immediate practical use. In August 1929 the German Junkers aircraft company launches a W33 Bremen-type floatplane from the river Elbe with the help of Sander rockets.

Also in 1929 another German by the name of Gottlob Espenlaub, an experienced glider designer and pilot, attempts to follow up on von Opel's rocket flights. The literature is confused with different sources giving different details on his flights; what follows is my reconstruction. On 22 October Espenlaub prepares his 'Espenlaub Rakete-1' rocket glider for its first flight at the Düsseldorf–Lohhausen airfield. Painted on the nose is the snout of an angry looking monster bearing large fangs, which is appropriate considering the dangerous nature of the machine. A pair of Sander's black powder rockets, each with 300 kg (660 pounds or 3,000 Newton) of thrust are installed, then the glider is towed into the air by a propeller plane. At a height of about 20 meters (65 feet) Espenlaub disconnects from the tow plane and ignites one of the rockets. A long stream of fire explodes from the back of the plane with a tremendous roar, giving it a powerful push. But the asbestos put around the tail of the plane turns out to be insufficient protection against the rocket's exhaust, and soon the rudder catches fire. Fortunately, he manages to land the burning plane safely and without injury. In May 1930 Espenlaub tries again at Bremerhaven, this time, wisely, with a tailless glider design, the Espenlaub-15. He equips the plane with two Sander solid propellant boosters each delivering 300 kg (660 pounds) of thrust and ten sustainer rockets, each with 20 kg (44 pounds) of thrust. For additional boost at take off, he places the E-15 on a catapult sled. When he hits the igniter switch, the combined power of the catapult and one of the large boosters launches him quickly to a height of about 30 meters (100 feet), whereupon he ignites the first of the small sustainer rockets to stay in the air. He decides to fire the second powerful booster to gain the speed required to make a turn, but the big rocket explodes, almost throwing Espenlaub out of his seat. As the little E-15 dives to the ground, he jumps out, strikes his head, and falls into a soft bog. Rescuers take him to a hospital, where he remains unconscious for two days. He never attempts another flight. After the war Espenlaub and his company go on to produce streamlined cars, which are much less dangerous than his high-powered rocket plane.

Inspired by the developments in Germany, daredevils in other countries also take to the skies in gliders fitted with black powder rockets. In June 1931 Ettore Cattaneo flies the 'RR' at the airport of Milan in Italy. A 280 kg (620 pound) rocket propelled glider built by the Italian company Piero Magni Aviazione, it resembles von Opel's RAK-1 with a single wing mounted above an enclosed fuselage and a tail with two vertical stabilizers set high enough that the exhaust of the rockets won't ignite them. It remains in the air for 34 seconds, covering a distance of about 1 km (0.6 mile). In 1928 the American aviation pioneer Augustus Post publishes a design for a rocket plane but it is never built. It was to have been propelled by liquid air. The air would be heated so that it would expand as a high-speed gas through nozzles in the tail of the aircraft to deliver thrust. Post designed the plane to operate in the stratosphere, where there would be little air to provide lift for flight and maneuvering. To generate lift at high altitude, some of the air would be expelled from a series of smaller outlets on the upper part of the wing's leading edge, delivering a stream of fast flowing air over the wing. Nozzles in the wingtips would help to steer the plane (an idea much later applied in the X-15). The air would be stored as a liquid rather than as a gas in order to minimize the size of the propellant tanks. The passengers would travel in a pressurized cabin, something now standard in airliners but in 1928 still a novelty (the first experimental plane to have a pressurized cabin, the German Junkers 49, flew in 1931). The first American rocket aircraft to actually fly is rather less ambitious than Post's design. William Swan's 'Steel Pier Rocket Plane' is a simple high-wing glider with an open-framework fuselage – a design that we would nowadays call an ultra-light. It takes off from Atlantic City, New Jersey, on 4 June 1931, powered by a single solid propellant rocket motor with about 23 kg (50 pounds) of thrust (nine more motors are installed but not fired for this test flight). The plane rises bumpily to an altitude of 30 meters (100 feet) and covers a distance of about 300 meters (1,000 feet). A day later "resort stunt flier" Swan employs twelve of the same motors. Now the little glider and its pilot are propelled to an altitude of 60 meters (200 feet) and remain in the air for eight minutes, demonstrating how rockets can launch a glider into the air without the need for a ground crew and a catapult or tow plane.

The experiments with rocket propelled gliders in the late 1920s and the first half of the 1930s show that rockets can indeed be used to propel an airplane: the 'instant'

William Swan flies the American rocketplane [*Modern Mechanics*].

high thrust rockets can certainly make a plane take off and ascend very rapidly, and in principle they can also enable a plane to fly very fast. Moreover, because rockets don't need oxygen from the atmosphere, rocket planes can potentially fly very high. In 1928, *Popular Mechanics Magazine* correctly tells is readers: "Opel's rocket-car and rocket-plane experiments are important because the rocket offers the one method yet discovered for navigating space at high altitudes, above the Earth's envelope of air. All existing motors depend on air for their operation, and all existing types of propellers screw their way through the selfsame air. The rocket, on the other hand, can be shot out into space and attain tremendous speed by escaping the resistance of the air." And rockets are much simpler than gas turbine engines, which existed only as concepts when the Ente and RAK planes were already flying (the first turbine jet aircraft, the Heinkel 178, only took off in 1939, also in Germany, which was at that time leading the world in advanced aircraft propulsion). Because of their simplicity, rocket motors can be added to existing aircraft, as demonstrated with the gliders that formed the basis of the Ente, RAK-1, Espenlaub E-15 and Swan's Steel Pier Rocket Plane, as well as the Junkers floatplane.

However, rocket engines are not very useful for sustained atmospheric flight. Jet engines scoop up air to collect the oxygen needed to burn the fuel, but rockets need to carry both their fuel and their oxidizer with them. As a result, rocket planes need relatively large tanks and are heavier than jet or propeller planes of a similar range. In 1931 the earlier-mentioned article 'Berlin to New York in less than One Hour!' remarks upon this: "With the present proportions between the weight of the fuel and that of the rest of the flyer, the former is so great that you need over 90 percent of the available space for fuel, and have left only 10 percent for cargo. This makes the venture economically unprofitable and, for that reason, no big machine has, as yet, been constructed."

Max Valier had already foreseen that most of the weight of his intercontinental rocket plane would be propellant, even when flying at the edge of the atmosphere in order to limit aerodynamic drag for an important part of the journey, and even when using liquid propellants which, by giving more thrust per unit of propellant weight, are much more efficient than solid powder rockets. Valier even suggested that in the denser atmosphere an intercontinental rocket aircraft could limit fuel consumption by using retractable propellers. It was self evident that major steps would be required in propulsion development, and not only in terms of efficiency but also in safety, since the primitive black powder rockets used in the early German rocket planes were not suitable for a trustworthy vehicle. Other major issues with powder rockets were that it was impossible to control their thrust during flight, let alone extinguish them, and they had the disturbing tendency to blow up or set the plane on fire. Rocket motors using liquid propellants would be much more controllable and therefore safer, and much better suited for propelling aircraft. Until these were available, it was hard to conceive of an aeronautical assignment that could not be performed more efficiently by conventional aircraft.

MONSTER ROCKETS

In the 1930s rocket motors, and especially those using liquid propellants, saw rapid development. In Russia the work of a deaf schoolteacher, Konstantin Tsiolkovsky, formed the basis for various experiments with rockets. He laid down the theory of rocket flight and derived the most important mathematical formula used in rocket design: the Tsiolkovsky Equation a.k.a. the Rocket Equation. In 1903, the same year as the Wright brother's first powered flight, this visionary man had already invented the method of 'staging', whereby a series of rockets are stacked on top of one another and discarded once their tanks run dry, in order that the remainder of the vehicle can fly on without their now useless empty weight; a method that has been applied to all the world's space launchers starting with the one which placed Sputnik into orbit in 1957.

Under the Russian Tsar's regime Tsiolkovsky never enjoyed any government support, but after the 1917 Revolution the Soviets saw in him a perfect example of the working class genius, and in 1919 he was even made Member of the prestigious Academy of Sciences. Apart from pure rockets, Tsiolkovsky also considered rocket propelled planes. Around 1930 he developed a fourteen-point plan to conquer space. The first step would be to build a low-altitude rocket plane that could fly to a height of about 5 km (3 miles). Step two would be a similar plane, but with higher thrust and shorter wings to limit the air drag at the higher velocities it would attain. Next a rocket plane capable of climbing to an altitude of 12 km (8 miles), equipped with a pressurized cabin, would be required. The subsequent steps would involve wingless rockets that could get into space, space suits, orbital stations, and even colonies on asteroids. Tsiolkovsky died in 1935 without personally building any rockets, but he inspired a number of young Russian rocket experimenters who, in the 1930s, began to build small experimental rockets based on liquid propellants. As a testimony to his vision, Tsiolkovsky's most famous words can be found on his grave: "The Earth is the cradle of mankind, but one cannot stay in the cradle forever."

In parallel, but on the other side of the ocean, the reclusive American Robert H. Goddard was designing, building and launching a series of ever larger rockets. By 1918 he had already constructed and flown a solid propellant rocket that was 1.70 meters (5.1 feet) in length and weighed 20 kg (44 pounds). In 1919 he published a now famous report called 'A Method of Reaching Extreme Altitudes', in which he described how a rocket could reach the Moon and signal its arrival by use of flash powder that ignited at impact. However, in spite of their scientifically sound basis,

The cover of the December 1931 issue of *Popular Science*, showing Goddard's rocketplane concept. Turbines in the rocket exhaust run the propellers [*Popular Science*].

Goddard's proposals were met with disbelief and ridicule. In response he became secretive about his plans, worked only with a small team of trusted people and hid his experiments from the public. Often not even his close associates knew exactly what he was up to. Like Tsiolkovsky, Goddard realized that rockets based on liquid propellants have several advantages over the simpler solid propellant (gunpowder) rockets used until then for artillery and fireworks. Liquid propellant rockets have a lower weight relative to their thrust and, unlike solids, can be throttled up or down, and even stopped entirely without blowing up. So Goddard began to experiment with liquid propellants and in 1926 flew the world's first liquid propellant rocket (running on gasoline and liquid oxygen); an event now recognized as historic even though at the time virtually no one knew about it.

Goddard also gave some thought to rocket planes, leading in 1931 to a patent for an especially innovative design. Observing that rocket motors are not very efficient in the lower atmosphere, he proposed that his rocket plane would start off by using the exhaust from its rocket engine to drive a pair of turbines each of which would be connected to a conventional propeller. At higher altitudes the turbine blades would be withdrawn from the path of the hot gas and the plane would fly on rocket thrust alone. The turbines would have had to be made from heat-resistant materials that far exceeded the capabilities of the time, but the general idea was sound and similar to what we call today a turboprop engine. The concept of dual-mode propulsion was also very innovative and would become a feature of advanced spaceplane concepts half a century later (although not involving propellers). In his book *The Conquest of Space*, the first non-fiction book on spaceflight in English, American David Lasser foresaw fleets of Goddard's planes connecting London, Paris and Berlin with New York by 1950. Flight time: 1 hour! The general public nevertheless kept ignoring Goddard's important work, probably wrongly associating it with the fanciful stories that featured in the popular science fiction pulp magazines of the time. After one of his rocket experiments in 1929, a mocking headline in a local Worcester newspaper read "Moon rocket misses target by 238,799 1/2 miles". By 1935 Goddard's rockets had exceeded the speed of sound and reached altitudes of 1.5 km (0.9 miles), but the US government did not seem to appreciate the possibilities of this novel technology. During the Second World War Goddard was only assigned a contract to develop rockets to assist propeller aircraft take off from carrier ships.

In Germany the situation was very different. The rocket experiments of a small club of hobbyists, inspired by their own rocket pioneer Hermann Oberth, attracted serious attention from the military. German Army leaders saw the development of powerful rockets as a means of circumventing the ban on the use of large cannon as stipulated in the Treaty of Versailles that Germany had been forced to sign after its defeat in the First World War. Under the technical leadership of the German Space Society's Wernher von Braun, who was only 20 years old at time, the Army began an enormous rocket development effort. It soon became the Third Reich's most expensive development project. A dedicated and huge development and launch center was built in Peenemünde, a remote and sparsely inhabited place on the Baltic shore, involving laboratories, wind tunnels, test stands, launch platforms and housing facilities for the 2,000 rocket scientists, 4,000 supporting workers and their

A captured German V-2 rocket fired by the British in 1945.

families. Peenemünde's vast resources enabled the team ultimately to develop the infamous A4 rocket; a 14 meter (46 feet) tall monster capable of throwing a 738 kg (1,630 pound) warhead a distance of 418 km (260 miles). In 1942, on its first flight, and A4 (without an explosive cargo) climbed to an altitude of over 80 km (50 miles) and thus became the first man-made vehicle to reach the edge of space. In a speech afterwards, Walter Dornberger, head of the rocket development program, observed: "This third day of October, 1942, is the first of a new era in transportation, that of space travel..."

But spaceflight was not what the rocket was developed for. It was a weapon and, renamed V2 for 'Vergeltungswaffe 2' (Retaliation weapon number 2), at least 3,000 were fired at London and Antwerp towards the end of the war, blowing away entire blocks of houses in an instant. Although impressive and scary, the A4 was however not a huge success as a weapon. For one, it was a very expensive means of dropping a 738 kg bomb on a city in a neighboring country, and not a very precise one at that. The 'terror' effect of the rocket striking without warning (because it flew faster than sound it could not be heard before impact) did not have the effect that Hitler hoped for either: Londoners did not panic en masse, and did not desperately demand their government negotiate an armistice with Nazi Germany. In retrospect it would have been better to use the money that went into developing and producing the V2 to buy advanced jet fighters like the Messerschmitt Me 262. Although the A4/V2 failed to change the outcome of the war, it demonstrated the future of big rockets as ballistic missiles. More importantly for von Braun and other space enthusiasts, it proved that a rocket could reach space and with further development would be able to launch a satellite into orbit in the foreseeable future.

By the end of the war rockets had reached such a high maturity that they could be used to propel operational planes. The rocket engines of the time were powerful and yet less complicated than turbojet engines, so if you needed to get an airplane up and high as fast as possible with a relatively simple power plant, a rocket engine was the obvious choice. In America the Air Force attached rocket pods to heavy cargo planes to help them to take off, but in Germany real rocket fighters were put into operation to counter high flying bombers. While it took a conventional high altitude propeller interceptor such as Focke Wulf's Fw 190D-9 some 17 minutes to climb to bombers at altitudes of up to 10 km (6 miles), the revolutionary Me 163 'Komet' (German for Comet) rocket plane could do it in just under 3 minutes! Even the Me 262 jet fighter took 10 minutes to climb that far. The Me 163 was the fastest aircraft of the Second World War: at top speed, the little rocket interceptor closed in on the Allied bombers at about 960 km per hour (590 miles per hour), allowing the defending gunners no chance to take aim at it. The Komet also outran the propeller-driven escort fighters, so there was essentially no defense against the little rocket plane during its powered attack. Only when it ran out of propellant and had to glide back to its base could the Me 163 be intercepted and destroyed.

Serious plans for launching rocket planes into space also began to be developed during the Second World War, starting with Eugen Sänger's design for a bomber to strike New York. This 'Silbervogel' (Silverbird) would begin its mission by riding a large rocket sled on a rail track over a distance of about 3 km (2 miles). On firing its own rocket engine it would climb into space, its altitude peaking at around 145 km (90 miles). It would fly below orbital speed, slowly descending into the stratosphere until the increasingly denser air would generate sufficient lift to 'bounce' the plane up again. Thus the Silbervogel would hop around the planet in the same manner as a stone skipping across a pond. It was calculated that the rocket plane would be able to take off from Germany, cross the Atlantic, drop a small bomb on the USA and land somewhere in Japanese occupied territory in the Pacific. It wouldn't be a real orbital spaceplane, but pretty close. Fortunately, Germany had no time to develop anything like this before the war ended, and even if they had it would not have worked: later analysis showed that the heat generated by re-entry into the atmosphere would have destroyed the Silbervogel in its original design. Additional heat shields might have saved the concept, but the associated extra weight would have cut the bomb load to zero. Certainly it would have been difficult to justify the vast effort and expense to develop the Silbervogel simply to drop the equivalent of a hand grenade on the US. Even its designer reckoned it would take decades to get the Silbervogel operational. But the idea of a rocket propelled space bomber would return later, even leading to concern in the Soviet Union that the NASA Space Shuttle was actually a disguised orbital bomber with heat shielding that would enable it to make shallow dives into the atmosphere to deploy nuclear bombs!

During the 1940s and 1950s many technological fields experienced radical and rapid advances, in particular aviation: jet aircraft replaced piston-engined propeller planes, and ballistic missiles quickly made strategic bombers all but obsolete. The amazingly fast progress in rocket and aircraft development led to series of rocket aircraft that repeatedly broke altitude and speed records. In October 1947 test pilot Chuck Yeager managed to get the little Bell X-1 rocket plane to fly faster than the speed of sound (Mach 1), thereby breaking the so-called 'sound barrier' (or rather, discovering there was no such barrier). In 1953 Scott Crossfield became the first to exceed Mach 2 in the Douglas D-558-2 Skyrocket. Three years later Milburn 'Mel' Apt pushed the Bell X-2 to the next magic number of Mach 3. Shortly before Apt's flight, Iven Kincheloe became the first pilot ever to climb above 100,000 feet, as he flew the X-2 to a peak altitude of 126,200 feet (38,5 km, or 23.9 miles).

The complexity of the aircraft had increased enormously in the quarter century that separated the first rocket propelled gliders to the Bell rocket research aircraft: while the Opel RAK-1 had required only a few simple readiness checks shortly in advance of the launch, the Bell X-1A pilot had a checklist of 197 points to tick off prior to flight. By 1962, test pilots were flying at speeds over Mach 6, and earning their 'astronaut wings' by reaching altitudes of 80 km (50 miles) and even higher in the incredible North American X-15; the US Air Force designates people who travel above an altitude of 50 miles as astronauts, although the International Aeronautical Federation (Fédération Aéronautique Internationale, FAI) defines the boundary of space at 100 km (62 miles). Many therefore saw the rocket spaceplane as the logical progression of aviation into space, as had Valier, Tsiolkovsky and Goddard thirty years earlier.

But based on the design of the A4/V2 and insight gained from the Peenemünde experts captured after the war, the US and the USSR had by then both developed an arsenal of intercontinental ballistic missiles. The multi-stage rocket technology used to lob a nuclear warhead halfway around the Earth had reached a very high level of maturity, and laid the basis for a much faster and easier road into orbit which all but suspended the development of rocket spaceplanes. Simply put, the warheads were removed and replaced with satellites and manned capsules. Of course these missiles were not reusable, but in the Moon Race of the 1960s money and sustainability were not the most important issues. Instead of the Silbervogel or similar rocket plane, the ultimate space machine thus became the enormous Saturn V rocket. It boosted Neil Armstrong and Buzz Aldrin to the Moon, in the process beating the Russians in the technological and political space race that had started with the launch of the world's first satellite by the Soviets in October 1957. Although the ultimate launch vehicle in terms of its capability, the Saturn V was also an extremely wasteful transport vehicle:

Saturn V on the launch pad [NASA].

excluding the actual payload, the total weight of the 111 meter (363 feet) rocket was some 2,896,000 kg (6,384,000 pounds), 94 percent of which consisted of propellant. At 47 tons (104,000 pounds), the Apollo spacecraft and lunar lander represented less than 2 percent of the entire vehicle that lifted off from the launch pad, and 187 tons (412,000 pounds) of precious hardware was lost by the time the astronauts arrived in lunar orbit; the Saturn V first stage fell into the ocean, the second stage burned up falling back into the atmosphere, and the third stage went into an orbit around the Sun or crashed on the Moon. Corrected for inflation, in 2011 economic conditions a Saturn V would have cost $2.9 billion per launch, which translates to nearly a billion dollars per astronaut. And for that they were only flying tourist class, with poor food and hardly any leg room. It got you to the Moon and made an impressive amount of noise and smoke doing so, but a Saturn V was clearly not an economical means of transportation. Good for winning a race, but not for the large-scale economic use of Earth orbit and beyond.

THE PERFECT SPACEPLANE

By the mid-1960s the age of the rocket fighter planes and experimental supersonic rocket aircraft had essentially ended, only some 30 years after it started. Jet engines could now also get a plane to high velocities and high altitudes in a short time, and most importantly were much more fuel-efficient than rocket engines. About the only advantage that rocket planes still had over jet aircraft was that they could operate at extreme altitudes and even in the vacuum of space.

In the 1970s the focus of spaceflight in the US turned to a vision of regular, easy and cheap access to low orbit, which seemed to mean good news for rocket plane development. The ideal launch vehicle would be completely reusable, reliable, safe, low-maintenance, efficient and require little work and time between flights (what aircraft operators call a short 'turn-around'). This sounds just like the description of a regular airliner, and so designs to address these requirements often resembled aircraft.

Normally rocket propulsion would need to be incorporated, because jet engines do not work in space due to a lack of oxygen. But in the 1970s the technology to build a single-stage orbital rocket plane did not exist. NASA therefore sought a multi-stage design. It initially envisaged a combination of two rocket planes in which a massive winged booster would release a smaller vehicle at sufficient altitude and speed for it to insert itself into orbit. The booster would fly back to the launch site. In due course the spaceplane would also land and be prepared for another mission. This two-stage design meant less severe vehicle empty-weight minimization challenges, while both stages of this combination would be fully reusable. But a combination of technical and budgetary constraints mandated a compromise in which the winged first stage vehicle was replaced by a pair of reusable solid propellant rocket boosters, and the orbital spaceplane got an external propellant tank that would be discarded on each flight. This became the Space Shuttle that NASA flew from 1981 to 2011. An ideal spaceplane it was not, because its long launch preparations, complex maintenance, partial reusability and relative fragility led to high costs and high risks. It turned out to be far less cost effective and a much more dangerous means of gaining access to low orbit than its designers envisioned. By the mid-1980s there was a sense in the US, Europe, Russia and Japan that it was time for a fully reusable spaceplane with aircraft-like operations. To limit the propellant load, and thus the overall size and weight, the vehicle would have to combine airbreathing and rocket engines: using oxygen in the atmosphere as oxidizer as long as practical, prior to switching to rocket propulsion when the air density fell below the minimum level required to operate a jet engine. It was believed that materials, flight control and propulsion technology were sufficiently matured to make the development of such a space plane possible.

In a perfect world, a trip in an ideal spaceplane that adheres to the constraints of known physics and near-future technical possibilities could go something like this: On waking up at home you put on your simple flight overall, have a pleasant and relaxed breakfast, then get into your (flying?) car and head for the spaceport. Upon arrival you check in, but don't need to put on a cumbersome spacesuit, you simply proceed to the plane. In contrast to the early days of spaceflight, there is no doctor checking your fitness, as the flight is extremely benign in terms of accelerations and shocks. Boarding the plane is similar to embarking a normal airliner. Soon after you have settled in, the vehicle starts to rumble down the runway. Just like any airliner, it takes off horizontally and initially ascends at a shallow angle. Apart from requiring a rather long runway there has been nothing extraordinary about the flight up to this point. Even when, two minutes after take-off, the pilot announces that the plane is going supersonic (faster than sound) you don't notice anything peculiar other than that noise of the engines diminishes because the spaceplane is now outrunning its own sound waves in the air outside. A bit more than another two minutes later you are at an altitude of 12 km (8 miles) and flying at twice the speed of sound. Normal airliners don't go higher than this, but your spaceplane keeps on climbing. Soon the sky goes dark and the stars become visible. The curvature of the horizon becomes noticeable, confirming the Earth to be a sphere rather than the flat surface it seems through the window of a normal airplane.

At over 28 km (17 miles) and a velocity exceeding five times the speed of sound the engines switch from their airbreathing to pure rocket propulsion mode, and the acceleration, which has been hardly noticeable, suddenly increases and pushes you deeper into your seat. The airplane has become a rocket spaceplane, independent of the oxygen and lift generating capabilities of the atmosphere. Sixteen minutes after take off you are 80 km (50 miles) high and accelerating more than three times more rapidly than a free-falling sky diver. At that moment the engines are shut down, and immediately all sensations of gravity and acceleration vanish and you feel yourself floating in your seat, held in place only by the safety harness. The spaceplane is now in an elliptical orbit around the Earth. You have not yet reached the highest point of your orbit, so you are still going up even although there is no sense of acceleration. Soon you cross the theoretical border of space at 100 km (62 miles) altitude. If you weren't one already, you're now officially an astronaut. At 400 km (240 miles) the spaceplane reaches the highest point and the rocket engines are reignited briefly for the boost required to achieve a nice circular orbit. By using small rocket engines, the spaceplane carefully maneuvers towards the space station that is your destination. A couple of hours later the vehicle docks at the large collection of cylindrical modules. You unstrap from your seat and simply float out of the cabin, through the docking tunnel into the space station. You stay there a couple of weeks to work in the biology experiments laboratory module.

When it is time to go home, you board a docked spaceplane of the same type you came up with. It is not the exact same vehicle, because spaceplanes arrive and leave every couple of days transporting people, food, oxygen and experiments to the space station. After undocking, the vehicle slowly drifts away. A half-minute burst of a set of relatively small tail engines in the direction the spaceplane is orbiting, slows the plane down just enough to change its orbit from a circle into an ellipse. The highest point of the orbit is at the altitude of the space station and its lowest point penetrates the atmosphere. As the altitude of the vehicle drops, the aerodynamic drag from the thin atmosphere slows the plane even more. Half an hour after the de-orbit burn you still appear to be in orbit as if nothing had changed, but you are actually falling slowly back to Earth.

The plane was flying backward when firing its rocket engines to break out of its circular orbit, which is okay in the emptiness of space, but it's not a healthy attitude for entering the thicker layers of the atmosphere. So the pilots rotate the spaceplane through 180 degrees to make it fly nose first. They also pull the nose up, to align its heat resistant belly with the wall of air it is about to encounter. This maneuvering is done using the small attitude control rocket thrusters, because in the near vacuum of space the rudders, elevators and ailerons on the wings and tail of the plane are totally ineffective. The heat shield protects the spaceplane from the extreme heat generated as it slams into the atmosphere. The edges of the wings glow red hot as they reach temperatures of 1,600 degrees Celsius (2,900 degrees Fahrenheit); greater than the melting point of steel. However, unlike the heat shields of the old space capsules the metallic shield on your spaceplane does not slowly burn up, and can thus be reused. It is also not as vulnerable as the old Space Shuttle's thermal protection tiles, which were reusable but also rather fragile: if it had to, the spaceplane could fly through a hailstorm without damaging its heat shield. And to minimize the vehicle's take-off weight no propellant was loaded as a reserve for the return flight, so the spaceplane glides back unpowered. As with the Space Shuttle, there is no option of aborting the landing and flying around to make another approach. This might appear to be risky, but actually it isn't because if the return conditions weren't perfect and for instance the weather at the airport was likely to be unfavorable, the spaceplane would have just waited in orbit or targeted another landing site well before the execution of the de-orbit burn. As usual the automatic flight system, supervised by the pilots, flies a perfect approach and landing. A bit shaky on your legs, because while you were in space your body adapted to weightlessness, you disembark from the plane. Even as you head home, the vehicle is being refueled, and after a short maintenance check it is declared ready for its next flight.

While this perfect reusable launcher does not yet exist, the above description is based on modern spaceplane concepts such as Skylon, currently under development

Skylon on the runway [Adrian Mann & Reaction Engines Limited].

at the British company Reaction Engines Ltd. This machine only exists on paper, but realistically illustrates what a near-future orbital rocket plane may look like. What you experience as a passenger may not be too different from what it is like to fly in a high-performance jet aircraft, and while the spaceplane does superficially look like one, there is a big difference in the amount of propellant that it has to carry. While 40% of a large airliner's take off weight may be fuel (51% in the case of the Concorde), a mixed-propulsion spaceplane such as Skylon will consist of 80% propellant. A pure rocket spaceplane that does not use any airbreathing engines at all would be over 90% propellant, a problem already accurately foreseen by rocket plane pioneer Max Valier in the late 1920s. Such percentages are similar to those of existing expendable launchers but are much more difficult to attain for aircraft with wings, wheels and a cockpit, that must also be able to survive atmospheric re-entry. In addition, the liquid hydrogen that Skylon uses for fuel has a density that is much lower than that of the kerosene used by normal airplanes: where 1 liter (0.3 gallons) of kerosene weighs 800 grams (1.80 pounds) the same volume of liquid hydrogen is 70 grams (0.15 pounds). The same amount of fuel thus takes much more room. The passenger cabin onboard a spaceplane will therefore be tiny compared to that of an airliner since most of the vehicle's volume will need to be filled with fuel (as well as additional liquid oxygen to burn with the hydrogen during the rocket propelled flight phase).

This difference between airplanes and spaceplanes is not so much a matter of their operating altitudes, rather it is the result of their vastly different velocities. Airliners fly at about 950 km per hour (590 miles per hour), whereas to achieve a low orbit a spaceplane will have to achieve a velocity of about 7.8 km per second (4.8 miles per second), which translates to 28,000 km per hour (17,400 miles per hour)! This means the spaceplane's velocity needs to be about 30 times greater than that of the average airliner. Furthermore, the energy needed to attain a given velocity increases with the square of the flight speed. This means a spaceplane needs some $30 \times 30 = 90$ times more energy than an airliner of the same weight. This energy must be gained by the engines converting the chemical energy of the propellant into kinetic (movement) energy. And this simplified calculation does not take into account the aerodynamic drag during the climb out of the atmosphere, which also increases quadratically with velocity.

When a spaceplane gets above the atmosphere and reaches orbital velocity, it can circle the Earth without any further need to burn propellant. An airliner however, needs to continuously compensate the drag of the atmosphere it is flying through to maintain its velocity. It does that using its engines, which consume fuel during the entire trip of often thousands of kilometers. This is why airliners in reality do not fly with 90 times less propellant than a spaceplane would require, which you would expect if taking into account speed alone. Nevertheless, whereas airliner designers achieve an optimum in terms of velocity, amount of propellant, cabin volume, cargo weight and ultimately cost, spaceplane designers are pretty much stuck with the need to cram as much propellant as possible into their vehicle, and hopefully in the end have some weight capacity left for the cargo that needs to be taken into orbit, which is, after all, the whole reason for the spaceplane's flight! In other words, the margin between success and failure is very small: if the spaceplane tank structure proves to be a little heavier than anticipated or the rocket engine yields just 1% less thrust than foreseen, you may end up with a very fast but useless suborbital rocket plane with zero payload, rather than a satellite-launching, money-making, orbital spaceplane.

This narrow margin, plus frustration with the Space Shuttle in terms of costs and risks, has made the world's space agencies and industries developing launchers extremely cautious with regard to spaceplanes. Since the development of the Space Shuttle hundreds of concepts for spaceplane (and other types of reusable launchers) have come and gone. Some hardware was build and tested, and some designs even got as far as flying a sub-scale test vehicle. But none of them has yet resulted in an operational vehicle, largely because the step to develop a full-scale spaceplane was deemed to be too risky and too expensive, and the benefits could not be sufficiently guaranteed.

In 2005 I attended an international conference on spaceplanes and hypersonic systems in Italy, where scientists and engineers from the US, Europe, China, Russia and Japan, and even Australia, India, South Korea and Saudi Arabia presented developments on exciting sounding topics such as pulsed detonation propulsion and aerospike engines, as well as highly specialized issues like 'Fluctuations of Mass Flux and Hydrogen Concentration in Supersonic Mixing' or 'Pseudo-Shock Wave Produced by Backpressure in Straight and Diverging Rectangular Ducts'. It seemed to me that half the world was involved in spaceplane technology development, and there was certainly no lack of concepts. However, at the time of writing none of the designs for large prototypes, let alone operational spaceplanes, have moved beyond the drawing board.

Whereas up until the 1970s advances in rocket plane technology were often soon incorporated in new experimental planes and high-altitude rocket interceptors, now engineers seem to be stuck in their laboratories, able to fly their innovations only on small-scale test models. This is due to the extreme complexity and enormous cost of developing modern high-performance airplanes in general and space launch vehicles in particular. In the Second World War it took the famous P-51 Mustang fighter only six months to progress from the conceptual design to its first flight, with just another 19 months until it entered combat service. At the time the US government purchased them at $50,000 per airplane, the equivalent of $600,000 today. The US Air Force's latest F-22 Raptor fighter took over 20 years to advance from concept to operational fighter at a development cost of $65 billion. Each of these sophisticated planes costs some $143 million: 238 times more than a Mustang! A future spaceplane will have a lot more in common with modern aircraft like the F-22 than with the nuts-and-bolts Mustang. There is no easy and cheap way to develop an operational spaceplane, so the technical and financial risks will be high. Hence, the benefits must also be high and more or less guaranteed.

Looking into the near future, it is clear that the preference for expendable launch vehicles is ongoing. Various new throw-away launchers or updates of existing ones are under development by a number of agencies, while research on reusable launch vehicles is continuing at a very slow pace with much lower levels of funding.

2

Crash course in rocket plane design

"Perfect as the wing of a bird may be, it will never enable the bird to fly if unsupported by the air. Facts are the air of science. Without them a man of science can never rise." – Ivan Pavlov (1849-1936)

To be able to understand the possibilities, limitations, history and evolution of rocket planes, we must look at how they work. We start with the 'rocket-' part, then explore the '-plane' element, and finally investigate the wonderful and dangerous things that happen when you combine them.

ROCKET ENGINES

The principle of a rocket engine is fairly simple: generate a gas at high pressure by burning propellant in an enclosed space and let it escape through a nozzle. The resulting thrust has nothing whatever to do with the rocket 'pushing against the air', but is purely a consequence of Isaac Newton's famous principle: for every action there is an equal and opposite reaction. If you stand on a skateboard and throw rocks away, then you will move in the other direction: the 'action' is throwing rocks backward and the 'reaction' is you moving forward. Pushing the rock away also means pushing yourself away from the rock. Another good example is a fire hose: as lots of water spews out at high speed you feel the thrust trying to push you back. The hose sprays water in one direction, and in reaction the hose itself is pushed in the opposite direction. In essence this is a rocket engine working on water, and if you stood on the skateboard holding the fire hose, then you would have basically created a rocket propelled vehicle. Instead of throwing rocks, you would be throwing out water continuously. The principle that Newton derived works because the rocks and the water have mass, and the greater the mass and the higher the velocity at which you throw them away, then the higher will be the velocity that you achieve in the opposite direction. You can imagine that throwing something with a small mass relative to yourself, such as a feather, won't have much effect. Throwing away a bowling ball with very little speed, essentially letting it fall out of your hands, will also not achieve much. Only if you throw away objects of substantial mass at a significant speed will the resulting thrust be enough to push you away on a skateboard. If the rock that you throw away is one-tenth of your own weight, then you will attain that proportion of the velocity at which the rock is flying out (ignoring the friction of the skateboard's wheels with the ground). If you want to go faster, you can either throw out a larger rock at the same velocity, or the same rock at a higher velocity, or indeed a smaller rock at even higher speed.

The thrust of a rocket engine is measured in 'Newton' in the metric system, and in 'pounds of thrust' in the US. One Newton is the force that a 0.1 kg mass exerts on a floor on the Earth's surface. Isaac Newton stated that a force (or thrust) is equal to mass times acceleration. On Earth, if you let something fall it will speed up by about 10 meters per second every second: i.e. after 1 second its velocity is 10 meters per second, after 2 seconds it is 20 meters per second, and so on. This means that on the Earth's surface the gravitational acceleration is 10 meters per second per second (or to be more precise 9.81 meters per second per second). If you stand on your skateboard again, and every second you throw away a 1 kg rock at 10 meters per second, then you will be creating a thrust of 1 kg times 10 meters per second per second = 10 Newton.

In real rocket engines the necessary high pressures are generated by combustion. The resulting gas expands out through a nozzle at tremendous velocity, and because it has mass this results in a powerful thrust. It is basically a continuous explosion: a single explosion gives a short kick, a series of explosions provides a series of kicks, and continuous combustion and expansion yields a steady thrust. For combustion to occur, a fuel and an oxidizer are required. An oxidizer is a substance that contains the oxygen that makes things burn. In the engine of a car the fuel is gasoline and the oxidizer is ordinary air, which contains some 21 percent of oxygen. Rockets don't use atmospheric air, but carry their own oxidizer.

In a liquid propellant rocket engine, the fuel and the oxidizer are in the form of liquids, for example alcohol and liquid oxygen. These are stored in separate tanks, from which they are fed into a combustion chamber using pressure in the tanks or powerful pumps. Such pumps are typically powered by a separate gas generator, in which some propellant (which can be the same as those used in the combustion chamber) is burned or decomposed to provide high-pressure gas. This gas is then fed

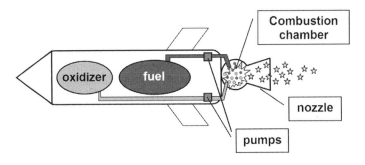

Rocket principle.

through a turbine that runs the pumps, and expelled through a separate exhaust; this arrangement is known as a turbopump. In more sophisticated engines some of the rocket's propellants are burned in a pre-burner and then used to run the turbopumps. However, instead of being dumped directly, the exhausted gas is then injected into the main combustion chamber along with more propellant, in order to complete the combustion; this is called a 'staged combustion cycle'. Turbopumps run at tremendous rates. For example, the turbines of the main engines of the Space Shuttle spin at 30,000 cycles per minute!

In the rocket combustion chamber the fuel and oxidizer are mixed and burned to create an extremely hot, high-pressure gas. This can only escape through an opening in the combustion chamber that is connected to the rocket nozzle. The gas flows out at high velocity through the nozzle, whose shape permits the gas to expand (and thus accelerate), and flow nicely in the right direction. For high-performance rocket engines the exhaust velocities are around 16,000 km per hour (10,000 miles per hour); much faster than throwing rocks! The nozzle is where the expanding gas exerts its forwards pressure, and the correct shape and length are critical in determining the achievable exhaust velocity. If the nozzle is too short or its shape does not allow the exhaust to expand properly, then lots of energy that could be used for generating thrust is lost. The nozzle first converges to a narrow throat so that the velocity of the gas stream is increased, just as occurs when water is passed through a narrowing channel. At the throat it reaches Mach one (the gas mixture's speed of sound) and creates a shock wave, after which the nozzle diverges to allow the high-pressure gas to expand and thereby flow out efficiently at speeds far beyond Mach one. The temperature of this gas stream can reach 3,000°C (5,400°F). The combustion chamber and the nozzle must be cooled to prevent them from melting. Their walls are often made hollow, so that rocket propellant can be pumped through in order to act as coolant before being burned inside the combustion chamber. It is also possible to make the nozzle so thick that it can be allowed to slowly erode during flight. These so-called ablative nozzles are relatively simple and cheap, but also heavy and obviously not reusable. They are usually applied in solid propellant boosters, which have no liquid propellants to use as coolants.

The efficiency of a rocket is indicated by its specific impulse, and is measured in seconds. It is one of the most important parameters in the equations that describe a rocket's performance. A

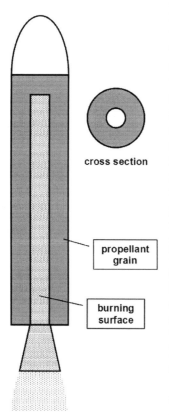

cross section

propellant grain

burning surface

Solid propellant rocket.

specific impulse of 400 seconds means that with 1 kg of onboard propellant the rocket can generate 1 kg of thrust (i.e. 10 Newton) for 400 seconds. (In the US, 1 pound of thrust from 1 pound of propellant for 400 seconds). You can think of the specific impulse of a rocket engine as the number of seconds a certain amount its propellant is able to generate the thrust required to keep itself in the air.

The simplest rockets combine the fuel and oxidizer into a solid propellant like for instance gunpowder. A solid propellant rocket can be viewed as a pipe stuffed with propellant. The propellant grain is usually hollow in order to expose a large burning area, and hence a high thrust due to the high pressure and the resulting large amount of gas flowing out. Firework rockets are of this type. The main advantage of solid propellant rockets is that they are simple, because they do not require any pumps, pipes and valves. As a result, they can provide a lot of thrust for relatively low cost. A big disadvantage however, is that they normally cannot be reused: the throats and nozzles of these motors typically burn away during firing, as there are no liquids to act as coolant. Other very important disadvantages are that, in contrast to liquid propellant engines, solid propellant motors cannot be stopped and you cannot actively control the thrust. Once ignited, the grain will burn away until it is all gone. If something goes wrong along the way, you cannot slow down or stop. In fact, if there is a problem with the motor itself, like a crack in the propellant grain or a piece of material blocking the nozzle, the propellant will keep on burning and the increasing pressure will result in a violent explosion. Solid rockets either work well or they blow up; there is no 'benign failure' which merely results in a loss of thrust, as is possible when using liquid propellant rocket engines. Nevertheless, by cleverly designing the shape of the propellant grain it is possible to vary the amount of thrust desired at certain times after ignition. Many solid propellant rockets used for launch vehicles have star-shaped cross sections. At first the burning surface will be large and the rocket will provide a maximum of thrust in order to get the vehicle off the ground quickly. As the pie-shaped sections burn away, the active surface is reduced and the thrust diminishes in order to limit the aerodynamic forces on the vehicle while it is flying at high speed through the atmosphere.

Another important disadvantage is that for any given mass of propellants, solids cannot provide as much thrust as liquids. A rocket engine that uses liquid hydrogen and liquid oxygen can have a specific impulse of 450 seconds but solid propellants can achieve no better than 290 seconds. On the other hand, because of their relative simplicity it is

**star-shaped propellant
grain cross section**

Solid propellant grain.

Space Shuttle [NASA].

easier to develop and build big powerful solid propellant rockets than large liquid propellant rocket engines, albeit they are much less efficient in terms of the amount of energy to be derived per kilogram of propellant. Also, the storage of solid propellants (in the rocket motor) is generally easier than of liquid propellants, which often require cooling and can be corrosive or toxic.

Probably the most famous solid propellant rockets are the Solid Rocket Boosters of the Space Shuttle. Each of these huge rockets provides a maximum thrust of 13,800,000 Newton, which means one booster could lift some 1,380 cars. Together the two boosters generate about 83% of the lift-off thrust of the Space Shuttle, with the more efficient but far less powerful liquid propellant engines of the Orbiter vehicle only providing the remaining 17%. The Space Shuttle's rocket boosters were actually recovered (by parachute, splashing into the ocean) and reused, but this required much cleaning and refurbishment, and was only economical because of the booster's huge size and production cost.

In contrast to a rocket engine used in an expendable missile, one that is meant to propel an aircraft has to comply with more requirements: it needs to be restartable, reusable, maintainable, and reasonably safe for both the pilot and the ground crew. Ablative cooling, where the rocket's throat and nozzle lose heat by slowly burning away, is not a viable solution for a reusable system; an aircraft rocket engine needs active cooling which pumps the propellant through cooling ducts in the combustion chamber and nozzle. The engine's igniter, which was an external piece of ground support equipment for the series of missiles of which the A4/V2 was part, must be incorporated into the rocket motor itself. On the other hand, the required reliability and safety means that performance may require to be sacrificed to improve safety margins, reduce wear and tear and simplify maintenance.

Overall, their non-reusability, lack of a throttle, poor efficiency and explosion hazards make solid rocket motors generally a poor choice for propelling manned aircraft, other than briefly for an assisted take-off. Liquid propellant rocket engines are more controllable, more efficient and can be made reusable, so are generally a better choice for aircraft propulsion, even if they are more complex.

The main benefit of rocket engines over jet engines and propellers, and the reason that they are used in spaceflight, is that they can operate outside the atmosphere. Jet engines uphold Newton's 'action equals reaction', just like rockets, but they depend on oxygen from the air to burn aircraft fuel. Propellers push air backwards to make a plane go forward, and are driven by engines that need atmospheric oxygen. Rockets carry both the necessary oxidizer as well as the fuel. Independence from the atmosphere is rather handy if you are flying at high altitudes or through the vacuum of space, where there is either insufficient or no oxygen available for your engines. Interestingly, rocket engines also offer an advantage in the thicker atmosphere because their operation is totally independent of velocity. Propellers lose efficiency because of aerodynamic shock waves which form when the rotation of their blades approaches the speed of sound. The same holds for the compressor fans in turbojet engines, although these can still be used at supersonic flight velocities because the airflow entering the engine can be slowed down to subsonic speed by the air intakes (but this does cost energy at the detriment of the plane's velocity, and becomes increasingly problematic at higher supersonic flight speeds).

Rocket engines are also intrinsically less complex than jet engines. This is not to say that rocket motors cannot reach very high levels of complexity (just take a look at a schematic of the Space Shuttle Main Engine) but it is generally easier to build a simple rocket motor than it is to construct a simple jet engine. Basic solid propellant rocket motors are much simpler than any piston engine and propeller combination; an alternative history in which the Wright brothers power their aircraft using not a primitive piston engine but a few simple solid rockets is not completely unrealistic. After all, solid propellant rockets had already been in use for several centuries when the piston engine and – even later – the jet engine were invented.

Rocket engines running on liquid propellants are much more efficient in terms of the amount of energy that can be obtained out of a certain amount of propellant, but the requirement for pumps, valves and cooling makes them more complex than their solid propellant counterparts. Nevertheless, without large compressors and turbines, complex air intakes, supersonic shock problems and intake drag, they were, at least initially, a simpler solution for high-speed, high-altitude aircraft than turbojets. This is why so many of the high-speed, high-altitude aircraft developed during the 1940s and 1950s were propelled by rockets rather than jet engines. For the same thrust, a liquid propellant rocket engine is also much less heavy than a turbojet engine: a large modern rocket engine can produce a thrust that is 70 (the Space Shuttle Main Engine) or even 138 times (the Russian NK-33) as great as its own mass (although this number is much lower for smaller rocket engines), whereas for a turbojet the thrust to weight ratio approaches eight at best. For the same thrust, rocket engines are also considerably smaller than jet engines. However, a rocket engine quickly loses this advantage if account is taken of the weight and volume of the propellants that must be carried. Considering just the fuel flow, since the oxidizer comes from outside, the specific impulse of a jet engine is some 20 times that of a rocket engine, and is thus much more efficient. For a plane that needs to have a considerable range it soon becomes more economical to use an air-breathing jet engine rather than rocket propulsion. In aviation rocket engines have therefore been mostly used for early high-speed, short range aircraft such as experimental planes and high altitude military interceptors.

ORBITS

A rocket can propel a vehicle to high speeds and high altitudes, both of which are required in order to achieve orbit. An orbit represents a delicate balance between a vehicle's velocity and the Earth's gravity. Imagine that you toss a ball away at low speed. You will see it follow a curved trajectory and hit the floor some meters away. If you want it to go further, you will need to throw the ball a bit faster. It will still fall and accelerate towards the ground at the same rate as before, because the force of gravity remains the same. However, because its initial horizontal speed is higher, it will cover a larger horizontal distance before landing. Now imagine that you shoot

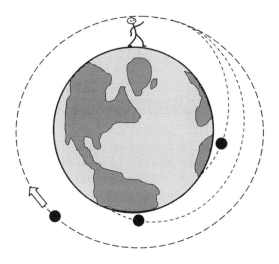

Orbit principle.

your ball away with a cannon: watch it fly, all the way over the horizon! Because the Earth is round, its surface drops away under the ball while it is falling. The result is that it takes the ball longer to reach the ground, and as a consequence it will manage to fly farther away than if we were living on a flat world. If you get the ball up to a velocity of about 8 km per second (5 miles per second), the curvature of the ball's trajectory under the pull of gravity is exactly the same as the curvature of the Earth. In effect, the ball is continuously falling around the world, without ever hitting the ground: it is in orbit! If you shoot the ball eastwards, it will circle around the planet and reappear from the west just under 1.5 hours later.

In reality, the atmosphere would slow the ball down so much that it would never come back, it would either burn up or fall short. However, at altitudes over 100 km (60 miles) there is hardly any atmosphere left to slow down a moving object. If you can get your ball up to speed there, it will be able to circle the Earth and become a satellite.

To launch a satellite, a rocket is initially fired straight up in order to rapidly climb above the thickest layers of the atmosphere and minimize aerodynamic drag. It then starts to pitch over, so that as the rocket keeps accelerating it increases both altitude and horizontal velocity. On the way up a conventional launcher drop stages to shed the dead weight of the empty tanks and engines that are no longer required. Without this 'staging' it would be too heavy to reach orbit. Most launch vehicles consist of two or three rocket stages, one on top of the other, plus sometimes two or more rocket boosters attached to the side of the first stage. Once the last stage reaches the proper orbital altitude and velocity, the satellite is released. The last rocket stage usually stays in orbit as well, but the other stages splash into the ocean, crash on land, or burn up in the atmosphere when falling at high velocity. In principle it would be possible to design these stages to be recovered and reused, but that would add an enormous amount of complexity and, most importantly, weight. A

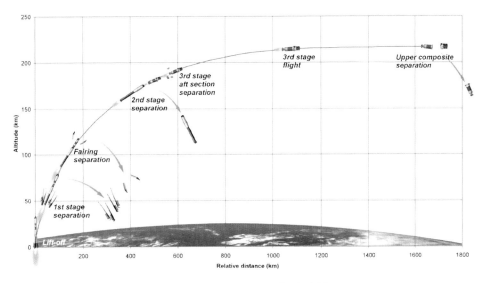

Soyuz rocket flight profile [ESA].

recoverable upper stage would need a heat shield to protect during its re-entry into the atmosphere, parachutes, and probably also airbags to cushion the impact on land or more likely at sea. Such a stage could easily become so heavy that it would not be of any use in a launcher. In addition, the recovery and refurbishment would be very expensive. All currently used satellite launch vehicles (except for the Space Shuttle, until recently) are therefore expendable, meaning they can only used once because everything except the satellite payload is discarded during the short flight up. As we will see later, in the case of the Space Shuttle the two Solid Rocket Boosters and the Orbiter itself were reused, but the large External Tank which held most of the liquid propellant for the Orbiter on the way up was still discarded. In operational terms, it would be best to use a fully reusable vehicle requiring only a single stage to get into orbit, just like you would not want to have a commercial airliner dropping off tanks and engines on the way to its destination. Expendable rocket stages are expensive to build, but usually it is more cost effective than providing a soft-landing capability and then retrieving, refurbishing and reintegrating reusable rocket stages. A Single Stage To Orbit (SSTO) vehicle must carry everything all the way up, including the large rocket engines and propellant tanks which will be used only for the ascent. It must also carry into orbit all the propellant, heat shielding, parachutes, wings and so on, that it will need to return to Earth. All the extra weight that an SSTO has to take into space diminishes the amount of cargo, or payload, it can transport, which is the actual reason for the launch. Each 100 kg of empty tankage that an SSTO takes into orbit is at the cost of 100 kg of payload. Since it takes about 30 kg of propellant and rocket hardware to place 1 kg of payload into a low orbit, and the payload typically represents only 3.5% of the total weight that leaves the launch pad, the design of an SSTO can easily result in a vehicle with a zero payload capability if the tank mass is a bit higher than expected or the rocket engine is slightly less performing. Such a launcher would only be able to put itself into orbit, if at all.

LIFT AND DRAG

Flying is all about balancing forces: thrust and drag in the horizontal direction, and lift and gravity in the vertical direction. The engines generate the thrust required to pull or push a plane through the air. For a steady speed, this thrust must be equal to the aerodynamic drag on the airplane. If the thrust is lower than the drag, the plane will slow down. If it is higher, the plane will accelerate. However, with increasing speed the drag will also increase, so at a certain moment the drag on the plane will again be equal the thrust. When that happens, the vehicle will continue to fly at a constant speed that is higher than it was at the lower thrust level.

In the vertical direction, the aerodynamic lift generated by the wings must be in balance with the force of gravity pulling the plane down. If the lift is too small, the airplane will descend; if the lift exceeds the plane's weight, it will gain altitude. To maintain a constant altitude, the lift must precisely balance the weight of the plane. When thrust equals drag, and lift equals weight constant velocity, straight and level flight is possible; but if any of the forces changes, the balance will be lost and the airplane will go up or down, accelerate or decelerate. These changes often occur in combination: for example in a dive, a plane will lose altitude and at the same time speed up.

Aerodynamic drag is a familiar thing: you feel it when you walk against the wind or when you stick you hand out of a car window. The amount of drag depends on the speed of the air (wind) or the speed of the car in the second example. Whether the air moves to you, or you move through the air doesn't matter: what is important is your relative velocity with respect to the air. Aerodynamic drag increases as a function of the square of the speed, so if you go twice as fast, the drag will increase by a factor of four. If you double your speed again, the drag will become sixteen times what it was originally. You can see where this goes: the drag increases at a much higher rate than your speed, so the higher your velocity, the harder it will be to go even faster.

Drag also depends on the size and shape of an object moving through the air. For similar shapes, an object with a large frontal surface will experience more drag than one with a smaller surface: a small hand out of a car window will feel less drag than a big hand, and a truck will suffer more drag than a small car. Aerodynamic drag on a vehicle can be decreased by using a good shape: the easier the air can flow around an object, the lower the drag will be. This is why sleek aircraft and racing cars have pointy noses. Lower drag means it takes less thrust to attain a certain speed, or that you can reach a higher velocity with the same thrust. Minimizing aerodynamic drag has therefore always been one of the driving issues in airplane design, and has led to continuous improvements in the shape of fuselages, the use of undercarriages which retract and the elimination of high-drag beams and cables.

The aerodynamic force that holds an airplane up is the lift is created by its wings.

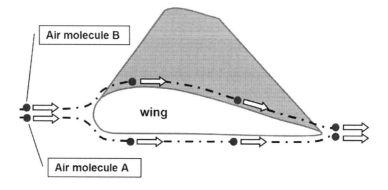

Lift principle 1.

The physics behind lift are complicated. I have a book written in 1909 called *Flying, The Why & Wherefore*, in which several theories on lift generation are presented; six years after the invention of the airplane, at a time when many types of planes were flying around, it was still not really understood what made those machines stay in the air! Even more surprising is that those same theories are all still employed to explain how lift is generated, even today, after over a century of flying experience. First there is the 'Longer Path' theory, which is also known as the 'Bernoulli' or 'Equal Transit Time' explanation. This theory is based on the assumption that air molecules that reach the leading edge of a wing at the same time, then flow either over or under it but all reach the trailing edge at the same moment. As the top surface of a typical asymmetrical wing is more curved than the underside, air molecules going above the wing have to travel a longer distance than those that pass under the wing, doing so in the same amount of time. Air flowing over the wing must therefore travel faster than that under the wing. Bernoulli's equation, a fundamental of fluid dynamics, states that as the speed of the air increases, its pressure decreases (the air molecules have less time to exert pressure on the surface). Hence the faster moving air on the top surface of the wing develops a low pressure area, while the slower moving air maintains a higher pressure on the underside. The low pressure essentially 'sucks' the wing upward (or the high pressure pushes it up, depending on your point of view). The weakness of this theory lies in its assumption that two molecules that become separated by the leading edge of the wing rejoin at the trailing edge at exactly the same moment. Even though you can measure that the air on top of a curved wing does indeed travel faster than under the wing, there is no fundamental reason why the molecules ought to meet again at the trailing edge and reach it at exactly the same time.

Another way to explain why wings generate lift is the 'Newtonian' explanation, based on the 'action equals reaction' idea which is also the working principle of the rocket engine. Air molecules hitting the bottom surface of an inclined wing bounce off and are deflected downward. In reaction, the wing is not only pushed up (lift) but also backwards (drag). You can ascertain that this is true by sticking your hand horizontally out of the window of a moving car, and slowly rotating it vertically. The

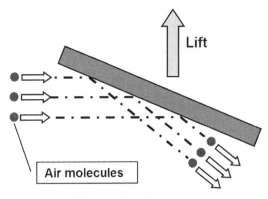

Lift principle 2.

greater the angle with respect to the airflow, the stronger will your hand be pushed up and in the direction of the airflow. This idea explains why airplanes with symmetrical airfoils (symmetrical wing cross sections) or even flat wings (such as those of paper airplanes) can fly, but it does not explain why an asymmetric airfoil with a strongly curved upper surface provides more lift. In fact, the Newtonian explanation leaves the top of the wing completely out of the picture. We also now know that molecules in the dense lower atmosphere, in which aircraft normally fly, do not act as individual particles, they actually interact and influence each other in complex ways. Nevertheless, air is indeed deflected downwards by an angled plate. The Newtonian explanation also correctly predicts that if you increase the inclination of the wing with respect to the airflow (its 'angle of attack') it will provide more lift but also experience more drag.

What happens in reality is a combination of these explanations, plus some more complex fluid dynamics. Air approaching the top surface of a wing is compressed into the air above it as it moves upward near the leading edge. The top surface then curves downward and away from the airflow, creating a low-pressure area that pulls the air above down toward the back of the wing. Simultaneously air approaching the bottom surface of the wing at the leading edge is slowed, compressed and directed downward. When this air nears the rear of the wing, its speed and pressure gradually match that of the air coming over the top. When you sum up all the pressures acting on the top and bottom of the wing, you end up with a net force that pushes the wing upward. However this force is not aimed straight up, it has a component in the backward direction. This is the aerodynamic drag described before. If the angle of attack is increased, the pressure differences between the bottom and top of the wing become larger, resulting in more lift as well as increased drag. There is of course a limit to how steep the angle of attack can be made, because beyond a certain angle the airflow over the wing is no longer able to nicely follow its curved contour; it no longer 'sticks' to the upper surface. This detached airflow creates a large turbulent wake that dramatically decreases the lift while increasing the drag. This is called a stall, and if the plane does not have enough engine thrust to compensate for the loss of lift it will fall out of the sky like a leaf falling from a tree. Normally it will accelerate going down, building sufficient speed to regain lift and control. However, if the plane is turning while falling it can enter what is called a spin. If the aircraft is forgiving and/or the pilot is lucky, this spin will be a normal one in which the nose points somewhat downwards and the corkscrewing descent provides sufficient control to achieve a recovery. The aircraft may be upside down, which is called an inverted spin. Much more serious is a flat spin, where the plane is falling straight down in a horizontal orientation and rotating on an axis perpendicular to its wings. In that case the airflow around the wings and tail is completely useless, and recovery often impossible.

The fact that the upper surface of an asymmetrical airfoil is curved means the pressure effects on the upper wing surface are more pronounced than those on the bottom of the wing. The upper part of the wing therefore contributes most to the generation of lift, which is one reason why most airplanes that use wing-mounted engines have them hanging under the wings rather than attached on the top of the wings: disturbing the air beneath a wing is less detrimental to flight than disturbing the air above a wing. This is also why the aforementioned 'liquid-air' rocket plane designed by Augustus Post in 1928 had air being expanded over the wing to create additional lift at high altitudes.

The wings of an airplane are optimized for a certain use in terms of speed and altitude. Airplanes that need a lot of lift at relatively low speeds have wings with strongly curved upper surfaces. This makes it possible to take off and land at low speeds and with heavy loads; very useful for military cargo carriers that have to be able to use short, improvised runways. The downside of such wings is that along with the powerful lift they also generate a lot of drag, which makes it difficult and uneconomical to fly at high speeds. If you want to fly really fast, you require small, thin wings designed to minimize drag while still generating sufficient lift to remain airborne at high speed. But such wings do not provide much lift at low speeds, and consequently planes that have such wings have higher take off and landing speeds and require long, smooth runways. If speed is of paramount importance, as it is for military fighter planes, you will go for the benefits of small, thin wings and accept the disadvantages that go with their use.

Lift not only depends on speed, but also on the density of the air. High up in the atmosphere the density of the air is much lower than at sea level, so that the lift that any given wing creates will diminish with increasing altitude (which is why aircraft have maximum operating altitudes). To be able to fly high, you must either employ larger wings or you must go very fast so that your wings generate more lift (as lift, just like drag, is a function of the square of the velocity of the air). But flying very fast increases aerodynamic drag and so requires large, powerful engines. Using big, slender wings is more economical in terms of engine power and fuel consumption. An example of the long-wing solution to reach extreme altitudes is the famous U2 spy plane, the modern version of which can reach an altitude of almost 26 km (16 miles) but has a maximum speed of only 800 km per hour (500 miles per hour). The SR-71 Blackbird spy plane can reach a similar altitude, but its short delta wings give it a top speed of no less than 3,530 km per hour (2,200 miles per hour). The penalty, of course, is that the SR-71 has a much more voracious fuel consumption.

U2 aircraft [US Air Force].

SR-71 aircraft [NASA].

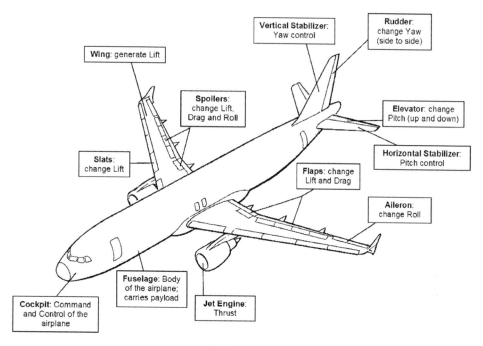

Airplane parts.

AEROPLANE ANATOMY

An airplane consists primarily of a fuselage, wings, stabilizers and engines. The fuselage is the body that connects all the other parts and holds the passengers, cargo and the pilots who control the vehicle from the cockpit. The required lift is provided by the wings. Besides these basic elements, much more is however needed to safely control an airplane.

Normal wings are designed to work optimally at the normal, cruise speed of the airplane. At take-off and landing, airplanes necessarily fly much slower, so ideally at those times you would rather have wings that give more lift at low speed. Airplanes are therefore often equipped with mechanical 'flaps' and 'slats' that can effectively change the shape of the wings. Flaps can be extended rearward and downward from the trailing edge to give the plane more lift at low speeds. Slats do a similar job, but on the front of the wing. In normal flight, when they are not needed, flaps and slats are retracted in order to minimize the aerodynamic drag of the wings. 'Spoilers' are door-like flaps on top of the wing. When moved up, they disturb (spoil) the airflow over the wing and thereby quickly diminish lift and increase drag. They are used to slow down and reduce altitude in landing, and also to assist with braking as well as to keep the plane firmly on the runway as it rolls after touchdown.

Like any object, an airplane tends to rotate around its center of mass. To steer an airplane, a pilot must control its movements around three axes that can be imagined

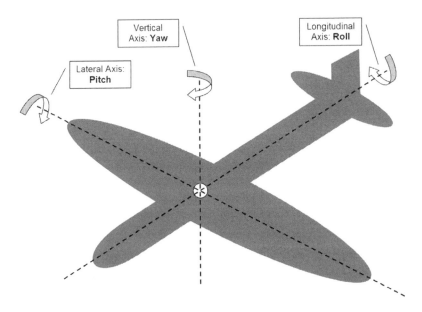

Rotational axes.

to radiate orthogonally from the center of mass, also known as the center of gravity. Rotation around the horizontal axis that runs from one wingtip to the other, making the nose go up or down, is called 'pitch'. The left-right movement of the nose, in other words the rotation around the vertical axis, is called 'yaw'. The third axis runs from the nose to the tail and is the line around which an airplane can 'roll' to make one wing rise and the other one drop. When driving your car you can only steer it to go left or right, but a pilot has two more rotation axes to take care of. In addition, your car can only go forward or backward, but a plane can also go up and down (although normally not backward). All this makes flying a plane much more complex than driving a ground vehicle.

Conventional airplanes have horizontal tail stabilizers, also called the tailplane. When normal wings are generating lift, they have the tendency to push the airplane's nose downwards, i.e. make the plane pitch down. To avoid this, the small horizontal stabilizers act as wings that provide a negative lift, pushing the tail down (and hence the nose up) and thereby counterbalance the pitching-down effect of the main wings. They are equipped with moveable flaps called 'elevators', which can increase or decrease the lift of the horizontal stabilizers and thereby make a plane pitch up or down. If a pilot pulls on his control stick, the elevators point up, pushing the tail down and therefore the airplane's nose up. This increases the angle of attack of the wings, increasing the lift and making the airplane gain altitude. Pushing the control stick forward has the opposite effect. By use of the elevators a pilot can control the altitude of the plane.

The vertical stabilizer on the tail gives stability in the horizontal direction, much like the keel of a boat. A moveable flap on its trailing edge, called the 'rudder', lets the pilot move the nose of the airplane left and right and thus provides yaw control. The rudder is operated with the pedals at the feet of the pilot: pushing the left foot forward moves the airplane's nose to the left and vice versa. Roll is controlled by 'ailerons' on the main wings. If the pilot pushes the stick to the right, the aileron on the left wing moves down and increases the lift of that wing. On the right wing the opposite happens. The result is the left wing goes up and the right wing goes down. From the pilot's point of view the airplane rolls clockwise. The spoilers can also be used to roll an airplane. Extending the spoilers on one of the wings will reduce the lift on that wing and make it drop, so that the plane will roll in the direction of the 'spoiled' wing.

Many planes also exhibit a so-called dihedral angle in their wings, which means the wings are canted slightly upward to form a weak V-shape. This helps to prevent unwanted roll, making the plane more stable. When a plane with wing dihedral rolls away from level flight, the lift force on its wings will no longer point straight up but somewhat to the side. As a result, the plane will sideslip, which means it is not only flying forward but also slightly sideways. Because of the wing dihedral the situation of the wings with respect to the airflow, which now comes slightly from the side, is asymmetrical. The upward tilted wing presents a less favorable angle to the airflow than the wing that is angled downwards, and hence produces less lift than the other wing. This unbalance in lift will automatically roll the plane back until both wings are again at the same angle to the horizon; in essence, the sideslip airflow 'pushes' the wings back to the horizontal level.

The undercarriage, normally fitted with wheels, enables a plane to move over the runway during take-off and landing, and to taxi around on the airport. At very low speeds the rudder on the tail of a jet or rocket plane cannot work, because of the low velocity of the air flowing over it. A rotating nose wheel can then be used to steer the airplane. The undercarriage creates a lot of aerodynamic drag during flight, so in most modern airplanes it is retracted when not needed. Although there still are airplanes that have fixed undercarriages with wheels, floats or skids, high-speed

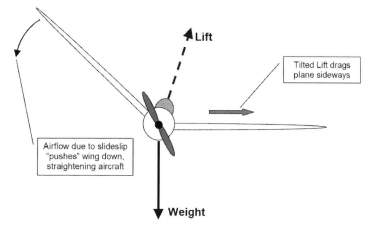

Dihedral wings.

airplanes must always get their undercarriage out of the way, either by retracting it into the wings and/or fuselage, or by dropping it altogether; the last option has the benefit that the plane does not have to drag the heavy undercarriage around in flight, but of course it still needs something to land on. The landing undercarriage can however be less robust and often smaller, because the weight of the plane at landing will be lower than at take-off by the amount of fuel consumed. This may be a good idea for experimental planes and spaceplanes with severe weight limitations, but it reduces the operational flexibility by requiring the airplane to be re-mated to the jettisonable undercarriage prior to each flight.

An important issue in the design of an airplane is whether it should be able to fly faster than the speed of sound. Sound has a speed limit that is easily noticeable: when you observe a flash of lightning in the distance the light reaches your eyes virtually instantaneously, but you may hear the thunder only several seconds later. Clearly, sound travels through the air at a speed that is much lower than that of light. In fact, every three seconds that elapse between lightning and thunder means that the event occurred about 1 km further away from you (or 1 mile every five seconds); if the flash and boom arrive at about the same time, you are in real danger. The speed of sound is defined as Mach 1, after the Czech/Austrian physicist Ernst Mach. The velocity of airplanes able to fly faster than the speed of sound is usually expressed in terms of Mach numbers, with Mach 2 meaning twice that speed. Flying significantly faster than the speed of sound is called supersonic. In contrast, subsonic flight means a speed of less than Mach 1. For example Mach 0.5 means half the speed of sound. Flying at precisely the speed of sound is called sonic flight. The Mach number not only depends on the actual speed of an airplane, but also on its altitude. The higher you go in the atmosphere, the lower the speed of sound because of the diminishing temperature of the air. At sea level and 20 degrees Celsius (68 degrees Fahrenheit), Mach 1 corresponds to a velocity of 340 meters per second (1,130 feet per second), or 1,240 km per hour (770 miles per hour). However, at 11 km (7 miles) altitude it means 1,060 km per hour (660 miles per hour). At altitudes from about 11 to 20 km (7 to 12 miles) the air temperature is constant, and so too is the speed of sound. But above this region up to about 50 km (30 miles) the air temperature and the speed of sound increase again. If it is mentioned somewhere that a plane is capable of flying at Mach 2, you will need to know at which altitude it can achieve this in order to be able to derive its real velocity.

The engines generate the thrust necessary to pull or push the plane through the air. For relatively slow aircraft like commercial airliners that do not exceed the speed of sound, the engines are often attached to the outside of the wings or the fuselage with struts. This makes it easier to maintain and replace the engines, and allows the use of different types of engines without the need to modify the rest of the aircraft. Aircraft that need to be more streamlined, for instance for supersonic flight or high speed at low altitudes, require their engines to be inside the fuselage or attached very closely to the wings, without struts. Typical examples are military fighter jets and the Concorde. Conventional airplanes which fly at relatively low speeds and low altitudes often use propellers, which, in simple terms, are spinning wings. The rotation provides the propeller blades with the necessary speed to generate lift in the horizontal direction. Helicopter blades operate on the same basic principle. Propellers work best in dense air, so are not suitable for high altitude flight. Another major disadvantage is that a propeller's performance drops quickly when the blade speed exceeds the speed of sound, because shock waves will form that dramatically increase the aerodynamic drag on the propeller whilst decreasing the thrust that it develops. As the speed of a propeller blade depends on both its rotational speed and the velocity of the aircraft, the blades will reach sonic speed long before the rest of the aircraft. Consequently, aircraft equipped with conventional propellers are only good for flight speeds up to about Mach 0.6. Propellers can be driven by combustion engines similar to those in cars, or by gas turbine engines.

Gas turbine engines (often simply called jet engines) all consist of the same basic components: an inlet for the air, a compressor, combustion chamber(s), a turbine and an exhaust. The compressor consists of rotating rows of fan blades that suck in air through the inlet and compresses it to increase the amount of oxygen available per liter of air. The high-pressure air then enters the combustion chamber(s), where it is mixed with fuel (typically kerosene) and burned. The resulting powerful stream of high-pressure exhaust gas then expands through (and therefore turns) another set of fans called the turbine. The turbine is connected to the compressor via a shaft, so the exhaust turns the turbines which turn the compressors to suck in and compress more air and thereby keep the engine running. In addition to the compressor, the shaft can be connected to a propeller, making the engine a so-called turboprop engine. Since the propeller needs to spin at a much lower speed than the compressor, it is linked to the shaft via a gearbox. Using the same principle a gas turbine engine, now referred to as a turboshaft engine, can drive the propellers of a ship, the wheels of a tank or the blades of a helicopter.

The momentum of the exhaust from a gas turbine engine can also be used more directly to provide thrust by Newton's 'action equals reaction', just like in a rocket engine. In such an engine, called a turbojet engine, it is the flow of gas that pushes

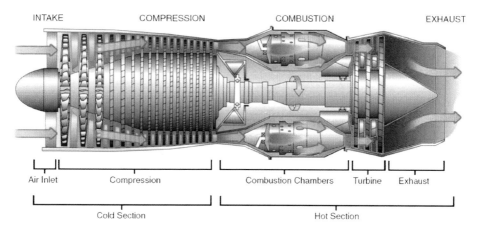

A turbojet engine [Federal Aviation Administration].

the aircraft through the air. This enables very high speeds and is therefore primarily used in military fighters. The principal difference between a turbojet engine and a rocket is that a turbojet uses atmospheric oxygen to burn with the fuel, rather than oxidizer drawn from a tank. The working principle is otherwise basically the same. The powerful thrust of a turbojet can be further increased by using an afterburner, a long tube that is installed behind the jet engine's exhaust, and into which additional fuel is injected and burned with the unused oxygen remaining in the hot gas coming out of the engine. As per the 'action equals reaction' principle, this provides a large boost in thrust. It is however a pretty inefficient means of propulsion and it greatly increases fuel consumption. Afterburners are typically used on fighter aircraft, and even then only for rapid acceleration during brief periods of time. Afterburners are however a great way to attract attention at air shows, because they provide lots of noise and long, dazzling exhaust plumes.

In the gas turbine engines that most modern jet aircraft employ, only part of the exhaust gas is used directly to provide thrust through a fast and hot exhaust jet. An important part of its energy is used to drive a large fan in front of the compressor. Instead of compressing air into the engine for combustion, this fan moves 'bypass air' around the actual engine, forming a cold jet that flows out at a lower velocity than the hot gas from the core. Rather than giving a small amount of air a very high velocity as occurs in the hot part of the engine, the fan gives a lot of air a relatively low velocity. Turbofan engines that have relatively large fans are typically used by commercial airliners, because they are ideal for the high but still subsonic speeds at which they fly. In addition, the low-speed air helps to cushion the noise of the hot and fast core exhaust, making the engine quieter than a pure turbojet. Modern

A Trent 1000 turbofan engine [Rolls Royce].

fighter planes also employ turbofan engines, but use smaller fans that are more optimal for supersonic flight.

Gas turbines configured as turboprop, turboshaft, turbofan or turbojet engines, deliver a lot of power compared to the weight of the engine. Their power-to-weight ratio is in fact much higher than for reciprocating engines such as those used in cars. In addition, gas turbine engines are relatively small for the power they provide. One disadvantage is that, because of the high rotation speeds of the compressors and the turbines, and the high temperatures employed, they are relatively complicated and expensive. Another limitation is that at flight speeds exceeding about Mach 3, the temperatures inside the engine can become so hot that the turbine blades melt and break apart.

How about removing the fragile turbine blades and fans? This is indeed feasible. At speeds over Mach 1, air is rammed into the engine at such high velocity that the shape of the inlet duct will satisfactorily compress the air. Such a 'ramjet' typically resembles a tube with a pointy core in the middle of the air intake; rather like a gas turbine engine without turbines and compressors. The shape of the tube and the core ensure that the incoming high-velocity air is squeezed into a small area and thereby compressed to a high enough density that it can be burned with the fuel (ramjets are therefore also known as 'stovepipe' jets). The good thing is that at around Mach 3, where gas turbine engines start to run into trouble, ramjets not only work well but actually perform more fuel-efficiently than turbojets. In addition, a ramjet is much less complex than a turbine engine, and therefore cheaper and more robust. These advantages make them ideal for use in long-range cruise missiles. A disadvantage is that a ramjet only produces thrust when it moves at high speeds, and only becomes reasonably efficient near supersonic speed; at low velocities the air does not rush into engine fast enough for proper compression and hence combustion. Cruise missiles are therefore usually accelerated up to high-subsonic or low-supersonic speeds using expendable rocket stages, whereupon the ramjet is engaged. But a manned high-speed plane used for reconnaissance or the interception of enemy planes needs to be able to switch from subsonic to high-supersonic velocities at any time. Expendable rocket engines that can only be used once are not very useful for that purpose. A reusable rocket engine or a gas turbine engine could be employed but at the penalty of extra weight. For the aforementioned SR-71 Blackbird, a solution was found by combining a gas turbine and a ramjet engine into a turboramjet. This hybrid engine essentially consists of a turbojet mounted inside a ramjet. For take-off and while climbing to altitude, flaps inside the SR-71 engines force the incoming air into the compressor of the turbojet part of the engine. Just short of Mach 1 the afterburners of both engines are ignited to accelerate the plane to supersonic speed. Then the bypass flaps are moved to block the flow into the turbojet and instead direct the air around the turbojet core and burn it with the fuel only in the afterburner part of the engine. At that moment the engine has been turned into a ramjet, with the air being compressed by the shock cones at the air inlets, without the need for fan compressors. This unique engine enabled the SR-71 to operate from zero speed to Mach 3+, and to fly at speeds between Mach 3 and 3.5 for long durations. As we will see later, the combination of different types of airbreathing engines with rocket engines may be the solution for future spaceplanes, something which Max Valier gave some thought to in the 1920s when he proposed designs for planes that would employ propellers in the lower atmosphere and rocket engines at extreme altitudes.

ROCKET + PLANE = ROCKET PLANE

Conventional planes depend a lot on the atmosphere: air is required for the wings to provide lift, to enable surfaces like rudders and ailerons to provide control, and as a supply of oxygen for the engines. However, flying through air also produces drag, which makes going very fast at low altitudes difficult. The higher the altitude, the lower the air density and thus the lower the drag at a certain speed. As a result, it is possible to fly faster at higher altitudes. But at the same time the lower air density limits the amount of available lift and thrust. Hence an aircraft's overall maximum speed is linked to an optimal altitude. Above that altitude the maximum attainable speed drops, because the engines are less efficient and the plane may require to fly with a higher angle of attack in order to produce enough lift, which increases drag. Furthermore, below the ideal altitude the greater air density will increase drag and hence also limit the plane's maximum velocity.

The use of rockets instead of airbreathing engines eliminates the reliance on the atmosphere for producing thrust. Without air obstructing the outflow of the exhaust jet, rockets can actually work more efficient in vacuum than within the atmosphere. With rocket engines, the maximum altitude of a plane is only limited by the lift it requires. And because lift is a function of the square of the velocity, rocket engines can accelerate the vehicle to such high speeds that even small wings can generate sufficient lift. Moreover, since rocket engines do not take in air from outside, they have no velocity constraints such as those that limit the speed at which airbreathing engines can be used. Unlike ramjets, rocket engines do not require a minimum flight velocity, and unlike gas turbine engines they aren't constrained by a maximum velocity beyond which the inrushing air heats up the engine so much that turbine blades are damaged. Rocket engines thus allow high speeds at high altitudes; in principle even virtually unlimited speeds at unlimited altitudes. Rockets can thus even be used to propel an aircraft into orbit, turning it into a spaceplane!

At high altitudes planes do experience control problems, however: because the efficiency of an aerodynamic control surface depends on the density of the air, it drops as the altitude increases. At very high altitudes and in the vacuum of space a control surface is completely ineffective, no matter how large it is. As a result, the vertical stabilizer and rudder can no longer ensure that the nose remains pointing in the direction of flight, and the elevators and ailerons can no longer control pitch and roll. Small misalignments in the rocket engine's thrust direction or weak residual aerodynamic forces on the wings and fuselage can cause a plane roll, pitch and yaw uncontrollably. If a plane falls in an uncontrolled orientation, maybe even sideways, the increasingly denser air could easily tear it apart. A solution is to equip the plane with reaction control thrusters: small rocket engines like those used on spacecraft for attitude control. When the aerodynamic control surfaces are rendered ineffective the thrusters can take over, pushing an aircraft around its axes to provide roll, pitch and yaw control.

In addition, the main rocket engine(s) can provide control during the powered phase of a flight by the use of gimbals. These enable a rocket engine to rotate in the horizontal and/or vertical direction in order to provide thrust vector control. When a rocket engine that is mounted in the rear of the fuselage is gimbaled up, its thrust will push the tail down. As the plane rotates around its center of mass, its nose will go up. Vertical gimballing of the engine thus provides pitch control. Likewise, rotating the nozzle in the horizontal direction provides yaw control. Roll control can be achieved if you have two rocket engines off-center from the rotational axis running from the nose to the tail of the plane, for instance one in each wing. In that case you can push one wing up and the other one down to make the plane roll. Rocket engine gimbals are pretty heavy however, and rocket planes are therefore normally controlled using aerodynamic control surfaces and small reaction control thrusters. The Space Shuttle Main Engines could gimbal to help to control the vehicle's attitude during the ascent through the atmosphere. When watching footage of a Space Shuttle launch, look for the test-gimballing of these three engines shortly prior to ignition. The Solid Rocket Boosters could also vector their thrust, but these massive rockets were discarded about two minutes into the flight.

Although rocket engines enable a plane to fly both higher and faster than any conventional airbreathing vehicle, a downside is of course that rocket planes, apart from fuel, need to carry oxidizer with them. This adds weight, requires additional tanks and takes up more internal volume. An added complication is that while most jet aircraft carry part of their fuel in their wings, this is generally not possible for a rocket plane. That is because rocket engine turbopumps require the propellant tanks to be pressurized to at least several times that of the atmosphere at sea level in order to ensure an efficient flow without cavitation (the formation of bubbles). Generally only tanks that are more or less cylindrical in shape can contain such pressure while remaining relatively light (like a balloon), but the resulting bulky shapes are impossible to fit into the thin wings of a fast plane. If you look at a cut-away drawing of for instance the X-1 or X-15 rocket planes, you will notice that indeed most of the available fuselage volume is taken up by propellant tanks.

The need to carry oxidizer means that the flight time of a rocket plane is a lot less than that of a similar jet-propelled airplane that can fill all its tanks with fuel and take its oxygen from the air. Rocket planes are therefore primarily useful for missions that require only relatively short-duration rocket-propelled boosts. Rocket aircraft usually return in an unpowered glide, what pilots call a 'deadstick landing', because the rocket propellant will typically have been spent earlier in the flight.

Many experimental rocket planes have been dropped from carrier planes. This saves the propellant that would have otherwise been needed to accelerate the rocket plane to take-off speed and to fly it to the planned test area. The function of a carrier plane is thus similar to that of the first stage of a conventional launch system. The rocket aircraft also has a shorter climb to attain its target altitude, because it begins its independent flight at the altitude of its carrier. Finally, at high altitude the air pressure at the rocket nozzle's exit is lower, enabling the exhaust gases to flow out more freely and expand further than at sea level. With a nozzle optimized for these conditions, the result is a higher specific impulse and a higher efficiency without any changes in the rocket engine itself. All this combined, saves considerable propellant weight. For example, the starting weight of the air-launched X-15 (about which you will read more later on) was almost 43% fuel and oxidizer. In other words, the propellant weight was equal to 'only' three quarters of the rocket plane's empty weight, but this was sufficient to enable the plane to climb to altitudes over 100 km (330,000 feet). A similar aircraft capable of doing that by taking off from the ground would have been about 70% of propellant: a propellant weight twice the vehicle's empty weight. The air launch enabled the X-15 to be a much smaller aircraft with lower structural weight constraints. Of course, the price paid is a large carrier aircraft, although for many applications an existing, slightly modified bomber suffices.

Apart from directly saving on propellant, dropping a rocket plane in the air means that it does not need a robust and heavy undercarriage that can handle the weight of the fully loaded plane prior to taking off. Wheels or skids will still be required to land the vehicle, but these can be small and light because they will only need to support a nearly empty plane, since half of the initial weight was propellant and that has been consumed. Indeed, the undercarriage and structure of the X-15 was only designed to carry the plane on the ground with empty tanks. On one mission in 1959 a small fire in the rocket engine forced pilot Scott Crossfield to make an emergency landing. He was unable to dump all of the propellant before he

An X-15 after hard landing [NASA].

touched down, so he landed at a tremendous speed and the heavy load snapped the vehicle in two! The Southern California Soaring Society awarded Crossfield the 'Order of the Streamlined Brick' for this flight, as it had set a record for the shortest descent time from 38,000 feet (11.6 km) to the ground as a glider. Fittingly, the trophy was a streamlined brick mounted on a piece of mahogany.

If, like the X-1 and the X-15, an air launched rocket plane is small enough it can be carried in the bomb bay or under a wing. When it is released it simply falls away from its mothership and can start its rocket engine once safely clear, so that there is little risk of a collision or damage to the carrier due to the rocket exhaust. A larger (space)plane will need to be put on top of its carrier, requiring it to have sufficiently large wings and initial thrust to quickly fly away without falling back down, yet not scorch its mothership with its super-hot exhaust. That this can be a risky procedure was demonstrated in 1966, when a unmanned experimental drone carried on top of an SR-71 suffered engine problems and struck the carrier's tail immediately after separation. Both planes were lost, and although the Blackbird's two crewmembers ejected and parachuted in the sea, one of them drowned.

The X-15 was what is called an ALHL: Air Launched & Horizontal Landing. There are a number of other possibilities, which aeronautical engineers designate with equally puzzling codes. Conventional aircraft are HTHL: Horizontal Take-off & Horizontal Landing. But if the thrust of the engines exceed its take-off weight, the plane could operate as a VTHL: Vertical Take-off & Horizontal Landing. At launch, the Space Shuttle System is a pure rocket and does not use its wings. Only on its way back to Earth does the Orbiter exploit its aerodynamics to fly and land like an airplane. One reason it takes off vertically is because it is attached to the huge External Tank and large Solid Rocket Boosters. It is hard to see how such a collection of bulky rocket stages and tanks could take off horizontally from a runway; imagine the enormous undercarriage required! Structurally, it is also much easier to design something that large and tall to be launched vertically. Just think of a long wooden pole: if you hold it vertically it will remain straight and you could put quite a load on top of it, but if you hold it horizontally its own weight may already be sufficient to bend or perhaps even break the pole.

A vertical launch also ensures that the Space Shuttle clears the denser part of the atmosphere as soon as possible, to limit aerodynamic drag and therefore the amount of propellant needed to achieve orbital speed: spending less time in the atmosphere and not requiring wings which create lift and also drag generally means that vertical take-off launchers need less propellant to get into orbit than a spaceplane taking off horizontally. Even if a rocket propelled spaceplane were to raise its nose sharply for a near-vertical climb immediately after takeoff, it would still use considerably more propellant than a vertical take-off vehicle because it must fire its rocket motor for a longer time due to its less efficient trajectory. But if the spaceplane combines jet engines with rocket propulsion, and spends a considerable time efficiently building up speed using its airbreathing propulsion within the atmosphere, this can more than compensate for the energy required to overcome aerodynamic drag. Moreover, a horizontally flying spaceplane can use its wings to stay in the air, while a vertically launched machine has only its thrust to keep it from falling back to Earth, and thus needs a more powerful engine. A winged launcher taking off horizontally can gently build up speed, while a vertical take-off vehicle needs to accelerate rapidly because otherwise the energy that it loses due to gravity will be too great (costing too much propellant to compensate). The result is that for a HTHL vehicle the acceleration during launch can be low, which is especially beneficial for passenger transport. Such accelerations are called 'G' forces. Standing on the ground you experience one G, resulting in your normal weight. In a fast accelerating sports car, the horizontal force with which you are pushed back into your chair may be 0.7 G, i.e. equivalent to 70% of normal gravity. A descending elevator initially causes a bit less than 1 G, making you feel lighter, but once the elevator achieves a constant velocity, only the normal 1 G gravity force remains. And while it is slowing down, you feel a bit more than 1 G. In a free fall you have no weight, i.e. 0 G. This is the situation in orbit, which is merely a continuous free fall around the Earth.

Another benefit of a HTHL is that it potentially has better abort characteristics: a horizontally flying plane that loses thrust can continue underpowered, even glide if necessary (provided it has sufficient speed). A vertically launched vehicle will simply fall out the sky unless it has a sufficient number of redundant engines ('engine-out' capability), but this imposes high weight and cost penalties. The sudden shut down of an engine in the first few seconds of a vertical launch is thus also likely to have catastrophic results. However a spaceplane starting horizontally may be able to stop before the end of the runway, or fly around on reduced thrust for an emergency landing in the same way as a normal multi-engined aircraft. The possibility to fly at less than full power also means that HTHL spaceplanes can be test flown, progressively increasing speed and altitude on successive missions. In contrast, the VTHL Space Shuttle's first powered flight had to be a full-blown orbital mission.

In contrast to vertical take-off vehicles, HTHL planes can, at least in principle, use existing runways and airports for take-off and landing (of which trillions of dollars worth of infrastructure already exists, spread all over the planet); launchers leaving the ground vertically need dedicated launch platforms and towers. On the other hand, horizontal take-off exposes a plane to failure modes which do not apply to vertically launched machines, such as collisions with obstacles or blown tires (for instance, out of the nineteen SR-71 Blackbirds that were lost in accidents, four were as a result of tire failures during take-off, and the only Concorde crash was caused by coming into contact with debris on the runway).

A Vertical Take-off & Vertical Landing (VTVL) vehicle is in principle possible, but generally does not make much sense for a winged plane that is optimized to fly horizontally. The Harrier 'jump jet' was equipped to take-off and land vertically in order to be able to operate from small clearings on a battlefield, but that capability adds a lot of complexity to the design of the plane and its engine (which uses four rotating nozzles), and was purely to satisfy the aircraft's military requirements. There are concepts for VTVL launch vehicles, but these do not have wings and can thus not be considered rocket planes.

If you have wings on your vehicle, it is generally best to exploit them as much as possible and thereby minimize complexity and the amount of thrust required. Rocket planes are therefore usually HTHL or ALHL vehicles. But HTVL does not really

An AV-8B Harrier lands vertically on an aircraft carrier [US Navy].

make much sense, since if you have the wings and sufficient runway to take off horizontally, it is hard to justify the need for a vertical landing when you get back.

So with this technical background in mind, let's now see how rocket planes have evolved since they first became a serious business shortly before the Second World War.

3

Germany's wonder weapons

"Science is one thing, wisdom is another. Science is an edged tool, with which men play like children, and cut their own fingers." – Sir Arthur Eddington (1882-1944)

GERMANY GETS SERIOUS

After the experiments of von Opel and Espenlaub with gliders equipped with simple powder rockets, work on rocket planes in Germany continued with the focus on more controllable motors using liquid propellants. Most spaceflight visionaries at the time believed that a reusable spaceplane would eventually be required to make launches into space routine and affordable. Wernher von Braun, the technical genius leading the development of the A-type rockets for the German Army at their Kummersdorf proving ground (the success of which ultimately led to the notorious A4/V2 missile), was no exception. But where others merely came up with ideas and published their hypothetical concepts, von Braun was actively trying to set up a practical program to evaluate an aircraft with a rocket motor propulsion system. The ultimate goal of all von Braun's efforts was space exploration, not the development of military missiles, but he recognized that in Germany in the 1930s the military was the only source of funding to develop rockets. He hoped later to use the technology for the exploration of outer space. To get the military, which was already funding his missiles, also to pay for rocket plane experiments he once again had to convince them of the novel technology's possibilities for making war.

The success of von Braun's rocket missile development program indeed managed to convince the Army High Command and the highest echelons of the Reichs-LuftfahrtMinisterium (RLM; Air Ministry) that a rocket propelled fast interceptor plane was feasible. In May 1935 Major Wolfram von Richthofen, in charge of developing and testing new aircraft for the German Air Force, the Luftwaffe, put forth a proposal to develop a rocket propelled interceptor for use against high flying bombers. He knew that the British were developing strategic bombing as a means of disabling enemy industry, and Richthofen (a fourth cousin of the First World War flying ace Manfred 'Red Baron' von Richthofen) proposed to defend German factories against this threat by equipping them with dedicated rocket propelled interceptors. Also, German airplane designer Ernst Heinkel, founder of the Heinkel airplane manufacturing company, decided to support von Braun. Heinkel was passionate about high-speed flight, and very interested in any form of aircraft propulsion which promised higher speeds than could be achieved using traditional piston engines with propellers. To get von Braun started, Heinkel sent Walter Künzel, one of his best engineers, to join the development team, and donated the wingless fuselage of a He 112 fighter aircraft for use in ground tests.

Work on the aircraft engine, which burned liquid oxygen and alcohol, began in Kummersdorf early in 1936. The development team came up with a system that was pressure fed, meaning that instead of using pumps the propellant was fed into the rocket motor by using a pressurized gas to force the fuel and oxidizer out of their tanks. This method would deliver less thrust than possible when using a turbopump, but in principle the engine's relative simplicity would make it easier and safer to operate and maintain. In the particular case of the engine that von Braun intended to use in the He 112 the pressure was created simply by letting the liquid oxygen propellant evaporate, eliminating the need for a separate pressurant gas and tank. To test the effects of acceleration on the propellant injection and combustion, a centrifuge was built consisting of an 8 meter long (26 feet) beam attached to an axis in its middle and with the rocket engine at one end. This allowed the engine to zoom around in circles of 4 meters (13 feet) radius, powered and accelerated by its own thrust. One day the engine didn't want to stop and the centrifuge brake failed, and the motor started to fly around out of control, faster and faster. The operator, who sat in the middle of the centrifuge, had to run for his life when the engine broke free. Much more than nowadays, rocket development was a dangerous business.

At the end of 1936, having gotten the engine to work as required, the engineers planned to install it in the He 112 fuselage. The tank with liquid oxygen was placed ahead of the cockpit and the tank of alcohol was placed behind the pilot. The engine sat in the tail of the airplane. Ground trials at Kummersdorf began in early 1937. The fuselage was secured to the ground with ropes and cables in order to prevent it from running off under the power of the engine, which had a maximum thrust of about 10,000 Newton. These experiments raised further interest in the RLM and later that year a secret rocket plane test program was established by Heinkel, von Braun and another rocket engine developer named Hellmuth Walter. The RLM also seconded to the program Erich Warsitz, one of their most experienced and technically proficient test pilots. Owing to the biography of Warsitz, *The First Jet Pilot* (written by his son Lutz), the details of his involvement in the rocket plane and jet plane programs of the late 1930s is now known. Warsitz did not know exactly what he had gotten himself into. He did know that it had to do with flying rocket planes and that his not being married was a major factor in his selection, so clearly it was going to be a dangerous job. Warsitz got some idea of how dangerous when he first arrived at Kummersdorf. He noticed a heap of torn and twisted, container-like metal objects. A mechanic told him they were combustion chambers that had violently exploded during previous tests, and that he, as the test pilot, might end up

A rocket-propelled He 112, probably the one with the Walter engine.

Ground testing a rocket engine in an He 112, probably the Walter engine.

among them if he were not careful! He then followed von Braun to the test stand where, kneeling on the wing root of the modified He 112, he watched von Braun confidently ignite the rocket engine from the cockpit. Warsitz was very impressed with the long exhaust flame and the ear-splitting noise that the engine produced. The force of the exhaust even managed to blow away several 1 cm thick metal plates that covered the ground some tens of meters behind the engine. Warsitz learned later on that the engine was normally ignited remotely from behind a thick concrete wall, and for safety reasons it had never before been started from the cockpit. Quite reasonably, von Braun and Künzel had been afraid that Warsitz would never get into the cockpit if he saw the engine being operated that way! Warsitz was impressed by von Braun's expertise and enthusiasm, and stuck with the hazardous project. However, tests did continue with the engine being ignited from the safety of the concrete wall's protection, and for good reason: one day and engine blew up during a demonstration for officials from the Army Weapons Office, totally destroying the He 112 fuselage. Fortunately Heinkel understood the risks inherent in testing revolutionary technologies and gave von Braun another He 112 fuselage to continue the experiments.

As an alternative to von Braun's liquid oxygen and alcohol engine, the RLM had also commissioned Hellmuth Walter's firm in Kiel to supply a rocket engine for the He 112. This engine ran on hydrogen peroxide which, under the influence of calcium permanganate as a catalyst, decomposed into hot oxygen gas and steam (a catalyst is a substance that facilitates a chemical reaction without being consumed by it; i.e. the calcium permanganate was itself not chemically affected by the decomposition of the hydrogen peroxide). Originally developed to turn turbine engines on submarines as "air independent propulsion" (contemporary submarines used diesel engines while on the surface and battery powered electric motors when submerged), the expanding gas could also be used directly to deliver rocket thrust. Experiments with a small Walter engine with a thrust of 1,500 Newton fitted to a Heinkel 72 propeller biplane in the autumn of 1936, and then a Focke Wulf 56 propeller plane with an engine of twice that power had been very successful (even the head of technical development for the Luftwaffe, Colonel Ernst Udet, had dared to make a flight in the latter aircraft). One benefit of the propellant combination used in Walter's rocket engine was that it did not require an igniter to get started; the fuel produced hot gases spontaneously upon contact with the calcium permanganate when they were simultaneously injected into the combustion chamber. There was thus less risk of the engine not starting, making it in principle simpler and more reliable. The combustion in the engine also occurred at about 480 degrees Celsius (890 degrees Fahrenheit), which was much lower than the 1,700 degrees Celsius (3,100 degrees Fahrenheit) of the rocket that von Braun's team was using for their project. The lower temperature reduced wear and tear of the engine. Another advantage that Walter's engine had over von Braun's rocket motor was that its propellants could be stored at normal temperatures, whereas liquid oxygen had to be chilled down to minus 183 degrees Celsius (minus 279 degrees Fahrenheit) because otherwise it would start to boil and rapidly evaporate. A missile or rocket plane using liquid oxygen could thus only be fueled shortly prior to launch, making it less suitable as a rapid response weapon.

An important disadvantage of the Walter engine was the low specific impulse inherent in decomposing hydrogen peroxide; von Braun's engine was much more efficient. But the main problem with Walter's engine was the hydrogen peroxide it consumed. In diluted form hydrogen peroxide (typically a 9% concentration) can be used to bleach hair, but the highly concentrated version used in the engine was able to make various materials ignite spontaneously and eat away human flesh. In 1934 von Braun had three members of his rocket development team killed by a hydrogen peroxide rocket engine explosion, so he was wary of it. Warsitz, who was to fly both types of engine gained firsthand experience of how dangerous Walter's engine was during a ground demonstration of a modified He 112 in Kiel. To show his confidence in the system in front of observers from the organization funding the program, the RLM, Warsitz planned to ignite the engine from the cockpit. Previous tests in which the engine was started remotely had gone well, but Walter's senior engineer Bartelsen nevertheless urged Warsitz not to operate the engine from the cockpit. It saved his life because the propulsion system blew up, spraying acid and metal fragments through the cockpit.

For the actual flight tests with modified He 112s (one equipped with von Braun's rocket engine and the other with Walter's engine) a suitable terrain had to be found. It had to provide enough room for emergency landings and be surrounded by open space so that crashes would not put populated areas at risk. A good location was found at the large Neuhardenberg reserve airfield, situated some 60 km (40 miles) east of Berlin. This site would enable flight experiments to be performed in secrecy, without locating the team far away from its important contacts in Berlin. However, because Neuhardenberg was only a backup field to be used in the event of war, it had no buildings or facilities. A number of tents were therefore erected to accommodate the technicians and their aircraft.

In addition to equipping He 112s with rocket engines, Heinkel and the RLM were interested in evaluating rocket engines as a way to get heavily laden bombers off the ground. Walter had designed an egg-shaped rocket pod that could be mounted on the He 111 medium bomber, one under each wing. Each pod would give the bomber an extra 3,000 Newton push for 30 seconds and enable it to get into the air with a load which would otherwise have prevented it from taking off within a reasonable ground run length. Flight tests with this system (what in the US would become known as a Rocket Assisted Take-Off or RATO system) begin early in the summer of 1937 with Warsitz at the controls. On his first flight he shows that the rockets work well and that the combined thrust of both rocket pods makes the plane climb very fast. Their exhaust jets disturb the airflow around the plane, but control remains manageable. He makes several further flights. On one of them a rocket pod comes loose just prior to take-off. Warsitz promptly stops the rocket engine and the plane. This demonstrates the advantage of a controllable liquid propellant rocket engine over a simpler solid propellant system that cannot be halted after ignition. Later he flies the He 111 with an overload of sand bags, concrete blocks and water tanks. The plane was too heavy to get off the ground unassisted, but with the RATO pods it manages to get airborne. He gets the plane rolling using the propeller engines alone, then ignites the boosters only after it has covered some 20 to 40 meters of runway. Because the boosters only work for 30 seconds, timing the ignition is crucial in order to gain the extra thrust at just the right time for take-off and the start of the climb. On another flight one of the rockets fails just when the plane gets airborne. The pod on the other wing, situated between the propeller engine and the wingtip, continues and the uncompensated leverage pushes the plane around in a rapid 180 degree turn. Warsitz is tempted to counteract the unbalanced thrust using the aircraft's rudder, but that would cause too much drag and make him lose altitude over the dense woodland that he is flying over. So he quickly extinguishes the other rocket engine as well. Now the plane has trouble staying in the air, lacking the speed and the thrust for a normal climb. With his wheels clipping the trees, Warsitz manages to stay airborne just long enough to make a safe landing. During a demonstration flight for RLM observers, Warsitz seeks to make a spectacular impression by taking off without any cargo, and with the thrust of both rocket pods the plane goes nearly straight up!

In the meantime work on von Braun's rocket engines are suffering problems. He too has developed rocket pods for the He 111 bomber. At 5,000 Newton these give even more thrust than the Walter rockets. However, they are not ready for flight tests. There are also problems with the rocket engine for the He 112: often the combustion chamber splits because of the high pressure inside. More ground tests are performed at the Neuhardenberg airfield. After some of these have been completed successfully, Warsitz presses to start the flight testing. He agrees to have one more standing test of the engine in the actual plane, and ignites it from the cockpit. The explosion not only destroys the engine, but rips the entire airplane fuselage apart! Warsitz is blown out of the cockpit and lands on the ground some 4 meters (13 feet) away, unharmed. It is another timely lesson that rocket propulsion is basically a controlled explosion, with the explosion easier to attain than the control.

Fortunately Heinkel agrees to give the team yet another He 112 to continue their tests; going through the official Luftwaffe channels to get a replacement might have ended the program prematurely because not all military officials are convinced there is any reason for rocket planes. On 3 June 1937 Erich Warsitz makes the first flight with a rocket powered He 112 fitted with von Braun's engine. The plane still has its standard propeller engine for take-off and landing; the rocket engine will be ignited once the plane is in the air. Planning of the flight is not easy, because von Braun's engine is pressurized by the natural boil off of some of its liquid oxygen propellant. Ten minutes after tanking, enough of the oxygen has evaporated to supply just the right amount of pressure in the tanks. If the engine is ignited too early, there will be insufficient pressure to force the propellants through the lines into the combustion chamber. But if the rocket is started late, the pressure may be so high that the engine explodes! Warsitz takes the He 112 up to 450 meters (1,500 feet) on propeller power alone, and flying at 300 km per hour (187 miles per hour) he waits until the pressure is just right and then he hits the ignition. Fortunately the engine starts correctly. The rocket gives a fixed thrust of some 3,000 Newton, quite modest for a plane with a weight of almost 2,000 kg (4,400 pounds), but sufficient for Warsitz to feel the kick. Within seconds the He 112 accelerates up to 400 km per hour (250 miles per hour). As Warsitz reported afterwards, at this point he noticed "a strong acrid odor of burning rubber and paint" and "clearly perceptible hot gases flowed under the pilot's seat". Because the gases irritate his eyes and hurt his lungs, he opens the canopy for ventilation and puts on his flight goggles to protect his eyes. Looking back, he sees flames in the fuselage! Unlike Walter's engine, von Braun's rocket cannot be turned off, and even although it has caught fire the plane continues to accelerate under the combined power of the propeller and the rocket. Warsitz cuts the propeller engine to slow down, and then simply waits for the rocket to consume its 30 seconds worth of propellant. Knowing the dangerous nature of the experimental propulsion unit on his plane he prepares to bail out and land by parachute but then realizes that his altitude has already fallen to about 200 meters (650 feet), which is too low to bail out. Side-slipping the plane to increase drag and hence lose altitude without increasing speed (which would happen if he simply pushed the nose down) he manages to get to the ground quickly. There is no time to deploy the wheels, so he belly-lands, scrambles out and runs for his life. The flames are quickly extinguished by the fire brigade, but the damage to the plane is significant.

Later the team finds out that the source of the accident are some ventilation slits that are fitted the wrong way around: instead of releasing the gases that build up due to the usual leaks in the rocket's propellant supply system, the slits were sucking the gases as well as jet exhaust forward into the cockpit. Analysis of the engine also shows that the combustion chamber has cracked. While the flight was not a complete success, it proves that rocket thrust from the tail of a plane can work and doesn't, as some critics had expected, make the aircraft flip over.

Warsitz continues flight tests with the other He 112 equipped with the fixed-thrust 3,000 Newton Walter engine, which performs several flights without blowing up. To protect himself against the dangerous acid fuel in case of a leak, Warsitz wears white clothing of a specially developed type of plastic; even his shoes and necktie are made of it. His normal clothes would act as a catalyst for the hydrogen peroxide, dissolving and burning when coming in contact with the angry substance.

The tests at Neuhardenberg using the Walter engine in the He 112 are completed satisfactory at the end of 1937 and the marquees are dismantled. But flights with the He 112 and the Walter engine are continued at the Luftwaffe's section of the secret Peenemünde center: Peenemünde West, on the other side from Peenemünde East where von Braun is developing the A4/V2 rocket for the Army. Eventually Warsitz dares to take off powered by both the propeller and the rocket engine: it makes the plane leap almost vertically into the sky. After this he starts it under rocket power alone, with the piston engine running in neutral and the propeller disengaged. This proves to be rather difficult because the rocket's thrust is never exactly in line with the central axis of the plane, making it veer to the side, while the rudder is not very effective at low speeds without the propeller blowing air over it. While accelerating for take-off he therefore has to steer using the wheel brakes, which costs much energy and speed. A better solution is found in adding a rudder just behind the nozzle, to deflect the rocket's exhaust jet and thereby help to steer the plane.

Even although von Braun's engines are in principle better performing, Walter's simpler hydrogen peroxide engines prove to be operationally more interesting for a rocket fighter plane: they are more reliable and do not depend on extremely cold liquid oxygen which is difficult to store and cannot be kept inside a plane for very long without a special cooling system. Interservice rivalry may also have played a role in the preference for a rocket plane with a Walter engine: the engines that von Braun was developing were primarily for use in the Army's A4/V2 rocket, and the Luftwaffe may simply have wanted an independent rocket propulsion system.

VON BRAUN'S VERTICAL TAKE-OFF ROCKET INTERCEPTOR

Von Braun became fully occupied with the development of the A4/V2 missile at the Army area of the Peenemünde center, but retained his interest in rocket planes. As with the A4, he again tried to sell impractical designs to those who didn't need them: in July 1939 he proposed to develop a rocket powered interceptor for the RLM. The first design had a cigar-shaped fuselage and straight, tapered wings. His trademark propellant combination of alcohol and liquid oxygen was stored in tanks behind the cockpit. The rocket engine was installed in the tail, and just as with his A3 and A4 rockets he placed four rudder-like jet vanes behind the nozzle to divert the exhaust jet and steer the vehicle by thrust vector control. Tilting the two opposing horizontal vanes up or down in the same direction would make the plane pitch, while the two vertical vanes would control the yaw. Rolling could be achieved by tilting opposite vanes in different directions. The pilot was to be seated in a pressurized cockpit that would be able to maintain a comfortable air pressure at high altitudes, and he would be protected from enemy bullets by armor plating. The vehicle would be armed with either two or four cannon mounted in the wing roots.

Whilst the plane was to land normally, von Braun designed his interceptor to take off vertically. That way the rocket plane could be launched straight up to the target and reach it in minimal time. A simple undercarriage would only require to be able to handle the empty weight of the vehicle at the end of the flight. The airplane was basically a rocket with wings. Von Braun envisaged large numbers of his planes would be stored vertically in a hangar/launch facility, hanging on the tips of their wings on two rails. When the air raid alarm rang, pilots would quickly board their interceptors via a removable bridge, then the plane and pilot would be rolled out of the building and launched straight off the rails. For the first minute or so, the plane would be remotely controlled from the ground and steered to the target by the help of radar (as was done with conventional air defense fighters). Then the pilot was to take manual control, switch off the main engine and start a smaller rocket motor that would enable the plane to engage enemy aircraft at sufficient speed while using its remaining propellant at a much lower rate. Spewing out rocket planes like a giant candy machine, a strategically placed launch facility would thus be able to quickly swarm enemy bomber formations with heavily armed interceptors. After

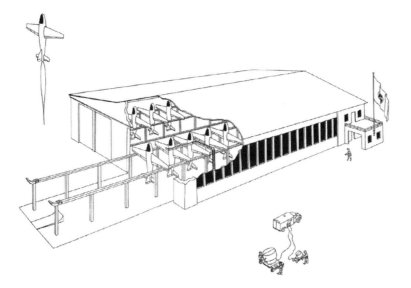

Design for von Braun's rocket interceptor launch facility.

completing his attack, the pilot was to glide his plane back to land on a grass field using a built-in skid.

However, the RLM considered von Braun's concept too impractical owing to the need for liquid oxygen, which was difficult to produce and store, and the specialized launch facilities that had to be constructed and maintained. Such facilities could also be easily identified by the enemy and destroyed by precision bombing. Several years later this fear was shown to be well founded when the elaborate bunkers constructed at the French coast to launch V1 and V2 missiles against England were destroyed by bombers, often before they became operational. The reluctance to use liquid oxygen in an operational military rocket system was also valid, as shown near the end of the war by the difficulties experienced in providing the mobile V2 launch systems with this propellant because of the bombing of the production facilities and transportation networks. Another reason that the RLM did not buy von Braun's proposal was that Germany expected to quickly win the upcoming war using its existing conventional weapons; in 1939 the prospect of large enemy bomber formations venturing far into Germany was not considered to be realistic. Unlike the other objections, however, this particular evaluation would soon be proven incorrect.

Von Braun reacted to the objections by producing a second version of his Vertical Take-Off (VTO) interceptor design. He switched to Visol and SV-Stoff as the rocket propellants because these are easier to store for lengthy periods and are hypergolic, meaning that they automatically ignite upon contact and thus do not need a separate ignition system, as does a hydrogen/oxygen rocket motor. SV-Stoff was mostly nitric acid, which is a very nasty substance; not something a pilot should feel comfortable sitting close to, and especially not in a combat aircraft whose tanks are quite likely to be punctured by enemy bullets, but, as we shall see, this was not a major concern in German rocket powered fighter design. Otherwise the new VTO plane was similar to its predecessor, with the vertical tail being a bit smaller and the wings now dihedral for improved flight stability.

Von Braun tackled the RLM's objections to the need for large ground facilities by proposing to launch his updated design from a mobile system based on a truck which hauled a trailer. These would first be used to transport the plane to wherever it would be needed. Once at the launch location, the truck and trailer would each be outfitted with a sort of tower structure and placed one wingspan apart from each other. A crane would hoist the rocket plane vertically between the two, and rest each of its wingtips on one of the support towers. A small flame deflector would be positioned beneath the rocket nozzle to avoid it burning up the ground or damaging the nearby equipment. In spite of the updated design, von Braun's VTO interceptor project was rejected by the RLM in 1941 because at that time the war was progressing well for Germany, with its forces continuously on the attack. Expecting the offensive war to finish soon, they saw no need for an interceptor which, because of its very limited range, was only suitable for local defense against intruding enemy planes that were in any case never expected to reach Germany in large numbers.

Undaunted, Von Braun retained his interest in rocket planes, and near the end of the war did launch two A4 rockets fitted with large swept-back wings. The military rationale was to develop a 'boost-glide' missile capable of reaching London when

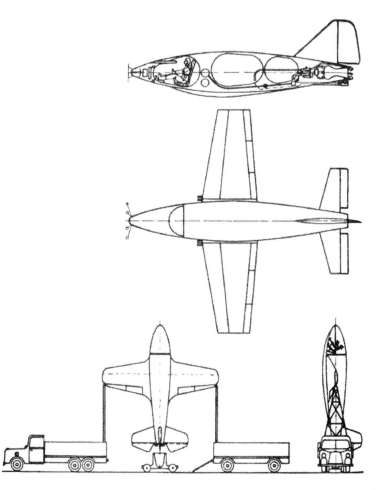

Original drawing showing the launch configuration of von Braun's updated rocket interceptor.

launched from inland, because at that time Germany was rapidly losing the coastal territory from which it had been launching its A4/V2 rockets. On 8 January 1945 a winged A4b left launch complex P7 at Peenemünde but failed in flight. The second attempt on the 24th was more successful: it reached an altitude of 80 km (260,000 feet) and then briefly performed a supersonic glide using its two swept-back wings until one of them broke off. The increasingly chaotic situation in Germany near the end of the war prevented any further flight tests.

The A4b launches were part of a plan to develop an A9/A10 two-stage rocket to attack the United States. This intercontinental ballistic missile was to have a winged, piloted upper stage (resembling the later X-15) to undertake an extended glide phase and accurate aiming. Once the A10 booster was jettisoned, the pilot/astronaut would steer the A9 to its target with the aid of radio positioning guidance from a network of U-boats along the flight path across the Atlantic. Once confirmed to be on course, the pilot was to use his ejection seat and land by parachute near an awaiting submarine if he was lucky.

Furthermore, Von Braun was planning the A6, which was basically a winged A4 with a pressurized cockpit instead of a warhead, plus landing gear and an auxiliary ramjet engine for continuing flight at extreme speed and altitude after the propellant for the main rocket engine was consumed. It would be launched vertically but land horizontally after gliding down to an airfield. To get funding for developing the A6, which von Braun saw as precursor to a real spaceplane, he offered it to the German military as a photographic reconnaissance aircraft. With an expected top speed of 2,900 km per hour (1,800 miles per hour) and a maximum altitude of 95 km (310,000 feet) he reckoned it would be impossible to intercept. But the Army did not see any urgent need for such an advanced, complicated and expensive machine, and it was rejected.

Von Braun's original concept for a vertically launched interceptor was also kept alive by Erich Bachem, at that time technical manager of the Fieseler aircraft plant. He proposed two designs for a Fieseler VTO rocket aircraft named the Fi 166-I high-altitude fighter. It initially involved a modified Messerschmitt Bf 109 from which the propeller and piston engine would be removed and replaced by an aerodynamic nose cover. It was to be launched with its aft belly affixed to a rocket stage with the same 250,000 Newton engine as von Braun's large A4 rocket, then under development at Peenemünde. Some sources say the engine of the smaller A5 rocket was to be used, but its 15,000 Newton thrust would not have been capable of lifting the engineless, empty Bf 109 of about 1,500 kg (3,300 pounds) together with a loaded rocket stage. At about 12 km (39,400 ft) the spent rocket would be discarded and parachute back down to be recovered and reused, while the engineless plane would attack enemy bombers during a gliding descent. A modification of this initial concept replaced the Bf 109 with a new, Bachem-designed aircraft which had two Jumo 004 jet engines installed beneath its wings to give the plane an extended flight capability. The RLM deemed the idea impractical. Undeterred, Bachem drafted a plan for a Fi 166-II. He deleted the rocket stage and designed the new two-seat aircraft (which looked very similar to Von Braun's VTO interceptor but was considerably larger) for a vertical take-off under its own rocket power. As before, the RLM was not convinced of the feasibility and the necessity for such a weapon. When the situation changed later in the war, Bachem revived the idea and developed the much smaller BP-349 'Natter' (discussed later in this chapter). It is also interesting that the concept of launching a plane vertically on top of a large liquid propellant rocket stage would much later be revisited many times for launching winged vehicles into orbit, and is of course the basic concept behind the Space Shuttle.

HEINKEL'S HE 176, THE FIRST REAL ROCKET PLANE

Whilst conducting the He 112 rocket plane tests at Neuhardenberg, Heinkel and the RLM decide to continue the development of a rocket plane interceptor. A secret

The He 176 with fixed nose wheel for taxi tests.

department at the Heinkel factory at Rostock-Marienehe is established to pursue this work. Whereas the rocket propelled He 112s were modifications of an existing type of airplane, the new machine is developed from the start as a true rocket plane. It is called the Heinkel He 176. Until recently it was unclear what the original prototype, the He 176 V1, looked like, and many books and websites include drawings of the proposed operational successor rather than the actual flying prototype (the fact that this improved version was also designated He 176 obviously caused the confusion). A recently discovered picture of the He 176 V1 indicates a configuration optimized for high speed flights: a tiny plane with a bullet-shaped fuselage and extremely thin, razor sharp wings in order to minimize aerodynamic drag. The cockpit is completely enclosed within the fuselage, with a flush upper glazing that can be removed for the pilot to gain entry to the plane. The picture shows two retractable main wheels and a fixed nose wheel that was fitted only for the initial taxi tests; for flights the plane was to land using the two main wheels and its tail.

The He 176 V1 has a Walter HWK RI-203 engine that uses the decomposition of hydrogen peroxide, as with the Walter engine for the He 112, but it is more powerful because the propellant is pumped into the engine rather than being pushed in (with a lesser force) by compressed air. Its maximum thrust is about 6,000 Newton, double that of the engine of the He 112. A second version of the design, the He 176 V2, will use an even more powerful von Braun engine to achieve the objective of a speed of 1,000 km per hour (620 miles per hour). To break speed records, the high thrust of the engine will be combined with a very lightweight fuselage and wing structure. In order to minimize the size of the cockpit, it is tailored closely around Erich Warsitz, the designated test pilot. It is so cramped that he can't even bend his elbows, and the controls that are to be operated by a particular hand have to be put on the opposite sides of the cockpit! To increase the stability of the plane, the wings have a positive dihedral. The design is quite a step in technology. The propellant tanks for the 82% hydrogen peroxide, for instance, are integrated into the thin elliptical wings and thus require to be welded by using a new process. In order to be able to handle the high accelerations, and also to minimize the frontal area of the cockpit and thus air drag, the pilot adopts an unconventional reclined position. There is no canopy bubble, so the entire nose section is made of Plexiglas for the requisite visibility. At the high speeds the He 176 is to fly, even the smallest movement of the aerodynamic control surfaces will have a big effect (because the generated lift forces are a function of the square of the velocity of the air flow) so these surfaces are kept small. However, at take-off and landing the pilot will have to make large movements with the stick and rudder pedals to produce some steering effect from the small rudder, elevators and ailerons. The sensitivity of these controls had therefore to be adjusted by the pilot to achieve sufficient control at all speeds. With a wingspan of 5 meters (16 feet) and a fuselage length of 5.5 meters (18 feet), the He 176 is very small: it would fit inside a modest living room. Looking at the picture of the tiny plane, you have to admire the bravery of Warsitz for volunteering to fly something so experimental which had such a dangerous engine in such a small package.

If anything were to go wrong in flight, even bailing out was going to be a novel experience. Jumping out of the cockpit in the traditional manner was expected to be extremely difficult at high speeds, if not impossible because the force of the air drag would be strong enough to rip the pilot's head off. Therefore the whole cockpit and nose formed a separate section that could be ejected from the rest of the plane by compressed air. A braking parachute would then slow it down sufficiently to enable the pilot to get out and land using his own parachute. Wooden mockups of the nose section with a dummy pilot inside (with weight distribution and body measurements reflecting those of Warsitz) were dropped from an He 111 bomber, and established that Warsitz would probably survive a parachute landing inside the cockpit if he did not manage to get out, and even without serious injury if he were lucky enough to set down on soft soil.

The first tests are performed by placing the actual prototype inside the huge wind tunnel of the Göttingen Test Institute. Once complete, the prototype is moved to the Luftwaffe area of Peenemünde which offers more secrecy than the Heinkel factory. Taxi trials in which the He 176 prototype is towed at speeds up to 155 km per hour (96 miles per hour) behind a 7.6 liter Mercedes car prove to be pretty useless, as the velocity is too low for the small rudder to become effective. Taxi runs are therefore continued on the plane's own rocket thrust, but all too often the wings hit the ground on the uneven grass airfield. Metal bumpers are therefore installed on the wingtips to prevent them from being damaged; something that can also be seen on the picture of the nose-wheeled He 176 V1. The tests show that the rudder only starts to be useful near the He 176's take-off speed, making it necessary to steer using the wheel brakes for most of the take-off run. As this costs a lot of energy and makes it difficult for the plane to get up to the desired speed, a rudder is installed in the engine's nozzle (this solution had also been implemented in the He 112 fitted with the Walter engine, for the same reason).

The first short rocket propelled hops into the air are made in March 1939 with very limited amounts of propellant in the aircraft for safety reasons. Over a hundred of these test runs are performed, with the plane getting no higher than 20 meters (70 feet) over a distance of about 100 meters (330 feet). Modifications are made to the plane. New concrete runways are built. In May 1939 a demo-hop is made for RLM officials including Ernst Udet, head of technical development for the Luftwaffe, and Erhard Milch, head of the RLM. The demonstration does not have the effect the He 176 team expects. Quite the opposite: Udet deems the plane to be too unstable, too small and too dangerous, and he grounds it! Nevertheless Warsitz talks the visitors into allowing the team to conduct the test flights. On 20 June the team prepares the He 176 for its real maiden flight. To prevent anyone from blocking the attempt, and to limit the repercussions of a failure, no officials are invited or notified – not even Heinkel; Warsitz assumes full responsibility for the historic flight. After take-off he quickly achieves a speed of 750 km per hour (470 miles per hour), then he makes a steep ascent and continues to fly a circuit at 800 km per hour (500 miles per hour): faster than any previous plane. After the 1 minute's worth of propellant is consumed, he glides back and makes a safe landing. Apart from the expected sensitivity to the controls, the He 176 proves to be a fine flying machine. News about the successful flight quickly gets out and the next day Warsitz performs a demonstration flight for Heinkel, Udet and miscellaneous other officials. On 3 July even Adolf Hitler and Hermann Göring, chief of the Luftwaffe, watch in amazement as it flies at a special air show of new Luftwaffe planes at Roggentin airfield. Coming in to land, Warsitz shuts off the engine too soon and almost flies into a brick wall; a last-second restart of the engine makes the plane suddenly rise some 50 meters and hop over the wall prior to landing safely. Most of the spectators think this spectacular maneuver is a part of the demonstration. At the same show, an impressive demonstration is made of the Walter take-off assist rocket pod with a pair of He 111 bombers, one with two 4,900 Newton thrust Walter RI-200 rockets and the other without. After starting at the same moment, by the time the standard He 111 leaves the ground the assisted one is already boosted to 200 meters (660 feet) altitude by the powerful Walter engines! The Walter rocket pods, which are dropped after burn out, deploy parachutes and are recovered for reuse, soon become standard equipment in the Luftwaffe's bomber squadrons in the form of the RI-202.

The more powerful He 176 V2 fitted with a von Braun engine is never built; on 12 September 1939 Hitler issues an order to halt all development work on weapons that cannot be made operational within one year, which is the time he expected Germany would to need to successfully conclude the recently started war. The He 176 V2 was to have had a rocket thrust exceeding the weight of the plane so that it could lift off vertically. It might even have been able to attain the magic number of 1,000 km per hour (620 miles per hour). Hitler's order also ended the test flights of the He 178, the world's first jet plane that was also flown by Warsitz. The He 176 V1 was put into a sealed container and sent to the Aviation Museum in Berlin to be displayed after the war but it was destroyed by an air raid in 1943. Sadly, no pictures or movies of the historic He 176 flights are available; according to Warsitz the Soviets obtained all the documentation when they captured Peenemünde at the end of the war and they kept everything secret.

MESSERSCHMITT'S LETHAL POWER EGG

The He 176 was by no means the end of rocket plane development in Nazi Germany. Alexander Lippisch, the brilliant aerodynamicist who had designed the Ente sailplane which formed the basis of von Opel's first rocket plane, had in the late 1930s been working on the design for the DFS 194. This was to be a delta-wing airplane without horizontal tail stabilizers, driven by a pusher propeller (at the rear rather than in the nose). The lack of a tail on an airplane not only decreases aerodynamic drag but also results in greater effective lift: a normal wing has the tendency to push the nose of a plane downward, which in conventional planes is compensated by the horizontal stabilizers in the tail. But for them to push the nose up they must push the tail down, which requires negative lift. The result is a balanced airplane but the negative lift of the tail must be made up by additional positive lift from the wings. A well-designed tailless airplane does not need such negative-lift stabilizers. The delta shape of the DFS 194 meant that the wing ailerons could be placed close enough to the rear of the plane to give sufficient leverage for pitch control when both were moved in the same direction, thus eliminating the need for separate horizontal stabilizers with elevators. When moving in opposite directions they would still be able to roll the plane. Due to their dual aileron and elevator functions these control surfaces are called elevons.

Although the DFS 194 is being developed at the Deutsche Forschungsanstalt für Segelflug (DFS; the German Institute for Sailplane Flight), the RLM considers it to be a good basis for a rocket propelled fighter plane. The work begins at the DFS in cooperation with Heinkel but apart from valuable tests using models in a wind tunnel and in the open air, progress is too slow for Lippisch's liking. When in late 1938 he is told that his group is to be disbanded because the DFS believes there is no need for a tailless military rocket plane, Lippisch and many of his staff decide to leave the DFS. After negotiations with both Heinkel and Messerschmitt, in January 1939 the project is transferred to the Messerschmitt factory in Augsburg. The decision to join Heinkel's main competitor may have reflected the fact that Willy Messerschmitt was a more diplomatic salesman than Ernst Heinkel and he had good ties with Hermann Göring.

At Messerschmitt the prototype DFS 194, which the DFS permits Lippisch to take with him, is further developed as 'Project X'. Late in 1939 the completed airframe is sent to Peenemünde West to be mated with its Walter RI-203 engine. This is similar to the one used earlier in the He 176 but has the advantage that its thrust can be adjusted by the pilot, whereas the previous engine had a fixed thrust. The first gliding flight is made in July 1939 piloted by Heini Dittmar, and by October the aircraft is undergoing engine ground tests. The successful engine tests are followed by the first powered flight in August 1940, when Dittmar flies the DFS 194 to a top speed of 550 km per hour (340 miles per hour). As he later recalled: "With nearly 1,500 pounds of thrust, and as I pulled up steeper and steeper without losing one knot of airspeed, I knew that we were beginning a new era in flight." Combining a graceful glider with the raw power of a rocket engine results in excellent flying characteristics, and the DFS 194 proves to be a good basis for an operational rocket fighter.

DSF 194.

The first powered flight of the DSF 194.

The RLM consents to the production of a series of five prototypes of an improved design, which is named the Messerschmitt Me 163A. The project name 'Me 163' had earlier applied to the development of a short-take-off, slow-speed aircraft; it is hoped that using the same name will deflect unwanted attention from the new, secret rocket plane project. In line with this subterfuge, the first prototype is named the Me 163A V4 to suggest a follow up to the three prototypes planned for the slow-speed plane. The Me 163A uses the Walter RII-203 motor in which 'T-Stoff' composed of 80% hydrogen peroxide with some stabilizing additives is decomposed by a liquid sodium or calcium permanganate solution catalyst called 'Z-Stoff'. It produces a maximum thrust of 7,400 Newton. The propellants are injected into the combustion chamber by pumps which run on T-Stoff sprayed over a cement-like solid Z-Stoff 'stone'. For stability of the aircraft, the bulk of the engine mass must be placed near the center of the airplane, close to the center of gravity of the airframe and the place where the lift from the wings is concentrated. The combustion chamber with the nozzle is placed in the tail, linked to the rest of the engine by a long thrust tube. The vivid violet color of the permanganate causes the Me 163A to leave a huge purple exhaust trail in flight.

The V4 prototype makes its first towed flight at Augsburg on 13 February 1941 under the control of Dittmar, but the first powered flight does not occur until August due to delays in the development of the Walter engine. When the plane finally flies under rocket power the results are spectacular: once airborne, Dittmar sets off on a horizontal trajectory close to the ground in order to gain speed, then pulls up the nose and makes an incredibly steep ascent: from the moment it leaves the runway the Me 163A needs only 55 seconds to climb to an altitude of 4.0 km (13,000 feet)! It also proves to have excellent flying characteristics. On 2 October Dittmar sets the incredible new world speed record of 1,003 km per hour (623 miles per hour) in level flight, making the Me 163A V4 prototype much faster than anything else in the sky at that time. For the occasion it was towed to an altitude of 4.1 km (13,500 feet) in order to maximize the amount of propellant available for the record attempt. Because of the project's top secret status, the flight was not officially recorded internationally and was disclosed only after the war. (It would not be surpassed until August 1947, well after the war, when the experimental Douglas D-558-1 Skystreak turbojet aircraft in the US flew slightly faster.)

However, the record flight of the V4 reveals that such high transonic velocities are pushing the limits of its aerodynamic design, because at its maximum speed the plane suddenly pitches nose down. Dittmar is subjected to a negative acceleration of 11 G (acceleration results in positive G, what pilots call 'eyeballs in'; negative G is deceleration, or 'eyeballs out') but he quickly manages to throttle back. The plane

Me 163A prototype number V4 [Bundesarchiv].

slows down and becomes controllable again. As the aircraft approached Mach 1, he had encountered a phenomenon called 'Mach tuck', which is a disruption of the air flow over the wing that causes a plane to nose over (as will be described in more detail later). The outer limit of the speed envelope of the Me 163 had been found.

The five Me 163A prototypes are followed by eight pre-production examples that are designated Me 163A-0, then by an improved and significantly larger operational combat version called the Me 163B. It is officially named the 'Komet' (Comet), but the pilots name it 'Kraftei' (Power Egg). The main advantage of the 'B' is its much more powerful engine. The He 176, DFS 194 and Me 163A all had engines in which hydrogen peroxide was decomposed to produce a reasonable thrust at relatively low temperatures; the Me 163A engine exhaust temperature was about 600 degrees Celsius (1,100 degrees Fahrenheit). These types are often referred to as the 'cold' Walter engines. In contrast, the Me 163B is powered by a 'hot' Walter HWK 109-509A (initially the 509A1 but later the improved A2 version) rocket that burns concentrated 'T-Stoff' (hydrogen peroxide) and 'C-Stoff' consisting of 57% methyl alcohol (made from potatoes), 30% hydrazine hydrate and 13% water. The T-Stoff decomposes under the influence of a dissolved copper salt catalyst mixed in with the C-Stoff, while the hydrate helps the alcohol fuel to ignite spontaneously with the hot oxygen produced by the T-Stoff decomposition. The exhaust temperature of the 'hot' engine is 1,800 degrees Celsius (3,300 degrees Fahrenheit). The new engine gives the Me 163B more than double the thrust of the Me 163A. It also produces less smoke, making it harder for enemy planes to spot the Komet.

The 16,000 Newton that the new 509A1 engine provides (17,000 Newton for the later 509A2) is high in comparison to the force of gravity on the plane when it is nearly empty (with a weight of 1,900 kg, or 4,200 pounds, corresponding to about 19,000 Newton) and therefore towards the end of its powered run the Me 163B can climb almost straight up. The maximum speed of the operational, armed fighter plane is 960 km per hour (600 miles per hour), which corresponds to Mach 0.91 at 12 km (40,000 feet) altitude; sufficient to enable it to "fly circles around any other fighter of its time", according to Me 163B chief test pilot Rudolf Opitz.

The thrust of the Me 163B's engine can be throttled in four stages (idle and then levels 1 to 3) by varying the number of propellant lines that are opened into the combustion chamber. The Walter engine is simple to build, simple to maintain and readily accessible because the entire tail section of the plane can be removed. And rapid replacement is possible because the engine is mounted to the thrust frame within the fuselage using only a few connections.

Two 163B prototypes are built for flight tests. In addition, a number of unpowered 163S glider trainers are built which have an additional instructor cockpit behind and above the usual cockpit. The first tow-glide with the engineless 163B airframe occurs on 26 June 1942 but there is then a long wait for the new engine to become available. The first powered flight is not until 24 June 1943. In part because of this the Komet is unable to enter squadron service until 1944, by which time large fleets of bombers are swarming over Germany and its revolutionary capabilities are desperately needed to stem the tide. Thirty Me 163B-0 are produced armed with two

20-mm Mauser MG 151/20 cannon and about 400 Me 163B-1s with two even more destructive 30-mm Rheinmetall Borsig MK 108 cannon.

The bat-shaped Me 163B is a remarkably small aircraft, with a wingspan of only 9.3 meters (31 feet), a length of 5.8 meters (19 feet) and a maximum take-off weight of 4,300 kg (9,500 pounds) of which about half is propellant. To save battery weight it has a little windmilling propeller on its nose that will turn in the air during flight and drive an electricity generator in order to power the onboard radio. In spite of the emphasis on weight minimization, the pilot is protected by armor plating behind him, a thick section of armored glass immediately above the instrument panel, and an armored nose cone (it was expected that the biggest risk for a Komet pilot was being hit by defensive fire from the bombers it attacked, rather than from an Allied fighter getting on his tail). Due to the lack of metals in Germany late in the war, parts of the wings, which blend beautifully with the fuselage, are made of wood. The short nose and the single-piece blown canopy, which blends aerodynamically with the aft fuselage, gives the pilot good visibility forwards, but he cannot really see what is at his six o'clock (straight behind him); on the other hand, nothing is able to chase him.

The plane takes off at a speed of about 280 km per hour (170 miles per hour) with a much longer run than the distance conventional fighters require. It remains just above the ground until it has achieved the optimal climbing speed of around 680 km per hour (420 miles per hour), whereupon it pulls up into a 70 degree climb and within 3 minutes reaches the bombers at 7 to 10 km (23,000 to 33,000 feet). Movies of its take-off show an amazingly abrupt change from level flight into a steep climb that appears spectacular even today. To spectators used to the gradual

Me 163B prototype.

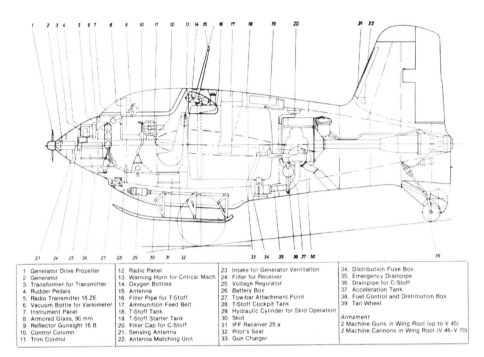

1. Generator Drive Propeller	12. Radio Panel	23. Intake for Generator Ventilation	34. Distribution Fuse Box
2. Generator	13. Warning Horn for Critical Mach	24. Filter for Receiver	35. Emergency Drainpipe
3. Transformer for Transmitter	14. Oxygen Bottles	25. Voltage Regulator	36. Drainpipe for C-Stoff
4. Rudder Pedals	15. Antenna	26. Battery Box	37. Acceleration Tank
5. Radio Transmitter 16 ZE	16. Filler Pipe for T-Stoff	27. Tow-bar Attachment Point	38. Fuel Control and Distribution Box
6. Vacuum Bottle for Variometer	17. Ammunition Feed Belt	28. T-Stoff Cockpit Tank	39. Tail Wheel
7. Instrument Panel	18. T-Stoff Tank	29. Hydraulic Cylinder for Skid Operation	
8. Armored Glass, 90 mm	19. T-Stoff Starter Tank	30. Skid	*Armament*
9. Reflector Gunsight 16 B	20. Filler Cap for C-Stoff	31. IFF Receiver 25 a	2 Machine Guns in Wing Root (up to V 45)
10. Control Column	21. Sensing Antenna	32. Pilot's Seat	2 Machine Cannons in Wing Root (V 46–V 70)
11. Trim Control	22. Antenna Matching Unit	33. Gun Charger	

Me 163B cutaway diagram.

Me 163B taking off.

climb of propeller fighter aircraft of that time, it must have been awe-inspiring. Also apparent in these movies is the large amount of smoke that escapes from the belly of the Me 163B: this is hot steam from a gas generator in which a small amount of T-Stoff is decomposed (according to the 'cold' principle) to drive the turbine which powers the pumps that feed T-Stoff and C-Stoff to the rocket's combustion chamber. (This same turbopump system was used by the 'cold' engine of the Me 163A.)

Famous German female test pilot Hanna Reitsch was mightily impressed by its power. In her memoirs (published in English with the title *The Sky My Kingdom*) she describes operating the plane during an engine check: "To sit in the machine when it was anchored to the ground and be surrounded suddenly with that hellish, flame-spewing din, was an experience unreal enough. Through the window of the cabin, I could see the ground crew start back with wide-open mouths and hands over their ears, while, for my part, it was all I could do to hold on as the machine rocked under a ceaseless succession of explosions. I felt as if I were in the grip of some savage power ascended from the Nether Pit. It seemed incredible that Man could control it." The mechanics near the plane had to keep their mouths open because otherwise the pressure differences caused by the rocket engine would rupture their eardrums.

Germany managed to equip only one operational fighter wing, *Jagdgeschwader* (JG) 400, with the Me 163 before the war was over. The first operational Komet mission was flown by commander Major Wolfgang Späte on 14 May 1944. For the occasion the ground crew painted his entire plane bright red, just like 'Red Baron' von Richthofen's famous aircraft of the previous war. It shows the high expectations for the Me 163 as an invulnerable wonder machine capable of outclassing all other aircraft in the sky. But Späte did not appreciate being made such a distinctive target and he had the color removed after the flight. In any case, the paint added 18 kg (40 pounds) of unwanted weight. Nevertheless, the red Komet has become a classic and many illustrations and scale models show the aircraft in this fiery livery. The planes also often had funny squadron artwork on their nose, including a depiction of Baron von Münchhausen flying through the air on a cannonball, or a drawing of a rocket propelled flea with the text "Wie ein Floh, aber oho!" (like a flea, but wow!)

The operational history of the different squadrons that belonged to JG 400 and the numbers of losses and successes is unclear, with different sources reporting different numbers since the claims of both Komet and Allied pilots about who shot down how many of whom are not always consistent. Komets often attacked bomber formations together with conventional Luftwaffe fighters and in the chaos of air battles mistakes were easily made about which planes were shot down by Komets and which by other aircraft. Allied bomber crews sometimes reported being under attack by Komets at times and places where none were flying. It appears that JG 400 Komets drew first blood on 16 August 1944, when they damaged several B-17 bombers and killed two American gunners. However, a gunner on another bomber managed to shoot down a Komet and a P-51 Mustang fighter claimed another. The pilot of the first Komet was able to bail out but the other pilot did not. Allied bomber formations were attacked intermittently up until May 1945, resulting in seven American B-17s and one British Lancaster being shot down. But several Komets were lost due to defending fighters and bomber gunners. In addition to attacking bombers, the Me 163B was also used to intercept the high altitude, fast reconnaissance planes that were usually hard to catch. By the time these were spotted it was normally too late to get conventional fighters airborne and up to the right altitude, but the extremely high climb rate of the Komet gave it a better chance of intercepting them. One Mosquito reconnaissance aircraft was indeed intercepted and shot down in November 1944. In March 1945 a Spitfire Mk XI photo reconnaissance plane was attacked by a pair of Me 163Bs but managed to escape by rapidly diving from 10.7 to 5.5 km (6.7 to 3.4 miles), and in the process reached a speed of about 800 km per hour (500 miles per hour). Later that month two Komets intercepted a Mosquito P.R. Mk XVI photo-reconnaissance aircraft near Leipzig. Nicknamed the 'Wooden Wonder' because it was mostly made of plywood, the Mosquito was an especially fast and rather agile two-engine plane, but it was not able to outrun an Me 163B. Although the Komet's cannon shot out the Mosquito's starboard engine, Canadian Pilot Officer Raymond Hayes managed to escape by very skillful maneuvering. Hayes and his navigator set for home in the crippled plane, but on the way were attacked by another fighter (not a Komet). Hayes managed to evade this second attacker but his aircraft sustained further damage and his navigator was badly injured in the leg. They finally crash-landed in Lille in France. Hayes received the Distinguished Flying Cross for his feat.

By the end of 1944 a total of 91 aircraft had been delivered to JG 400, but due to a severe propellant shortage most never left the ground: the two main factories which produced its special propellant had been demolished in bombing raids in September of that year. Many Komets were destroyed when their bases were bombed. Attacks from the air, and the approach of Allied ground troops, repeatedly obliged the JG 400 pilots to change airfields and this further hampered their operations. When they did manage to find an intact Komet and sufficient propellant, a typical tactic was to shoot upward through a bomber formation in a 70 degree climb, fire a few rounds at the enemy planes, and then turn and dive through the formation at zero thrust while shooting again. Sometimes this could be done twice before running out of propellant. The plane was remarkably agile and docile even at these high velocities, so it could easily both outrun and outmaneuver Allied fighter aircraft. Moreover, it would zoom by so swiftly that the gunners on the slower bombers hardly had time to swing their guns towards the small Komets.

However, the huge difference in airspeed also made hitting the enemy difficult for the Me 163 pilots, even though a bomber was a larger target to aim for and just a few hits from the powerful cannon could cripple even a large multi-engine plane such as a B-17 or a Lancaster. The Komet had to close within a range of about 600 meters (2,000 feet) to have any chance of hitting its target, but had to break off when at 180 meters (590 feet) to avoid a collision. That left less than 3 seconds for shooting the twin MG 151 cannon. Although powerful, these had a rather slow firing rate of 650 rounds per minute. Per attack, the Komet's two cannon could thus fire only some 65 rounds at a particular bomber. This may sound like a lot, but within that same time a conventional Fw 190D fighter could fire a total of 165 rounds using its four (smaller caliber) cannon. Furthermore, the slower fighter could hold its target for far longer. Another issue was that if the Komet fired its cannon while it was maneuvering, the ammunition belts often jammed. JG 400 First Lieutenant Adolf Niemeyer came up with the idea of installing 24 R4M 'Orkan' (Hurricane) missiles under the Komet's wings. When fired in brief barrages these would be able to knock out a bomber much faster and far more effectively than the twin cannon. Nobody knew how the Me 163 would react to the disturbance of the airflow around its unusually shaped wing while carrying rockets but because the situation for Germany was desperate there was no time for wind tunnel testing and careful analysis. Niemeyer took off in an Me 163A test vehicle with a set of dummy missiles under its wings to find out. Apparently the flight went reasonably well because he survived the experiment (as well as the war). Had Niemeyer been able to mount real rockets on his Komet, he would have become the first man in history to fly a missile-equipped rocket propelled airplane. Such an Me 163B would have been a very lethal weapon indeed, but no Komet ever carried missiles in combat.

One Lancaster bomber did however fall victim to an Me 163B with an experimental SG500 'Jägerfaust' (Hunter's Fist) weapon system consisting of a set of eight upward-firing, short-barreled guns triggered by a photocell able to detect a bomber's shadow as the Komet flew underneath it. The system was designed to make it easier for Me 163B pilots to shoot down targets because it did not require careful aiming and timing. The guns were built into the wings, four on each side, and they had to be placed rather far out towards the wingtips and be fired with minute intervals to preclude the explosive shock waves from shattering the canopy. To prevent the recoil from disturbing the aircraft's flight, the barrel of each gun was ejected downwards as the shell was shot upwards.

Sitting in an Airbus 320 reading about the exploits of the Komet pilots in Stephen Ransom and Hans-Hermann Cammann's *Jagdgeschwader 400*, it dawns on me that I am comfortably flying at a speed and altitude at which Me 163B and Allied bombers were slugging it out over northern Europe some 65 years ago. Pilots and crews did not have pressurized cabins, and the engines used were either reliable but slow as in the case of the bomber's piston engines, or dangerous but fast rocket motors. What at the time were rather extreme flight speeds and altitudes have been standard cruising performance for commercial airliners since the late 1950s: a Boeing 747 typically flies at around 900 km per hour (560 miles per hour) at 11 km (35,000 feet) altitude. The very rapid developments in post-war jet aircraft technology meant that what in 1945 was attainable only by an experimental and extremely dangerous fighter aircraft was perfectly normal for the average tourist-class airline passenger a mere 15 years later. It is also interesting to note that since then up to the Airbus that I am flying in, airliner operating speeds and altitudes have hardly changed apart from the fantastic Concorde, which is unfortunately no longer in service.

The Me 163B's performance was revolutionary, but the design was not perfect for operational military use. From the long list of in-flight explosions, engines quitting, weapons jamming, crash landings and aircraft being shot down while gliding back to base, it is clear that the Komet was an experimental aircraft rushed into operational service. On combat missions it was expected that one-third of the aircraft and pilots would not make it back to base. Of those lost, some 80% crashed during take-off or landing, 15% burst into flames or went out of control in flight, and the remaining 5% were shot down. Abysmal numbers for what was supposed to be an operational fighter plane. Another major issue was that the Komet guzzled its propellant at such a rapid rate that it had a total of just 7.5 minutes of powered flight. This meant the fighter could only operate as what we now call a point-defense interceptor, needing to be stationed close to where the enemy bombers were expected to pass over. A German animation movie of the time shows how the plane would take off when a bomber formation came within a range of 42 km (26 miles), flying towards the Me 163B airbase at an altitude of 7 km (4.4 miles). By the time the bombers had halved that distance, the Komet was in the right position to attack.

Once it ran out of propellant, the Me 163B had to glide back home. Allied pilots noticed this and tried to attack them during this unpowered descent. This was still a difficult task because even whilst gliding back down the little rocket plane was much more maneuverable than any Allied fighter, and with an unpowered diving speed of over 700 km per hour (440 miles per hour) it was very fast. Shortly before landing, however, the Me 163 was an easy target since it had to fly straight and slow. Allied fighters would circle just outside the defensive perimeter of the airfield's anti-aircraft guns and from there mount quick attacks on Komets making the final approaches. To counter this tactic, experienced Me 163 pilots would rapidly dive into the protected area at about 800 km per hour (500 miles per hour), then fly circles within the range of the anti-aircraft guns while they bled off excess speed prior to lining up to make a landing approach.

To keep the plane light, it was not equipped with a proper undercarriage. For take-off it used a jettisonable dolly with two large wheels which, if dropped prematurely, could bounce back against the belly. The plane landed on grass fields using a single extendable skid which was light and easy to fit into the plane without disrupting its clean aerodynamic shape. It slowed the aircraft down after landing simply by friction with the ground. The skid had hydraulic shock dampeners but these would not work when the skid activation lever in the cockpit was not properly set after deployment. The resulting lack of cushioning in combination with bumpy fields and high landing velocities caused some pilots to suffer back injuries, including experienced test pilot Heini Dittmar when he landed an Me 163A in November 1943. And Hanna Reitsch was badly hurt when the take-off dolly did not separate on her fifth towed training flight in an Me 163B, forcing her to discontinue and land with the wheels attached. Because the dolly made it impossible to deploy the skid and its shock dampener, the plane hit the ground hard, somersaulted and threw her head against the gunsight. She spent several months in hospital and lost her chance to fly the Me 163B under power.

The Me 163B was aerodynamically very clean, and showed its glider heritage by its very flat gliding angle: while unpowered, it would only lose one meter of altitude per 20 meters of distance. Whilst this was great for gliding back to base, it made the plane rather difficult to land: just above the ground the air would become compressed between the wings and the field, forming an air cushion. This 'ground effect' could keep the Me 163B floating just above the ground for a long distance, with the merest updraft being sufficient to make it ascend back into the air again. As a consequence, pilots found it difficult to put the Komet down quickly. Since on approach the plane would be unpowered there was no opportunity to circle around for another landing attempt, and a delayed touchdown could easily result in a crash amongst the trees at the periphery of the airfield. And even after touching down the plane needed a run of about 370 meters (1,200 feet) on dry grass to halt, and nearly

Deployed landing skid of the Me 163 displayed in the Science Museum in London [Michel van Pelt].

twice that on wet grass. With the high landing speed of around 200 km per hour (120 miles per hour) this made landing an Me 163B very challenging. A set of landing flaps, to increase drag while still allowing sufficient lift, provided a modicum of control but it was almost impossible to put an Me 163B down at a precise landing point. Even if the landing were successful, the pilot was still not safe because the Komet had no power to taxi, and was anyway not easy to move on its single skid. It had to be retrieved by a small tractor pulling a trailer equipped with two lifting arms which fitted under the wings, and while waiting out in the open the aircraft was very vulnerable to attack by enemy planes.

Take-off could also be dangerous for an inexperienced pilot because the absence of a propeller-driven airflow over aircraft meant that the aerodynamic controls only became effective at about 130 km per hour (80 miles per hour). Steering during the taxi run was therefore done using the tail wheel, which was coupled to the rudder and so could be operated using the foot pedals. However, it meant that the position of the stick (and thus elevons) was initially immaterial and a pilot might accidentally put it in a turn during the take-off run which could seriously ruin his day when the control surfaces suddenly became functional. A crash soon after an aborted take-off would inevitably lead to a violent explosion of the volatile propellants. The long take-off run of the Komet was also an issue because it required a lot of room and, moreover, used a lot of propellant. A rocket propelled launch rail system was developed to help the plane on its way. This was successfully demonstrated using an Me 163 mockup, but the rails would mean losing the operational flexibility of taking off in any direction from any sufficiently large and reasonably level grassy field.

Unlike modern airplanes, the cockpit of the Me 163 was unpressurized, exposing its pilot to a very quick drop in atmospheric pressure during the fast climb to altitude. Pilots had to eat special low-fiber food to prevent any gas in their gastrointestinal tract from rapidly expanding on the way up. They were also given altitude chamber training to familiarize them with operating in the thin air of the stratosphere without a pressure suit and while breathing through an oxygen mask.

Another major hazard for the Me 163 pilots and their ground crews was that the propellant was very corrosive. The T-Stoff and C-Stoff would explode violently if they came into contact with each other prematurely; even a missed drop could set off a massive explosion of all the propellant in the aircraft. The filling caps on the plane were therefore very clearly marked with a 'T' or a 'C', and trucks transporting these liquids to the planes were never allowed to get close to one another. After each flight the propellant tanks and rocket engine had to be flushed with water to remove any residual fluids, and the entire aircraft's surface was washed just to be sure. The pilots wore protective asbestos-mipolam-fiber clothing that would not burn on coming into contact with the concentrated hydrogen peroxide T-Stoff, but the suit was not perfect and if this fluid seeped through a seam it could rapidly react with the pilot's skin in a chemical reaction that released oxygen, the concentration of which would soon rise to that for spontaneous combustion. In short, the concentrated peroxide would burn a pilot alive, and several pilots were injured by it when propellant lines broke in crash landings. Part of the dangerous liquid was actually put in two small tanks located on each side of the seat in the cramped cockpit, so even a small leak there would have grave consequences for a pilot. The tanks for the T-Stoff had to be kept meticulously clean because even a fingerprint would be enough to cause it to decompose violently into oxygen and hot steam. Because of the dangerous propellant vapors, a pilot had to breathe through his oxygen mask even while the plane was still on the ground, and always had to wear flight goggles to protect his eyes. The Me 163B was not equipped with an ejection seat (the now standard means for escaping from a supersonic or transonic airplane) as this equipment was still very experimental at the time. However, the Me 163B designers did manage to give the pilots some chance of bailing out of a Komet in trouble at 900 km per hour (560 miles per hour): a small drogue chute could be deployed behind the plane to slow it to a velocity at which it was safe to get out of the cockpit, and at which the pilot's own parachute would not be torn to shreds. But this was not of much use if the engine suddenly exploded upon being ignited for take-off, which frequently happened.

The Komet was the first airplane to really enter the domain of transonic flight near to Mach 1, where the compressibility of the air starts to be noticeable. When a plane flies at the speed of sound, air no longer flows nicely around its nose and over the wings. Wherever the supersonic air hits a significant obstacle, like the blunt nose of an airplane, or a region in which air is flowing at a lower velocity, it is slowed

An Me 163B pilot getting ready for flight.

down. But the air molecules are no longer able to move out of the way in order to ensure a smooth flow. Instead they collide with each other and are compressed into a shock wave. While the Me 163 could not reach Mach 1, it still encountered compressibility effects because the airflow over part of its airframe would actually go supersonic. We have seen earlier that air is accelerated when flowing over the wings. This means it locally flows faster than the plane itself is flying through the atmosphere, and thus shock waves can occur even if the plane's airspeed is just below the speed of sound. The flight Mach number at which the local flow over the wing first goes supersonic is called the 'critical Mach number' of the airplane. When flying above the critical Mach number but below the speed of sound, the airplane is in the transonic flight regime.

The thicker the wing, the more the air passing over it will accelerate and thus the lower the critical Mach number of that airplane. Several fast conventional propeller planes of the Second World War experienced air compressibility issues in a fast dive because their wings were relatively thick. The P-38 Lightning had a critical Mach number of 0.68, and thus already started to experience transonic aerodynamic effects when flying at 68% of the speed of sound. Pilots of early versions of the Lightning therefore often ran into trouble when they put the plane into a high speed dive, with the controls suddenly freezing, the tail shaking violently, and the plane nosing into an ever steeper dive beyond the pilot's control.

When the airflow over the surface of a wing goes supersonic, the point where the sum of the lift forces across the wing's surface can be thought to originate (called the plane's center of pressure) moves aft. Basically what happens is that the speed of the airflow over the aft part of the wing is faster (supersonic) than the subsonic flow over the front part of the wing. The aft part thus gives relatively more lift and causes the center of pressure to migrate further aft in the airstream. As a result, the aircraft is no longer balanced along its pitch axis. As the plane enters the transonic regime, the lift force will try to push the nose down.

Another nasty consequence of transonic flight can be a sudden loss of lift. When the airflow over an area of a wing goes supersonic, a shock wave can form at the aft boundary of the supersonic-flow region, beyond which is subsonic flow. This shock wave can grow so strong as to cause the airflow behind it to separate from the wing and no longer nicely follow the wing contours. This leads to a serious loss of lift, a condition called 'shock stall'. On the P-38 this also meant that the airflow behind the wings went more straight towards the tail and gave the horizontal stabilizer more lift, which only worsened the pitch-down problem. If a P-38 flew faster than its critical Mach number of 0.68 the shift in the center of pressure, in combination with shock stall, would inevitably induce 'Mach tuck'. The pilot would naturally try to get the nose back up by pulling on the stick, deflecting the tail elevators upward. This would actually give the airflow at the elevator hinge line on the underside of the stabilizer more room, making it expand and hence accelerate. The resulting shock wave near the hinge line would make the elevators shock-stall as well, become ineffective, and cause the plane to uncontrollably nose down into an ever-steepening dive. Since the shock stall happens irregularly and is subject to constant change, the aircraft would also shake violently as it plunged towards the ground. Not surprisingly, a pilot would like to bail out at that point, but without an ejection seat to boost him clear of the airplane this would mean certain death. But if he remained in the plane he actually had a chance of survival because when the diving plane encountered the denser, warmer air at lower altitudes the local atmosphere's speed of sound would increase. As a result the Mach number would diminish, the shock waves would lose their strength, and the elevators would regain their effectiveness. With luck, the pilot would be able to pull up before creating a hole in the ground. A more effective way to solve the Mach tuck problem was to equip the P-38 with dive flaps that could quickly be extended downwards and influence the center of pressure on the plane in a manner that would prevent it from nosing down. In late 1943 these flaps became standard on all new P-38s, and the company issued kits so that they could be retrofitted to already operational planes.

Alexander Lippisch knew about compressibility and Mach tuck, and he designed the Komet with relatively thin delta wings. The thin wings (at least in comparison to other contemporary planes) resulted in less acceleration of the air flowing over them, which contributed greatly to a high critical Mach number. The delta shape, or rather the swept wing in general, was another revolutionary good idea for high-speed flight. The most important factor in creating shocks on a wing is the velocity of the airflow perpendicular to the leading edge. If a wing is swept backwards, the component of the velocity perpendicular to the leading edge is diminished even if the total velocity remains the same. Or in other words, the air is presented with a more gradual change in wing thickness as the wing is swept back. This means that for the same thickness, shock waves will build up at higher speeds and thus the critical Mach number of the wing is increased. Another benefit of swept-back wings for supersonic flight is that they remain out of the V-shaped shock wave that forms at the nose of a plane when it exceeds the speed of sound. Straight wings partly project through this shock wave and this produces high drag. The downside is that the more a wing is swept back the less lift if will generate at any given velocity, reducing its effectiveness, especially at low speeds. The triangular shape of a delta wing combines wing sweep with a stronger structure that is supported by a longer section of the plane's fuselage. As a result, the wing is more capable of resisting the strong force of high-speed flight. Moreover, at high angles of attack a delta wing creates a vortex of air which is able to stick to its surface much better than a normal straight airflow, thereby delaying wing stall. This is very handy at low flight speeds, where a delta-winged airplane can fly at a higher angle of attack and create more lift than a plane with normal swept wings. A delta wing is therefore efficient for high-speed flight as well as take-off and landing, as demonstrated by the excellent flying characteristics of the Me 163.

The combination of a relatively thin wing and its delta shaped planform gave the Komet a critical Mach number of about 0.84, so that it experienced transonic effects only at a much higher velocity than the P-38 or even the P-51, the latter of which had a critical Mach number of 0.78. This was very important, since even in level flight the Komet could fly much faster than a diving Lightning or Mustang and could more readily run into trouble. In addition a delta-winged aircraft needs no horizontal stabilizers, which saves weight and eliminates a source of drag. But the need for a high critical Mach number meant that the Me 163 had very little 'washout'. Most wings on conventional airplanes are twisted so that the tip encounters the air at a lower angle of attack than the root. This is done to make sure that when the wing stalls it does so at the root first, leaving sufficient effectiveness for the ailerons near the wingtip (where they have the biggest leverage) to control the plane. The wing's twist is called 'washout', and it is particularly important for swept wings since they more readily stall at the tip, causing the wing to drop and potentially result in a spin. Sufficient washout on the Me 163 would have caused the lower surface of its wingtips to shock stall at high speeds and bring on Mach tuck. To keep the tips from stalling at low speeds and high angles of attack in spite of the lack of washout, Lippisch installed fixed leading edge slots that would suck air from below the wing and pass it over the wingtips, thus delaying their stall.

Because the Me 163 flew much faster than propeller fighters and could surpass Mach 0.84 even in level flight, compressibility and Mach tuck remained a danger in spite of Lippisch's revolutionary wing design. In comparison to modern supersonic planes, the wings of the Me 163 were still relatively thick. When flying the Me 163A Heini Dittmar was the first pilot to encounter the now familiar problems associated with supersonic flight. In his book *Raketenjager Me-163*, former Me 163 pilot Mano Ziegler quotes Dittmar regarding his record-setting flight of October 1941: "When I looked at the instruments again, I had gone over the 1,000 km per hour mark, but the airspeed indicator was unstable, the elevator started to vibrate, and at the same time the aircraft plummeted out of the sky, gathering speed. I could not do anything. I immediately turned off the engine and was certain that the end was near when I

Leading edge slot of the Science Museum's Me 163 in London [Michel van Pelt].

suddenly felt the steering responding and managed to get the Me out of her nose-dive relatively easily." To prevent pilots from exceeding the critical Mach number, a red warning light was installed in the cockpit to tell them when it was time to throttle down. However, while accelerating horizontally just before the sharp climb, or when leveling off in preparation for attacking enemy bombers, pilots tended not to look at their cockpit display very often, so later an alert horn was positioned right behind the pilot's head to gain their attention.

After the war, other high-speed tailless airplanes with relatively thick wings had similar Mach tuck problems. The British de Havilland D.H. 108 Swallow research aircraft killed three pilots, including Geoffrey de Havilland, son of the company's founder, because of its nasty Mach tuck characteristics. The American Northrop X-4 experienced similar difficulties during flight tests. Later tailless combat aircraft had thinner wings and better designed elevons that were shaped to prevent the loss of control power.

In his book, Mano Ziegler claims that during a flight in July 1944 Heini Dittmar actually broke the sound barrier in an Me 163B and reached a flight speed of 1,130 km per hour (702 miles per hour) by placing his Komet into a steep powered dive. Several observers on the ground are reported to have heard the sonic boom. If they indeed heard this, it would be positive proof of the plane flying faster than sound. A sonic boom is created by the pressure waves a plane causes by moving through the

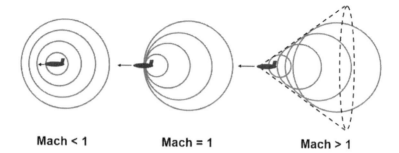

Mach < 1   **Mach = 1**   **Mach > 1**

Sound waves from a plane flying faster than the speed of sound create a shock wave that causes the 'sonic boom'.

air, similar to the bow and stern waves of a boat. These waves travel away from the airplane at the speed of sound, so when the plane flies at Mach 1 or higher the waves are forced together and merge into a single shock wave in the shape of a cone which has the aircraft at its apex. When that wave reaches your ear as a sharp increase and following decrease in air pressure, you hear a bang. However, modern analysis of the Me 163 design shows that it was incapable of surpassing the speed of sound because its wings were too thick and its fuselage not slim enough. As the Komet got close to Mach 1, the resulting strong shock waves would create so much drag that the aircraft would not be able to accelerate further, even in a powered dive (shock waves require energy to form, and this is taken from the aircraft's speed and creates a form of aerodynamic drag). The speed registered by Dittmar was very likely inaccurate, as indicators designed for subsonic airspeeds cannot measure transonic velocities very well. So the noise heard by the spectators was probably not the fully developed sonic boom associated with faster-than-sound flight.

As with all German 'wonder weapons', the Me 163B Komet was too little and too late to change the outcome of the war. Due to enemy raids on their airfields, the lack of rocket propellant, and the constant need to retreat, Komet pilots managed to shoot down only eight bombers and a single Mosquito reconnaissance aircraft. The cost to JG 400 was high, with many aircraft and pilots lost due to exploding engines and rough landings ending in deadly crashes, or because they were shot down. Overall, the Komet represented more of a threat to its own pilots than to the enemy. Nevertheless the Me 163B was the fastest plane in the sky during the Second World War, and had it been deployed earlier it would have made a much bigger impact than it did. The Komet was the first, and last, pure rocket propelled airplane ever to fly in combat.

As the war neared its conclusion, a more advanced rocket fighter, the Me 263, was already under development. Although based on the Me 163B it would have a proper retractable undercarriage with a nose wheel, a larger and better shaped fuselage, a greater propellant load and a pressurized cockpit with a bubble canopy for all-around vision. It would initially be powered by a new HWK 109-509C Walter rocket engine with two combustion chambers and nozzles, mounted one above the other. The larger chamber would provide power for take-off and climbing (and at 20,000

Newton was much more powerful than the engine of the Me 163B) whilst the smaller chamber would provide a lower thrust of 4,000 Newton for cruising and the return to base. In that way the plane would be able to fly under power and spend significantly longer attacking bombers than was possible for the Me 163B. The smaller chamber was also to enable the Me 263 to taxi to a safe shelter, making it less vulnerable on the ground (this smaller part of the engine was tested on two experimental Me 163Bs). The new plane would have the same wings and tail as the Komet, and so have a similar critical Mach number. One vital difference however, was that some of the propellant would be stored in tanks fitted in an empty space in the wings.

Because Messerschmitt was totally overloaded with production demands for its conventional fighter planes, the project was assigned to the Junkers aircraft company, which renamed it the Ju 248 and developed it under the leadership of Dr. Heinrich Hertel. Three prototypes were built, and the first one made several unpowered flights under tow by Messerschmitt Bf 110 twin-engined fighter aircraft before testing was halted by a shortage of fuel for the tow plane. It appears no powered flights with the Me 263 were made before the war ended.

After Germany's capitulation, many Komets were removed to Britain, the US and Russia for analysis and tests. None of these were ever flown under power, mostly due to unfamiliarity with the rocket engine and its dangerous propellants. It is likely that the revolutionary airplane simply scared most of the potential test pilots. Ten of the expropriated planes survive (most of which were once part of JG 400) and are on display in museums in Britain, Germany, the US, Canada and Australia. During the restoration of one in Canada in the late 1990s it was discovered the aircraft had been sabotaged by the French laborers who had been forced to build it. They had wedged a small stone in between the fuselage fuel tank and a supporting strap in an attempt to cause a dangerous leak, and they had also weakened the wooden wing structure by using contaminated glue. On the inside of the fuselage they had written in French "Plant Closed" and "My heart is not occupied". This particular Me 163B can now be seen in the National Museum of the United States Air Force near Dayton, Ohio. In the mid-1990s a former Luftwaffe pilot built an unpowered glider reproduction of the Komet which flies very well and from a distance looks very much like the real thing (its internal construction and the materials used are very different). The builder, Josef Kurz, had been in training to fly the Me 163B near the end of the war but never flew it. He made the first flight in his replica in 1996, being towed into the air by another plane. He later sold his Komet reproduction to the large European aircraft company EADS, one institutional strand of which could trace its roots back to none other than Willy Messerschmitt.

The original Komet hanging from the ceiling in London's Science Museum now looks somewhat dusty, silent and inert: no longer advanced and top secret, merely an artefact of a war which was fought long ago. Just like a stuffed jaguar, it is initially hard to imagine that this display item was once an agile and terrifying hunter; a killer that could leap up to attack its victims high in the sky in an unprecedented manner, digesting toxic liquids and spitting fire. Suspended next to the museum's Hurricane and an early Spitfire fighter, it looks rather diminutive, but it could readily out-climb, outrun and outmaneuver these conventional propeller planes that entered

Me 163B of the National Museum of the US Air Force [National Museum of the US Air Force].

service only a few years before the Me 163B took to the sky. But if you look more carefully you cannot help but admire its sleek shape, the audacity of its design, and the pilots who dared to strap themselves into those dangerous little machines. They zoomed through the air at what were then utterly amazing speeds, in machines powered by incredibly dangerous propellants. A punctured propellant tank from a bullet or a rough landing would probably be fatal. On the one hand these machines were part of the jet age in terms of aerodynamic sophistication and propulsion technology, yet the construction technology and primitive landing skid clearly belonged to the realm of Second World War propeller aircraft. Its wooden wings and fabric-covered control surfaces would normally even be associated with planes of the First World War! The Komet pilots were the kings of the sky while they had propellant in their tanks (and provided their plane did not spontaneously explode), flying and climbing even faster than the early jets, but once out of power they found themselves in a glider that could be shot down by conventional fighters.

The Me 163's legacy of advanced aerodynamics based on the delta-wing, tailless design led to such fast vehicles as the Concorde and the Space Shuttle, and it lives on in today's advanced fighter aircraft. After the war Lippisch, the 163's chief designer, was moved to the US as part of Operation Paperclip (along with von Braun and his A4/V2 team). There he continued his work at the Convair airplane company, which subsequently produced delta-wing airplanes such as the XF-92, F-102 Delta Dagger, F-106 Delta Dart, B-58 Hustler and the Navy's F2Y Sea Dart. One Me 263 prototype was taken to the US. After Soviet forces captured the Junkers factory another ended up in Russia, along with some technical staff and engineering documentation. It formed the basis for the Soviet MiG I-270 (to be discussed later). Walter and his technology were captured by the British, and later gave rise to the powerful engine that pushed the Black Arrow rocket into space, as well as the Spectre engine of the Saunders-Roe SR.53 rocket interceptor (also to be discussed later). And his 'cold' rocket pods were developed into the Sprite engine that de Havilland designed to assist aircraft take off from airports located in hot places and at high elevations where the low air pressure prevented the wings from generating sufficient lift. A pair of Sprites were tested on a Comet jet airliner in 1951.

The only Allied pilot ever to fly the Komet under power was the famous British test pilot Eric Brown, who assessed the astonishing capabilities during a single flight of a captured plane in Germany shortly after the surrender. He later commented: "I was struck by how small it was, and yet how elegant it looked, and at the same time how very lethal. Lethal not only to the enemy, but to those who flew it." Brown flew no fewer than 487 types of aircraft, not even counting different versions of the same basic type, so his judgment can be trusted.

BACHEM'S BUDGET INTERCEPTOR

Another German rocket interceptor that posed an extreme risk to its own pilots was the Bachem Ba 349. This was really a surface-to-air anti-aircraft missile with a pilot on board to compensate for the absence of sufficiently advanced electronics, and was designed in response to a request by the Luftwaffe's Emergency Fighter Program in early 1944 for a 'Verschleissflugzeug' (literally 'Wear-and-tear Aircraft', meaning a short-life aircraft). The situation had changed dramatically since the start of the war. Gone were 'Blitzkrieg' (Lightning War) and aerial supremacy for the Luftwaffe. The war had become one of attrition, and Germany, its heartland now under attack, was unable to compete with the industrial might of the US. The Luftwaffe no longer ruled the skies over the occupied countries, and was rapidly losing control of the airspace above Germany. Some radical defense system was needed to stop the Allied bombers from destroying the German industries, cities and transportation infrastructure. This new weapon had to be introduced as soon as possible, not require much scarce raw materials such as airplane-grade metals, and be easy to operate by pilots with little training (since by then most of the experienced pilots had either been killed or were prisoners of war). Moreover, the new aircraft would preferably not need runways or other vulnerable ground facilities that could be easily found and destroyed by enemy bombers.

Outsider Erich Bachem, who had left Fieseler in order to set up a small company to manufacture parts for aircraft, devised a truly radical design which fitted the need. His 'Projekt BP 20' was a rocket interceptor to be made mostly using cheap wooden parts, glued and screwed together. Its configuration was extremely simple, having a tube-shaped fuselage, two straight stubby wings and a T-shaped tail. Roll control was by differential use of the rudders on the lower and upper vertical fin (moving one to the left and the other to the right) so the wings required no moving parts. It would use few 'war-essential' materials reserved for other aircraft and be easy to build in volume by semi-skilled labor: a **BP 20** could be constructed in only about 1,000 man-hours. Moreover, owing to the design's simplicity most of the parts could be made in small woodworking shops distributed throughout Germany without interfering with aircraft production in factories which were already at peak capacity. Because of its distributed nature, production would also be relatively safe from sudden destruction by enemy bombers.

The BP 20 would be launched straight up while being controlled from the ground, and the pilot would bail out at the end of the flight. That way the pilots would not require training in the most difficult aspects of flying, which are take-off and landing. Bachem proposed to build large numbers of launch platforms around key industrial targets so that his interceptors would deter Allied bombers sufficiently for them to leave those targets alone. The RLM's response was less than enthusiastic so Bachem showed his plan to Heinrich Himmler, chief of the security forces. Himmler, eager to increase his influence in the Third Reich, agreed to fund the development and placed an order for 150 machines paid with SS funds. Himmler, of course, intended that the interceptors be flown by SS men. By saving Germany from the "bomber menace" he would increase his own power in the Nazi government. Not wanting to be left out, the RLM later placed an order for another 50 interceptors for the Luftwaffe. The highly classified project was assigned top priority and the operational vehicle became the Ba 349, code-name 'Natter' (Viper).

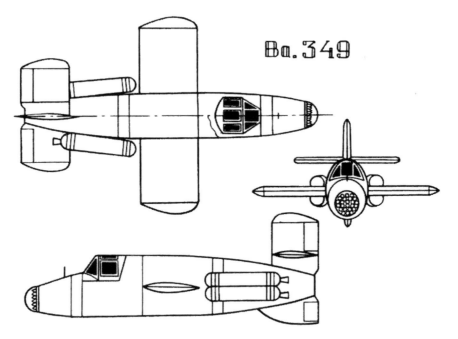

Layout of the Bachem Natter.

If flying the Me 163 was a pretty brave thing to do, getting into a Natter should be considered truly heroic. Or rather, suicidal. The flight would start with the wooden contraption standing upright against a 15 meter (50 feet) tall, open-structure launch tower (or even just a modified telephone pole) with three vertical tracks engaging the wingtips and the edge of the ventral fin. The pilot would climb in, lie on his back in a seat angled at 90 degrees, and wait for the bombers to come within range. The entire gantry could turn around its vertical axis to enable the Natter to be oriented correctly depending on the direction the bombers were coming from. On command, the ground crew would take cover and the pilot would test the flight controls. Then he would start the central Walter HWK 109-509A engine (similar to that of the Me 163B) and verify its functionality. Next he would ignite the four solid propellant Schmidding rocket boosters, two strapped to each side. These delivered a total thrust of 47,000 Newton and because the vehicle weighed only 2,200 kg (4,850 pounds) they could readily lift it off the ground. Riding the tower, which provided initial stabilization, the machine would soon build up enough speed for the aerodynamic surfaces to keep it flying straight. The flight controls would remain locked in the neutral position until the solid boosters burned out, some 10 seconds into the flight. The initial acceleration was nearly 2 G, meaning twice as strong as the acceleration of gravity (i.e. the Natter would accelerate upwards about twice as fast as it would fall, because the total thrust was twice its total weight at lift-off). Hence at that moment the pilot would feel three times his normal weight, a combination of one G force of gravity plus the two G of the acceleration. When the strap-on boosters burned out they would be jettisoned by explosive bolts, the flight controls would unlock, and the autopilot would begin to steer to its assigned target using guidance commands received from the ground by radio. The vehicle would continue to climb under power of the Walter engine alone, accelerating at a rate of about 0.7 G to a maximum velocity of 700 to 800 km per hour (435 to 500 miles per hour). It would be guided towards a position in front of the bombers at an altitude of 9 km (30,000 feet), and only then would the pilot take over for the final phase of the attack. During the automatic phase of the flight the pilot was required to hold on to hand grips to prevent him from accidentally pulling on the control column while under the force of acceleration.

Closing in on his victim, the pilot would jettison the plastic nose cone to reveal a battery of missiles in front of the cockpit: either 24 R4M 'Orkan' (Hurricane) missiles of 3.5 kg (8.0 pounds) each, or 33 Henschel Hs 217 'Föhn' (Warm Wind) rockets of 2.6 kg (6 pounds) each. Pulling the trigger would fire all of the unguided anti-aircraft missiles at the target in a single devastating salvo. Since the Natter would have little time for the attack, it made more sense to pack all its firepower into a single punch than to use conventional cannon more suitable for repeated precision strikes. Shortly after the missiles were fired, the engine would exhaust its propellant and the pilot would quickly glide at high speed down to an altitude of about 3 km (10,000 ft), where he would jettison the entire nose section, release his safety harness and fold the control column forward, an action that would release a braking parachute from the rear of the vehicle. The pilot would be thrown out by the sudden deceleration and subsequently deploy his own parachute. The tail with the engine, the most precious part of the aircraft, would land under the braking parachute to be refurbished and used again; only the nose section would be lost.

Development commenced with scientists of the Technical University of Aachen calculating the Natter's aerodynamics using a large analog computer. The Deutsche Versuchsanstalt für Luftfahrt (DVL) in Braunschweig then tested models in a wind tunnel. The results showed that the Natter would behave "satisfactorily" up to speeds of about Mach 0.95. In the meantime the solid propellant boosters were tested at the Bachem-Werke factory in Waldsee. Due to the high priority given by the SS to the development, less than 4 months after Bachem made his initial design sketches three full sized Natter prototypes were completed.

On 3 November 1944 the M1 (first) prototype, which took off from a trolley, was towed to an altitude of 3 km (10,000 feet) by a Heinkel He 111 bomber. Pilot Erich Klöckner made various tests of the control and stability of the machine, discovering it to be rather sensitive and to react pretty violently to small movements of the stick. Due to the plane's bucking and shaking he could not finish the test flight as planned with a realistic drop of the forward section, bail-out and retrieval of the tail section, so he decided simply to eject the cabin roof and jump out. The roof got caught at the hinge and did not fall away entirely, but Klöckner managed to bail out from the side and landed safely by parachute. The Heinkel tow plane tried to land with the pilotless prototype, but without active control and lacking an undercarriage it hit the ground hard, tumbled end over end and was destroyed. Klöckner went on to make several flights with the M3, which was fitted with a fixed undercarriage, each time bailing out successfully while the tow plane landed with the complete Natter behind it. The Natter design clearly did fly, but the tests showed that the dynamics of the tow cable and the aerodynamic influence of the fixed undercarriage had adverse effects on its stability. A free glide without the undercarriage would be required to unambiguously determine the Natter's flight characteristics. Klöckner refused, convinced that more tests and modifications were needed before such a flight could be made. The SS did not approve of his criticism and he was kicked out of the project.

The SS found another guinea pig, and on 14 February 1945 Luftwaffe test pilot Hans Zübert took off in M8 towed by an He 111 bomber. Like the M1, the M8 took off from a jettisoned trolley and once in the air was as aerodynamically 'clean' as the envisioned operational Natter. At an altitude of 5.5 km (18,000 feet) Zübert released himself from the tow plane and pushed the plane into a steep (75 degree) nose dive to gain speed, then leveled off to execute a series of flight control tests. He found that the plane handled well at the attained maximum speed of about 600 km per hour (370 miles per hour) and that its stall velocity was about 200 km per hour (125 miles per hour). As per the test plan he then tried to deploy the braking chute, only to find that the release lever was stuck. Next he tried to eject the nose by activating the explosive bolts that held it to the rest of the vehicle but again nothing happened. It was time to get out: Zübert ejected the cabin hood, put the plane into a steep climb to bleed off speed and struggled to escape. When just 1,200 meters (3,900 feet) above the ground and flying at 300 km per hour (190 miles per hour) the Natter went into a spiral dive. Zübert hung on for three full turns, then dived over the left wing, passing beneath the tail unit. He landed safely by parachute but the M8 smashed into the ground at high speed. Of this flight Zübert reported: "My general impression of the machine is very good and I can describe its flying characteristics as very benign." The reason for the braking parachute failing to release was that a screw was blocking the release lever. The screw had been inserted in order to prevent inadvertent deployment of the chute on the ground, and the technicians had forgotten to remove it just before flight.

In parallel with these gliding flights a series of unmanned vehicle launches tested the rocket boosters, the Walter motor and the autopilot. Beginning on 18 December 1944, fifteen of these were shot into the sky from the Lager Heuberg military training area near Stetten am kalten Markt. These tests led to the abandonment of the plan to reuse the Walter engine, because it invariably got damaged by the rather hard impact of the parachute landing. The tests also proved that the thrust of the boosters varied by as much as 50% either way, making the stability of the vehicle very unpredictable during its initial boosted ascent.

Preparations on the launch pad for the flight of an unmanned test version of the Natter.

Launch of an unmanned Natter prototype.

By early 1945 Germany had all but lost the power to defend its own airspace and, driven by desperation, the SS ordered a manned Natter launch even though many of the unmanned test flights had ended in crashes and explosions. The SS found a pilot willing to risk his life in Lothar Sieber, a Luftwaffe Second Lieutenant who had been demoted to the rank of Private after drinking alcohol on duty, and who hoped that his flight would restore his previous rank. On 1 March Sieber got into Ba 349A1 number M23 at the Lager Heuberg military area. The booster rockets were ignited and the wooden contraption raced along its launch rail and shot straight up, thereby making Sieber the first person to be launched vertically from the ground by rocket power. He did not have much time to enjoy this accomplishment though, because after climbing about 100 meters (330 feet) the machine suddenly turned over on its back and lost its cockpit cover. Since this cover included the headrest it is likely that the acceleration snapped Sieber's head back against the fuselage, at least rendering him unconscious and perhaps even snapping his neck. The Natter reached a peak altitude of about 450 meter (1,500 feet) some 15 seconds into the flight, then it dived essentially vertically into the ground. Whether the cover had not been installed correctly or had simply not been sufficiently closed was never discovered. Sieber was posthumously promoted to First Lieutenant. Development and testing continued

US soldiers with a Natter.

using unmanned prototypes but it was clear that the Natter would not be ready for service before the Allies overran Germany. And of course its use would be limited owing to the scarcity of the special propellants required for the Walter engine. Thus no Natter was ever launched against an Allied bomber. After the war, US forces found many Natters in various states of completion. One of those that they took back to America for evaluation is currently awaiting restoration at the National Air and Space Museum's Steven F. Udvar-Hazy Center close to Washington Dulles International Airport.

A high-fidelity Natter replica is on display at the Fantasy of Flight Museum in Florida. Looking at this crude wooden tube with its small wings, you cannot imagine anyone volunteering to pilot it, to be shot up at close to 800 km per hour (500 miles per hour) to attack a fleet of heavily armed bombers. The Planes of Fame Air

Museum's Chino location in California has a full scale model that was built by hobbyist George Lucas (not to be confused with the creator of *Star Wars*), who also made the full-scale model of the Me 163B that is on display in the same museum. In Europe a Ba 349A replica is displayed in the Deutsches Museum in Munich, Germany, featuring realistic colors and markings of one of the unmanned test aircraft. On its horizontal stabilizers there is an authentic-looking text promising a reward to anyone who finds the tail section and returns it to the military, evidently because some units were expected to drift out of the test area while descending on their parachute. Another German museum, the Stettener museum located near the place where the Natters made their test flights, shows a replica of the one flown by Sieber. It is depicted standing against the launch tower with mannequins of Bachem and Sieber in front of it in discussion about the upcoming flight. At the time of writing, a group in Speyer, Germany, is also building a detailed replica of a Natter according to the original plans. This is intended to be flightworthy except for the rocket propulsion, but it is not actually meant ever to take to the sky. Eventually the team hopes to build a small series of ten aircraft.

The Natter was basically a guided surface-to-air missile with the pilot acting as its control system; a job that was soon taken over by electronics in the new air defense missiles that became operational after the war. These were smaller and lighter than a human, did not require training, and were cheap enough not to require retrieval after use. The Ba 349 remains unique as the first, and last, piloted air defense missile. It is interesting to note that the Natter bears similarities to the Space Shuttle in that both were crewed, vertically launched, liquid propellant rocket planes that had jettisonable solid propellant boosters for take-off, with the liquid propellant motors being started and checked (and stopped, if needed) prior to igniting the boosters, at

Natter replica in the Fantasy of Flight Museum in Florida [Rogier Schonenborg].

which moment there was no way to prevent the vehicle from leaving the ground. In 1952 Bachem noted, "within a few months we had to track down, go through and solve numerous problems associated with vertical take-off, problems which the designers of future spacecraft will also have to look at", and also, "The attainment of great goals is not possible by a single leap, it is the result of an arduous climb up a steep ladder, step by step! Perhaps through our labor we have constructed one rung on that ladder!" Although true, these words appear to impart too much philosophical meaning to what was basically a last-ditch, near-suicidal weapon system.

Erich Bachem remained in Germany after the war and resumed his innocent pre-war business of designing and building camping trailers and motor homes. He joined the business of a neighbor, Erwin Hymer, whose company is still manufacturing recreational vehicles and camping trailers in the same buildings in which Bachem's rocket interceptors were made. One of Hymer's popular lines of camping trailers is called 'Eriba', Bachem's nickname as a student derived from ERIch BAchem. You can thus still ride a Bachem, not at transonic speeds but certainly at less risk to your life.

LAST STAND ROCKET PLANES

The need for a revolutionary air defense fighter led to many other German rocket interceptor designs near the end of the war, as well as an even more bewildering array of advanced propeller planes, jet fighters and intercontinental bombers that are beyond the scope of this book. Even limiting the focus to rocket airplanes results in a lengthy list of concepts.

Junkers came up with the EF127 'Walli', a rather whimsical code name for what was to have been a relatively large, horizontal take-off rocket fighter equipped with a retractable tricycle undercarriage (i.e. one with wheels under the wings and the nose), a take-off weight of 2,960 kg (6,530 pounds), a wingspan of 6.7 meters (22 feet) and an armament comprising a pair of MG 151/20 20-mm cannon and a dozen air-to-air rockets. With two solid propellant take-off assist boosters each delivering a thrust of 10,000 Newton and an HWK 105-509C dual-chamber rocket engine it was predicted to be able to climb to an altitude of 10 km (6 miles) in a mere 75 seconds, which was significantly faster than either the Me 163B or the Me 263. Apart from the boosters it had many similarities with the Me 263, but had a more conventional shape with two straight wings and a tail with horizontal stabilizers.

Then there was the Arado Ar E381 minifighter, of which the first design had a length of just 4.7 meters (15 feet), a wingspan of 4.4 meters (14 feet) and a weight of only 1,200 kg (2,650 pounds). It was to have been carried to altitude slung under an Ar 234 jet bomber, a plane that was actually operational in August 1944. Because the E381 was so small and simple, heating and electrical power would be supplied by the host during the ascent. Communication between the E381 pilot and his colleague in the Ar 234 would be via a telephone link. The E381 was to be released at an altitude of about 1 km (0.6 miles) above the enemy bomber formation to execute its attack in a shallow dive at close to 900 km per hour (560 miles per hour), firing either a single MK 108 cannon or air-to-air missiles. The Walter HWK 109-509C dual-chamber engine would subsequently be ignited to climb for a second attack pass. At that point the E381 would dive and glide to the ground unpowered. It would land using a skid, similar to the Me 163. Then the little plane was designed to be easily dismantled into wing, fuselage and tail units that could be manually loaded onto a truck for return to base.

Due to the limited ground clearance of the E381 when suspended under its carrier plane whilst on the ground, its pilot would lie on his stomach in the cramped cockpit and view through a bubble window in the nose. This prone position would also help him to sustain the high G-forces when making sharp maneuvers at high speeds. Pilots in a conventional sitting position could experience tunnel vision, black-outs (loss of vision) and even loss of consciousness when sharply starting a steep ascent or a tight turn because the centrifugal force would push blood into the lower part of their body, depriving the brain of sufficient oxygen. Similarly, when pilots would suddenly push their planes into a dive the negative G-forces would push blood into the upper part of the body, causing a reddening of his vision known as 'red-out'. These phenomena were much less trouble when the pilot was in a prone, horizontal position, since there was little vertical distance between his upper and lower body, making it easier for the heart to keep pumping sufficient blood to the brain. Also, the prone position led to a more aerodynamic fuselage shape with a smaller frontal area, reducing drag.

The second design of the E381 was slightly larger and had a weight of 1,270 kg (2,790 pounds). What appears to be the Ar E381 III had a length of 5.7 meters (19 feet), a wingspan of 5.1 meters (17 feet) and a weight of 1,500 kg (3,300 pounds). Instead of a single MK 108 cannon, the third version was armed with six RZ-65 air-to-air rockets that would be fired from the leading edges of the wings. But the E381 project never materialized. After the war, the prone pilot position was tested in the UK using a second, experimental cockpit attached to the nose of a Gloster Meteor F8 jet fighter. For safety, the conventional cockpit was also manned. The tests showed that the prone pilot could endure slightly higher G-levels, but this position proved to be very uncomfortable, to lead to vertigo (a sensation of motion when one is actually stationary) and to seriously limit the pilot's rear view; the latter being a very serious disadvantage in a dogfight. Furthermore, it would be very difficult to devise a good ejection seat escape mechanism for this awkward position: ejecting forward was not an option because the plane would catch up with the pilot and collide with him. After 55 hours of flight testing, the idea was abandoned. The prone position soon became obsolete after the war anyway thanks to special anti-G suits with inflatable bladders in the trousers: when they experienced negative G-forces in a sitting position, these bladders would quickly inflate to firmly press against the abdomen and legs in order to restrict the draining of blood away from the brain.

Arado also proposed to develop the TEW 16/43-13, a low-wing rocket interceptor that had a wingspan of 8.9 meters (29 feet), a Walter HWK 109-509A engine, and an armament of two MG 151/20 20-mm cannon and two MK 108 30-mm cannon set in its nose. The wings and tailplane were moderately swept back, and the pilot would sit in a conventional upright position. The company also devised the TEW 16/43-15, a combined jet/rocket powered variant in which a jet engine was mounted on top of the (lengthened) fuselage.

In his 1929 book *Wege zur Raumschiffahrt* (*Ways to Spaceflight*) Hermann Oberth had contemplated a rocket aircraft built like a tank to fly into an enemy air fleet. It could then use conventional gunfire to destroy targeted planes but he suggested that collision would also be an effective means of attack. By 1943 this dramatic concept started to look like a valid idea to stop Allied bombers, leading Lippisch to suggest a 'ram rocket'. This would be launched vertically using a liquid propellant engine and solid propellant boosters (like the Natter) and be equipped with a sharp steel nose to slice through a bomber without damaging the ram rocket (or its pilot). In November 1944 the Zeppelin works, famous for building giant airships, had the same idea and proposed the 'Rammjäger' (Rammer Fighter). Taking off from a jettisonable tricycle carriage, their little plane would be towed aloft by a fighter aircraft and be released near the bombers. The pilot would ignite a single solid propellant Schmidding rocket (similar to the Natter boosters) to accelerate to 970 km per hour (600 miles per hour) and launch his 14 R4M rockets. He was then expected to make a second pass and actually ram his plane through the tail section of a bomber. It was calculated that the Rammjäger would slice cleanly through without great loss of speed and stability, and survive the collision thanks to its armored cockpit and reinforced fuselage and wings. It would then glide back unpowered and land on a retractable skid. By virtue of not having a liquid propellant rocket engine, the Rammjäger was even simpler than the Natter. After towed glide flights in January 1945 an order for sixteen pre-production prototypes was placed. However, US bombers destroyed the Zeppelin factory before their construction could commence.

Another rather ludicrous proposal by Zeppelin was for the 'Fliegende Panzerfaust' (Flying Armored Fist, or Flying Bazooka). This was to be an aircraft filled with high explosives to ram and blow up an enemy bomber; the pilot was supposed either to eject just before impact or sacrifice himself to ensure an accurate hit. If possible, the plane would also be equipped with a cannon to shoot down other bombers in advance of the final ramming attack. In line with the need for a very simple and cheap design, it would be powered by six solid propellant rocket motors only. The machine would be towed to altitude, for which it was equipped with a hooked nose connecting to the tail of a regular fighter aircraft. It would have been some 6 meters (20 feet) in length with a wingspan of 4.5 meters (15 feet). The project never got any further than a full-scale mockup that was presented to the SS in January 1945.

Then there was the DFS 'Eber' (Boar), a tiny rocket fighter that either an Fw 190 propeller fighter or an Me 262 jet fighter would tow using a long pole, then release some 300 meters (1,000 feet) above the bombers. The Eber would engage a target in a gliding dive utilizing either a single Mk 108 cannon or air-to-air rockets, and then ignite two solid propellant motors that would deliver a thrust of 15,000 Newton for 6 seconds to make a second attack in which it would ram a bomber. The prone pilot would be protected by an armored cockpit and use a special spring-dampened sliding seat to soften the 100 G deceleration shock that would result from the collision with a large enemy plane. After the final attack, a parachute would pull the pilot and his seat from the cockpit. At sufficiently low speed, he would jettison the seat and land using his own parachute. The airplane itself was expendable.

The fact that the idea of ramming enemy planes was popular near the end of the war reflects a level of panic in German military planning and an increasing fanatical demand for pilots to sacrifice themselves in defense of their doomed motherland. The Sombold So 344 'Rammschussjäger' (Ram-shoot Fighter) was another approach to the ramming plane concept. Rather than ramming a bomber itself, this rocket aircraft would carry a 400 kg (880 pound) warhead on its nose and fire it into the middle of a bomber group using a solid rocket motor built into the bomb. The warhead was fitted with four stabilizing fins and a proximity fuse for automatic detonation when close to an enemy plane. Allied bombers flew in tight formations so that the onboard gunners could provide cover to each other's planes and work together to bring down attacking fighter planes. The explosion of the warhead released by the So 344 was expected to destroy three of four bombers and open the formation sufficiently for other German fighters to engage individual bombers. The So 344 was expected to join this carnage employing a pair of machine guns.

The So 344 was to be powered by either a single Walter HWK 109-509 engine or a number of solid propellant rocket motors, have a wingspan of 5.7 meters (19 feet) and a take-off weight of 1,350 kg (2,980 pounds). Its cockpit was located well behind the wings, just in front of the vertical and horizontal stabilizers. It was designed to be carried into the air on top of another aircraft, to land employing a skid, and then be dismantled into two sections for transportation by truck back to its base.

Messerschmitt also produced designs for simple towed ramming rocket aircraft. The basic design for the Me P.1103 of July 1944 had its pilot lying prone inside an armored cylinder cockpit. After release from a conventional Bf 109 fighter or even an Me 262 jet fighter tow plane, it would attack at 700 km per hour (435 miles per hour) shooting a single MK 108 30-mm cannon before ramming one of the bombers. A parachute would then pull the pilot and his seat out of the aircraft. Another chute would soft-land the reusable assembly consisting of the armored cockpit, the cannon, and the floor-mounted Walter RI-202 'cold' rocket engine (the same as in the RATO pods used on German bombers) for retrieval. The tail section would be lost, but that was to have been only a simple wooden empennage derived from the V1 unmanned flying bomb. Another interesting fact about the Me P.1103 is that it was to have a rearward firing rocket launcher to defend itself from Allied fighters. A design for an alternative P.1103 version of September 1944 shows a sitting pilot looking through a conventional bubble canopy. The Me P.1104 was another Messerschmitt design for a simple towed, rocket propelled defensive interceptor. This one would have had an HWK-109-509A rocket engine (as on the Me 163B) and been equipped with a single MK 108 cannon. The total weight of the fully loaded plane was to be 2,540 kg (5,600 pounds), so its rocket engine should have provided a maximum speed of 800 km per hour (500 miles per hour) and a range of some 90 km (60 miles).

All these concepts for small and cheap rocket propelled fighters were consigned to the archives by the Luftwaffe, which was undoubtedly better news for the pilots who would have been required to fly these rickety death traps than it was for the crews of the Allied bombers. However, some other rocket fighter designs did progress a little further through the development process.

The Focke Wulf 'Volksjäger' (the People's Fighter; not to be confused with the Heinkel He 162 'Volksjäger' jet fighter) was based on the design for the Ta 183 jet fighter. It was to be a defense interceptor with a similar role and mode of operation to the Me 163. It also looked a lot like the Komet, with a Walter HWK 109-509A2 and sharply swept wings mounted mid-fuselage which were to be mostly made of wood. The wingspan would be 4.8 meters (16 feet) and the total length a mere 5.3 meters (17 feet). It would take off from a jettisonable dolly, climb to the enemy bombers and attack using two MK 108 30-mm cannon located in the lower fuselage sides, then glide back unpowered and land on a retractable skid. Unlike the Me 163 (but like the Ta 183) it had a horizontal stabilizer mounted on the vertical fin to form a 'T' shaped tail. With the addition of four solid propellant strap-on boosters, the Volksjäger was intended to reach an altitude of 5.9 km (19,400 feet) in just 60 seconds, and 16.5 km (54,100 feet) in 100 seconds. This was a significantly faster climbing speed than the Komet, which did not have boosters. The maximum speed was expected to be about 1,000 km per hour (620 miles per hour). Three aircraft were under construction when the war ended but none ever flew. As was the case for the He 162 jet that was also called the 'People's Fighter', the Focke Wulf was probably meant to be flown by relatively untrained pilots recruited from the Hitlerjugend, the Nazi youth movement. But the He 162 proved too difficult for amateurs to fly, and this would probably also have been the case for the Focke Wulf rocket interceptor.

Another Focke Wulf design was the Ta 283, an airplane with a slender fuselage, wings swept back at 45 degrees, and two large ramjet engines. Since ramjets do not work at low speeds, for take-off the aircraft was to have a single Walter HWK 109-509A rocket engine and about 30 seconds worth of rocket propellant. To prevent the big ramjets from disturbing the airflow over the wings, they were to be mounted on the sharply swept tailplane. The armament was envisaged to comprise two MK 108 cannon. When this concept was judged lacking in performance an improved design was proposed in which the fuselage intended for the Fw 252 jet fighter would have a small jet engine emplaced in the central fuselage, two ramjets on the tailplane and a Walter HWK 109-509A just beneath the tail boom. This design was so preliminary that it didn't even have a proper designation, but after the war it came to be known as the Fw 252 'Super Lorin' (after the inventor of the ramjet concept, Frenchman René Lorin).

Heinkel's contribution to the mixed bag of radical rocket fighter concepts was the P.1077 'Julia'. As designed in August 1944, the wooden Julia would have been 6.98 meters (22.9 feet) long, had straight shoulder-mounted wings with a total span of 4.6 meters (15 feet) and a tail with twin vertical stabilizers. Its propulsion was to have been one Walter HWK 109-509C engine with separate combustion chambers for the initial climb and subsequent cruise, and two booster rockets mounted on either side of the fuselage. It would have had two MK 108 cannon housed in blisters on the sides of the forward fuselage. It was initially envisaged that the pilot would fly in a prone position but an alternative design was prepared in which he would sit in a conventional manner. Like its Bachem Natter competitor, the Julia would have been launched vertically off the ground, but after the attack the pilot would have landed it on a retractable skid rather than bail out. In September 1944 the RLM ordered 20 Julia prototypes and then two weeks later demanded the production of 300 operational aircraft per month. Nothing came of this, of course. Only towing trials with a full-scale mockup were performed before it was destroyed when the Vienna woodworks was bombed, and the Schäffer company in Linz had time only to start two unpowered prototypes, neither of which was completed.

The DFS 346 of the German Institute for Sailplane Flight was not intended for combat, it was designed purely as an experimental research plane. Its chief designer, Felix Kracht, gave it wings swept back at 45 degrees, a streamlined fuselage with a prone pilot, and an HWK 109-509C rocket engine. It was to be taken to high altitude by a Dornier Do 217 bomber and then air-launched, a novel concept which, it was hoped, would save the DFS 346 sufficient propellant (otherwise needed for take-off and ascent) to break the sound barrier. The altitude provided by the carrier plane also resulted in lower aerodynamic drag due to the thinner atmosphere and meant Mach 1 could be reached at a lower absolute flight speed since the speed of sound is lower in the colder high-altitude air. The pilot was in a pressurized section that formed a self-contained escape capsule which could be separated from the plane in an emergency. Stabilized by a small parachute, the capsule would fall to an altitude of about 3 km (10,000 feet) where the air pressure is safe, the Plexiglas nose section would separate and the pilot would slide out and land using his parachute. The unfinished DFS 346 prototype was captured by the Soviets at the end of the war, taken to Russia, rebuilt and actually flown (we will discuss this further in a later chapter).

The concept of air-launching was later adopted in the USA for the world's first supersonic rocket plane, the post-war X-1. The highly swept wings of the DFS 346 were actually much more advanced than the conventional straight wings of the X-1 and make the DFS 346 look more like a modern supersonic fighter than the famous

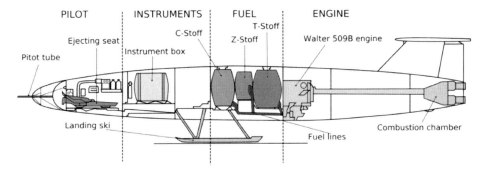

Overview of the internal design of the DFS 346 [Mothmolevna].

American rocket plane which relied upon brute rocket power and sturdy wings rather than sophisticated aerodynamics to blast through the sound barrier.

Felix Kracht also designed the DFS 228 rocket propelled reconnaissance aircraft, a fairly conventional sailplane design with long, slender but straight wings spanning 17.6 meters (57.7 feet). It was to have a dry weight of 1,350 kg (9,280 pounds) and a take-off weight of 4,210 kg (2,980 pounds). It would be air-launched from on top of a Do 217, be powered by a Walter rocket motor, and land using a belly-mounted skid. The pilot was to be prone in a pressurized escape capsule. The long wings were designed to enable the plane to achieve a cruise altitude of 24 km (80,000 feet), safe from interception by Allied fighters. Its maximum speed was estimated to be about 900 km per hour (560 miles per hour). The mission profile was for a powered climb followed by a slow, unpowered descending glide, then reigniting the rocket motor to climb back to altitude, and so on until the propellant was finished. The resulting saw-tooth flight pattern that would give the aircraft the relatively long range of 1,050 km (655 miles) also prompted its nickname of 'Sagefish' (Sawfish).

Walter designed a new rocket motor for the plane because the requirement for a very streamlined fuselage in combination with weight balance made it necessary to place the motor near the plane's center of gravity, which required fitting a long thrust tube to the tail. The resulting motor was an elongated version of the single-chamber HWK 109-509A2 named the HWK 109-509D. The DFS built three prototypes: V1, V2 and V3. The DFS 228 V1 had a conventional seat but it was slightly inclined to the rear to accommodate the fuselage streamlining and to assist the pilot in handling high G-loads (a similar idea was introduced in the F-16 fighter jet several decades later). The V1 made several unpowered glide flights after being carried to altitude on top of a Do 217, and was later fitted with the HWK 109-509D for ground tests. This prototype was captured by the Americans at the end of the war and in 1946 was sent to Britain for study, but it arrived in very bad condition and was apparently scrapped. The V2 had a cockpit with a prone pilot position and apparently made several glide flights before being damaged in a landing accident. The V3 was never finished. None of these aircraft ever flew under rocket power.

Rather than develop a completely new aircraft, in mid-July 1943 Arado proposed to produce a high-altitude reconnaissance rocket plane by modifying the Ar 234 jet, which had made its first flight in June. In the initial design the two jet engines were to be replaced by two HWK 109-509A rocket engine pods. The resulting Ar 234R 'Höhenaufklärer' (High-altitude Scout) would be able to take off by itself, ascend to an altitude of 16.5 km (54,000 feet) and photograph its targets while descending in a shallow glide. It would be equipped with the pressurized cockpit already created for the Ar 234C reconnaissance jet. A later design had a dual-chamber HKW 109-509C built into the tail instead of pods slung under the wings. This version was to be towed to an altitude of 8 km (26,000 feet) by a Heinkel He 177 bomber (a large, long-range bomber then under development) and upon being released it would boost itself up to 18 km (59,000 feet). The Ar 234R would be able to photograph areas some 250 km (155 miles) from its base, being towed 200 km (125 km) towards its target and flying a further 50 km (30 miles) on rocket power. The plane would subsequently glide over its targets, igniting the rocket engine at intervals to maintain sufficient altitude.

Freed from the need to mount engines under the wings, a special highly efficient wing was designed to give the aircraft a glide angle of 1 to 14 (in which it would lose 1 meter of altitude for each 14 meters of horizontal flight), sufficient to enable it to glide the entire 250 km (155 miles) back to base.

Another Ar 234 variant fitted with a rocket engine was actually under construction when the Allies captured the factory in which it was being assembled. It consisted of an Ar 234B fuselage equipped with a new concave-curved swept wing optimized for high-speed flight. It was powered by a pair of BMW 003R engines, each consisting of a BMW 003 turbojet with a BMW 718 liquid propellant rocket engine mounted on top. The fact that the pumps for the rocket propellant were powered by the jet engine resulted in a nicely compact assembly. The rocket engine could deliver an additional 10,000 Newton of thrust for 3 minutes during take-off and ascent. Unfortunately the prototype was scrapped.

The sole surviving Ar 234B can be seen in the National Air and Space Museum's Steven F. Udvar-Hazy Center near Washington Dulles International Airport. It is the basic jet type with a pair of Walter RI-202 'cold' RATO pods under its wings. The parachute pack intended to retrieve the pods following jettisoning soon after take-off can be clearly seen on these examples.

Another pragmatic design enhanced the standard Me 262 jet fighter (which had a jet engine under each wing) with an additional rocket engine. To test the principle an Me 262 was adapted to accommodate a redesigned HWK 109-509A2 motor with the combustion chamber installed in the tail; a portion of the lower part of the rudder was cut away to make room for the nozzle and the rocket exhaust. This Me 262 C-1a 'Heimatschützer I' (Home Defender I) prototype made its first rocket assisted take-off on 27 February 1945. The combined thrust from the two jets and the single rocket reduced the take-off run by at least 200 meters (660 feet) and pushed the plane to an altitude of 8 km (26,000 feet) in about 3 minutes (about half the time required by a standard Me 262 and similar to the Me 163 Komet's performance). Major Heinz Bar later managed to intercept and shoot down an American P-47 Thunderbolt fighter in the rocket propelled test plane by climbing to about 9 km (30,000 feet) in a little over 3 minutes.

A similar concept was the Me 262 C-1b 'Heimatschützer II' which was provided with two BMW 003R combined jet-rocket engines for boosted thrust. Only a single prototype was built and its only flight with the combined propulsion was made on 26 March 1945. This engine was also to have been mounted on an He 162 'Volksjäger' (People's Fighter), the standard version of which had a single BMW 003 jet engine on its back. However, the creation of this He 162E prototype could not be achieved before the war ended.

Despite the success of the 'Heimatschützer I' it was judged that the rocket engine added too much complexity to the jet fighter, making it more difficult to maintain. In addition, the tanks needed for the less propellant-economic rocket severely reduced the available space for fuel for the jet engines, which limited the range and flight duration of the fighter in comparison to a standard Me 262 (issues that would also plague the post-war mixed-power interceptor designs that we will discuss later). These considerations led to the design of the 'Heimatschützer III' which had a rocket engine bolted onto the belly of the aircraft and its propellant in drop tanks that would be jettisoned once they were empty so that the plane would not be burdened by their useless weight and aerodynamic drag. The advantages of this arrangement were that it required fewer modifications of the standard Me 262, and the modified Me 262 could take off without the rocket engine if it were not needed. Another plus was that maintenance of the rocket system could be separated from maintenance of the aircraft: a fighter would not be grounded by a malfunctioning rocket engine, but could be quickly fitted with a replacement engine while the faulty unit was taken to the repair shop (an idea which, after the war, would be applied to the French Mirage rocket-motor equipped fighters).

Walter designed a powerful 'hot' engine based on his HWK 109-509 series for the 'Heimatschützer IV'. This HWK 109-509 S2 gave a thrust of 20,000 Newton and weighed just 140 kg (310 pounds). The engine was to be mounted on the belly of the plane, just behind the roots of the wings. Two jettisonable tanks would be carried externally on bomb attachment points under the nose, with flexible hoses delivering a total of 1,200 liters of T-Stoff to the Walter motor. The C-Stoff would be carried in the rearmost converted fuselage tank. This was designated the Me 262 C-3 and work on the prototypes started in January 1945. Tests on the T-Stoff tanks apparently revealed a problem with feeding fuel when the drop tanks were mounted lower than the rocket motor. In April the Allies overran the factory at Jenbach before a single prototype could be completed.

Eugen Sänger's design for the 'Silbervogel' (Silverbird) intercontinental bomber surely ranks as the most ambitious of Nazi Germany's rocket plane concepts. It was the first, more or less realistic design for an aircraft capable of flying on the edge of space and it is therefore described in more detail in a following chapter.

Germany hoped that its superior rocket and jet airplane technology would counter the vast numbers of conventional planes that were being manufactured by Americans in their secure homeland. Given a bit more time, this might have worked at least for a while and greatly prolonged the war in Europe. The Luftwaffe might have been able to put some really revolutionary aircraft into the skies, and a German rocket aircraft would likely have been the first to exceed the speed of sound. However, by the time the 'wonder weapon' airplanes received proper support it was already too late, and in any case the developments were too scattered, rivalry between the military services too disruptive, and the constraints in terms of material and propellant availability too great. Some of the small interceptors relied on Arado jet carrier planes that were not themselves available, and those meant to be towed behind conventional bombers and fighters would have been vulnerable to enemy fighters during the first phase of their mission due to the low speed of the towing operation. Many of the RLM designs now seem to be more the results of desperation and wishful (albeit creative) thinking than the reality of military operations. Nevertheless many post-war airplane designs were strongly influenced by some of the more brilliant German developments, particularly the Me 163 and Me 263.

4

Non-German wartime rocket fighters

"Make everything as simple as possible, but not simpler" – *Albert Einstein*

Also during the Second World War the Japanese and the Russians were actively pursuing rocket and rocket aircraft technology. In both cases this was to make up for ineffective conventional air power, but for the Japanese this necessity grew towards the end of the war whereas for the Russians the need was much more urgent during the disastrous early years of the German invasion. The accomplishments of both of these countries have been largely overshadowed by the well-known developments in Germany but are nevertheless impressive. However, as with many of the German concepts, the rush to get rocket aircraft with spectacular performance into operation led to hasty designs that lacked many of the refinements required to make the aircraft safe and of military value. Uniquely, the Japanese created a rocket vehicle that was actually intended *not* to be safe: a manned suicide missile. The pilots of some of the planned German rocket planes would have had little chance of surviving a mission but at least they were not specifically required to die.

Rather than losing themselves in complex technological adventures, the US and the British instead focused on the mass production and overwhelming deployment of conventional aircraft. This proved to be a devastatingly successful strategy, but it left them initially lagging considerably behind in the new military field of rocketry and rocket planes.

JAPAN: TOO LITTLE TOO LATE

Just like its axis partner, Japan was seeking a 'wonder weapon' that could counter the overwhelming power of the Allied forces. When Japanese military attachés first saw the Me 163 on a visit to Peenemünde West in 1943 they were mightily impressed and recognized in the revolutionary aircraft a means of halting the expected onslaught of American bombers. The attachés had seen the devastation bombing had caused in Germany and knew that soon Japan would be facing a similar fate. The high-flying Boeing B-29 Superfortress bomber, already in production, would surely soon darken the skies over Tokyo. Still struggling to develop turbo-superchargers to enable their conventional propeller airplanes to reach the altitude at which the B-29 would cruise, Japan was desperate to find an alternative defense measure (turbo-superchargers compress the thin air at high altitude, so that sufficient oxygen gets into an aircraft's piston engine for proper combustion).

Commander Eiichi Iwaya (who went on to play a crucial role in this story) saw a Komet demonstration at the Rechlin airfield in April 1944, then wrote in his diary: "The roar, and the blue-green flame from the Walter motor, with its immense 1,500 kg of thrust, was evidence that German technology was still alive here, even if the combat situation was steadily getting worse."

In early 1943 the Japanese had negotiated licences to manufacture the Me 163B and its engine, although this did not come cheap: the rights and information to build the HWK 109-509A rocket motor alone cost them 20 million Reichsmark, which is equivalent to something like $100 million today. The Japanese were allowed to study the aircraft's production in Germany, as well as the operational procedures of the Luftwaffe. Documentation on the Komet was sent to Japan by the German submarine U-511, which left Lorient in occupied France on 10 May 1943. It was then put into service with the Japanese navy under the name RO-500. To enable Japan to build its own forms of the aircraft, Germany later agrees to supply complete blueprints for the Me 163B and the HWK 109-509A, as well as a complete Komet, two sets of sub-assemblies and components, and three complete rocket engines.

This equipment and documentation is loaded onto two Japanese submarines, one of which leaves the German harbor of Kiel on 30 March 1944 and the other leaves from Lorient just over a fortnight later. The first, RO-501, is sunk in the mid-Atlantic by the destroyer USS *Francis M. Robinson*. The second boat, the I-29, reaches the harbor at Singapore. Soon thereafter, Allied intelligence intercepts a message from Berlin to Tokyo that lists all the strategic cargo carried by the I-29. Before leaving Singapore the I-29 commits the mistake of radioing a detailed itinerary for the final part of its journey to Japan; the message is intercepted and decoded by US Navy's Fleet Radio Unit. The Navy sends three submarines to hunt the I-29 and on 26 July the USS *Sawfish* finds the enemy boat running on the surface near the Philippines. The Navy submarine launches four torpedoes: three hit and sink the I-29. With all the Me 163 parts and plans on the ocean floor, the Japanese would not have received any of the material sent from Germany in 1944 were it not for a Japanese naval mission member who left the I-29 in Singapore soon after it docked. Unaware that the I-29 is doomed by sloppy communications security, Commander Iwaya flies to Japan with a briefcase containing only twenty pages of the design manual, a photo of the Me 163B and another of its wing, a document describing the production and handling of the dangerous propellants, and data for several types of valves used on the plane. It is by no means sufficient to build a copy of the Komet and its advanced rocket engine, but this is all that reaches Tokyo.

For the Japanese aeronautical engineers the scarce information, together with what they received from the U-511 the previous year and radio telegraph communications with the air ministry in Berlin, actually turns out to be sufficient to develop their own forms of the Komet. The project is set up as a joint effort by the Imperial Japanese Army Air Service and the Navy Air Service, but they argue about how to proceed. The Army proposes to develop a new, completely Japanese rocket interceptor that can carry more propellant than the Me 163B in order to increase the duration of its powered flight. The Navy seems to better understand the urgency of Japan's gloomy military situation, and proposes to stick as closely as possible to the already proven design of the German Me 163B. The Navy Air Service wins the dispute, and in July 1944 specifications are issued. Aircraft manufacturer Mitsubishi Jukogyo KK wins the contract to develop and produce two models of the new plane: the J8M1 'Syūsui' (Autumn Water) for the Navy and the almost identical Ki-200 for the Army.

However, the Army secretly decides to develop its own concept, independently of Mitsubishi, at its aerotechnical institute Rikugun Kokugijitsu Kenkyujo at Tachikawa (as in Germany there is considerable rivalry between the different military services, even although by now the desperate situation of the war leaves little time for such a divergence of effort). This Ki-202 is to be a better plane than the Me 163B, the J8M1 and the Ki-200, but it never leaves the drawing board. It was to have had a stretched fuselage to carry more propellant and used a KR20/Toku-Ro 3 dual-chamber rocket engine similar to the HWK 109-509C, with the large chamber and nozzle producing 20,000 Newton of thrust for take-off and climb, and the smaller one 4,000 Newton for cruise flight.

Work on the J8M1 and Ki-200 rocket plane versions quickly gets underway with Mitsubishi manufacturing the Japanese form of the HWK 109-509A engine, which is known variously as the KR10 and the Toku-Ro 2, and joining forces with Nissan and Fuji to develop the airframe. The Naval Air Technical Arsenal in Yokosuka develops the MXY8 glider 'Akigusa' (Autumn Grass), which will initially be used to study the handling characteristics of the Syūsui and later to train its pilots. On 8 December, Lieutenant-Commander Toyohiko Inuzuka makes the first flight in an MXY8 glider after being towed into the air from Hyakurigahara airfield, and finds that it closely matches the described handling characteristics of the German Komet. Two additional gliders are made in Yokosuka, one of which is delivered to the Army for evaluation at its own aerotechnical institute.

The resulting J8M1 rocket plane looks very similar to the Komet (as intended) but it has a take-off weight some 430 kg (940 pounds) lighter due to a reduced propellant volume, less ammunition, and the deletion of the armored cockpit glass (the lack of armor protection for pilots and engines was a common feature of Japanese fighter aircraft, resulting in a weight reduction and hence increased agility at the cost of a higher vulnerability to enemy bullets). However, the weight reduction does not fully compensate for the fact that the KR10 engine produces less thrust than the original Walter engine: 14,700 instead of 17,000 Newton. The Syūsui will therefore not be as fast as the Komet and will have a slower rate of climb. Nevertheless, its top speed of 900 km per hour (560 miles per hour) will still be more than sufficient to outrun any Allied propeller fighter aircraft, and its rate of climb of 48 meters per second (156 feet per second) is still phenomenal. The German Komet could achieve 960 km per hour (600 miles per hour) and had a rate of climb of 61 meters per second (199 feet per second). The distinctive power-generation propeller of the Komet is omitted from the J8M1, which instead has a longer nose section housing a battery. It is

The Japanese J8M rocket interceptor.

armed with two 30-mm Type 5 cannon that can each spit out 500 rounds per minute. A J8M2 is planned that will differ only in that it sacrifices one cannon for a small increase in the propellant capacity for slightly longer endurance. The Army's Ki-200 is very similar to the J8M1, the most important difference being that it is equipped with two 20-mm Ho-5 cannon or two 30-mm Ho-155 cannon, each of which can fire as many as 600 rounds per minute.

A production plan is put together that should lead to having at least 3,600 rocket interceptors in operation by March 1946. The first J8M1 prototype is used for load testing on the ground. The next two are for flight testing, and on 8 January 1945 one of these is towed into the air by a Nakajima B6N1 (Allied designation 'Jill') bomber from Hyakurigahara airfield. The aim is to test the low-speed aerodynamics, so the plane has no engine or propellant. Water is used as ballast to obtain a realistic total weight and mass balance. The test shows that the design is very good for gliding and should handle well under rocket power at high speeds.

Development of the engine is not going well. The first KR10 prototype explodes immediately upon being started, as does a modified engine known as the KR12 (the KR12 design does not offer any real advantages over the KR10 and its development is halted). Since the Japanese engineers have little experience with liquid propellant rocket engines and have only a limited amount of design and production information on the German technology, they are having difficulty designing the equivalent of the Walter engine and especially the small, high-speed turbopump. The resulting delays mean that by mid-1945 the engine is still not available and the J8M1 still cannot be flight tested under power. Time is running out because the Allied forces are virtually banging on the door of the Japanese home islands. Captain Shibata, commander of the Navy air group which is to be the first to operate the J8M1, tries to speed up the development by agreeing with the development team that the engine will be deemed ready for flight if it can operate for at least 2 minutes without failures; desperation is clearly becoming a driving factor in the plane's development.

In the meantime, another submarine is dispatched from Germany to deliver Komet documentation and equipment to Japan. Also on board are Messerschmitt engineers Rolf von Chlingensperg and Riclef Schomerus. The U-864 leaves Bergen in Norway on 5 February 1945 but a couple of days out is obliged to return due to trouble with one of its two diesel engines. On the way back her periscope is spotted by a British submarine, HMS *Venturer*, which has been dispatched to intercept the U-864 and its cargo in response to the interception of German radio transmissions. On 9 February the submerged *Venturer* fires four torpedoes in a spread pattern at the U-864, which crash dives but suffers a hit by one torpedo and breaks in two. With it the last load of Komet equipment sent to Japan disappears into the depths.

The date for the first Syūsui powered flight slips further when another engine prototype blows up. In addition, relocating the KR10 and Syūsui development teams lest they be bombed by B-29s results in even more delays. Only in June 1945 does the engine meet the 2-minute thrust requirement. In fact the KR10 development team working in the Yamakita factory runs an engine for 4 minutes. The Mitsubishi J8M1 group at the Matsumoto research center runs another engine for 3 minutes. The striped pattern in the exhaust flame caused by the shock waves in the supersonic gas flow leads the engineers to dub the thrust plume the "tail of the tiger". Unpowered glide tests of one of the Syūsui prototypes with an engine installed, and tests running the engine within the plane's fuselage, are quickly organized and successfully completed. On 7 July the first J8M1 with an engine of doubtful trustworthiness finally stands ready at Yokoku airfield for its first flight under rocket power. Following the procedure established for the Komet, Lieutenant Commander Toyohiko Inuzuka rolls 320 meters (1,050 feet) down the runway and takes to the air after only 11 seconds, successfully jettisoning the dolly. He then flies horizontally to gain speed prior to climbing at a 45 degree angle. All is well up to this point, but at an altitude of some 350 meters (1,150 feet) the engine sputters, puffs black smoke and quits. The speed of the plane carries it up a further 150 meters (500 feet) whereupon Inuzuka levels off, dumps the remaining propellant and banks to the right intending to glide back to the airfield. However, the maneuver makes the plane lose speed at an alarming rate and causes it to drift off in the direction of a small building. Inuzuka pulls the nose up in a desperate attempt to avoid it but nevertheless clips the side of the structure. The aircraft comes down in a terrible crash. Inuzuka is extracted from the wreck severely injured, and dies shortly after.

It was soon realized why the engine had cut out. For the test flight the propellant tanks had been only half-filled for safety reasons and when Inuzuka started his steep climb the liquids sloshed to the back of the tanks. But the feeds to the engine were at the front of the tanks. This deprived the engine of propellant. The reason for the feed points being at the front of the tanks was that during actual combat operations it was expected that the Syūsui would climb above the enemy bombers and then attack in a powered dive. Such propellant as was left in the tanks would then slosh forward, and this had prompted the designers to put the feed lines there. The incompatibility of this system with a powered climb instead of a dive with half-filled tanks had not been noticed before the test flight. It is decided that the propellant supply system from the tanks to the KR10 must be modified, and in the meantime all powered flight testing is stopped. While the crash investigations are underway, two new KR10s blow up on the test stand indicating that the engine still has other issues in need of resolution.

However, less than a month after the disastrous test flight the first atomic bomb is dropped on Hiroshima. Another incinerates Nagasaki several days later. In response, Japan surrenders unconditionally.

Component production for the new airplanes had already begun in preparation for mass production of J8M1s and Ki-200s. After the three prototypes, four machines had been produced (one of them the first Ki-200). Training courses for Army and Navy pilots were also being organized using the Ku-53, an engineless glider with the same configuration and flight characteristics as the actual plane. However, even more than in Germany, the revolutionary rocket interceptor came too late to make a difference and was never used in combat.

Only two examples of the J8M1 survive today. One is on display at the Planes of Fame Museum in Chino, California. It is one of two aircraft taken to the US aboard the USS *Barnes* in November 1945 and which, after evaluations, were sold for scrap. One was apparently turned into pots and pans but 19-year-old Ed Maloney found the other one in a southern California storage yard. The owner of the facility thought it was some kind of boat but Maloney recognized it for what it was and bought it for the cost of its unpaid storage charges. It became the first plane in the museum that he was planning and is now the Planes of Fame Museum. Another Syūsui fuselage was discovered in 1961 in a cave in the Yokosuka area south of Tokyo, badly damaged and incomplete. Until 1999 it was on display at a Japanese Air Force Base near Gifu, then it was restored and completed with replica parts by Mitsubishi, whose archives contain 80% of the original blueprints (additional information for the restoration was obtained by studying the Planes of Fame Museum example). This plane can now be viewed at the company's Komaki Plant Museum, bearing its original bright yellow-green paint.

Towards the end of the war the Japanese Navy were also working on a completely home-grown rocket attack aircraft called the Mizuno 'Shinryū' (Divine Dragon). The project started as a fairly conventional-looking design for a simple kamikaze glider that could be launched from the shore using a trio of 1,300 Newton Toku-Ro Type 1 solid propellant rockets with a 10 second burn time. This plane was to crash into and thereby destroy Allied ships, or even tanks if these managed to get onto the beaches of Japan. A prototype for the glider was tested without the rocket engines, and these flights showed that the rocket propelled version could be expected to be difficult to fly. But experienced and well-trained pilots were too valuable to sacrifice in suicide attacks. A revised design was therefore proposed for a nimble attack aircraft powered by four 1,500 Newton thrust Toku-Ro Type 2 solid propellant rockets burning for 30 seconds. Armed with eight unguided air-to-ground rockets it would be able to attack ships and tanks, or even intercept B-29 bombers at a top speed of 300 km per hour (190 miles per hour) without the need for a suicidal collision. The Mizuno Shinryū was a canard design with large swept wings, small vertical stabilizer wings on the nose, and a vertical tail fin; the rocket boosters were to be fitted inside the tail part of the fuselage.

As explained earlier, the vertical stabilizers behind the wings of a conventional airplane push the tail down and compensate for the tendency of the main wings to push the nose down. An obvious disadvantage of this balancing technique is that the vertical stabilizers effectively provide negative lift, which has to be compensated by additional lift from the main wings. On a classical canard design the stabilizers are put in front of the wings, creating balance by pushing the nose up rather than the tail down; both the wings and the horizontal stabilizers provide positive lift, resulting an aerodynamically more optimal design. However if the plane pitches up, the angle of attack on the canard stabilizers increases and this gives them more lift, which in turn makes the plane's nose rise even faster. A canard airplane can thus be rather unstable in pitch. In the hands of a skilled pilot this translates into high maneuverability but for an inexperienced pilot tends to produce a crash. There are several other pros and cons regarding canard airplanes and different effects depending upon the location of the center of gravity relative to the center of pressure (lift), but it is interesting that a number of highly maneuverable modern jet fighters have shapes remarkably similar to the Shinryū; notably the Eurofighter Typhoon, the Sukhoi Su-30 and the Dassault Rafale.

To enable it to attack B-29 bombers flying at high altitude a pressurized cockpit was proposed for the Shinryū, but a pressure suit for the pilot in combination with an unpressurised cockpit was also possible. The plane would take off with the help of a two-wheeled jettisonable dolly, and it would land using fixed skids under the wings and nose. It seems that other ways of getting the plane into the air were considered, likely involving towing or carrying by other aircraft in order to extend the range and preserve the rocket motors for the actual attack (as was proposed for several German rocket interceptors). Although some sources claim the vehicle was also designed to be used for kamikaze attacks with a warhead fitted into the nose, the complexity of the aircraft, its innovative aerodynamics for high maneuverability, and the presence of landing skids make this improbable. The plane never even reached the prototype stage before Japan surrendered, and even the earlier glider design never flew under power due to problems in developing the required rocket motors.

The only Japanese rocket airplane that actually did see action near the end of the war was the Yokosuka MXY-7 Type 11, which was more of a manned suicide anti-ship missile than a plane. It had a torpedo-shaped fuselage fitted with small, straight wooden wings and a horizontal wooden stabilizer that had a vertical fin at each end; there was also one experimental version built, the Type 21, that had thin steel wings manufactured by Nakajima instead of the standard wooden wings. The cockpit was positioned behind the wings and just behind the 1,200 kg (2,650 pound) bomb which occupied most of the fuselage. The pilot was protected by armor plating beneath as well as behind his seat because it was expected that the plane would encounter anti-aircraft fire from ships as well as from interceptor fighters. Instrumentation inside the cockpit provided the bare minimum of information required for the short flight into oblivion. Five fuses where installed to ensure the warhead would go off after impact. At least one of these was expected to be triggered by the impact shock, then detonate the bomb 1.5 seconds later so that the explosion would occur inside the target ship and cause maximum damage.

An Ohka suicide missile found by US soldiers.

The kamikaze piloted bomb was carried partly inside the modified, doorless bomb bay of a Mitsubishi G4M ('Betty') bomber, and dropped within striking range of its target (interestingly, the later X-1 in the US would start its mission in a very similar manner, also with the pilot entering the rocket plane inside the bomber shortly prior to the drop). The pilot would start by gliding towards his target and then ignite a trio of Type 4 Model 1 Mark 20 solid propellant rocket motors, each delivering a thrust of 2,600 Newton. He could choose to fire the motors sequentially or simultaneously depending on the range and speed required. In a sharp dive firing the rockets together in combination with gravity could smash the aircraft into a ship at about 930 km per hour (580 miles per hour). This tremendous velocity made it virtually impossible for the ship to aim its anti-aircraft guns and shoot the little plane down; something the gunners often succeeded in when attacked by much slower, conventional kamikaze planes.

The Japanese called their kamikaze weapon the 'Ohka' (Cherry Blossom) but the Americans referred to it as the 'Baka' (Idiot). The plane was conceived by Ensign Mitsuo Ohta, a transport pilot of the Japanese navy who made the first design aided by professor Taichiro Ogawa and students of the Aeronautical Research Institute at the University of Tokyo. Within weeks of contacting the university, Ohta was able to send plans to the Naval Air Technical Arsenal in Yokosuka complete with drawings, wind tunnel model test results and performance estimates. The Navy was sufficiently impressed to tell the Yokosuka engineers of its First Naval Air Technical Bureau in

Ohka dropped by a Betty bomber.

August 1944 to develop the design into an operational machine. The first variant, and the only that was put into service, was the Type 11. As it would not need to take off by itself, land or fly at low speeds, its wings were kept very small to minimize drag and thereby maximize attack speed. In part because of this, the maximum range of the rocket propelled Ohka when dropped from an altitude of 6 to 8 km (20,000 to 27,000 feet) was only 36 km (23 miles). The slow and vulnerable transport bombers laden with their heavy Ohkas were therefore obliged to fly close to the targets (which often included well-defended aircraft carriers, themselves prize targets) before they could release their Ohkas. Many bombers were shot down by defending fighters long before they could get near enough to the enemy fleet to deploy their Ohka.

Solid propellant rocket motors are simple, cheap and expendable, and therefore a natural choice for a single-mission kamikaze plane. However, the disadvantage of the short range led to several Ohka proposals powered by jet engines for longer flights. None of these alternative designs could be put into operation before the war ended.

Unmanned tests of unpowered and powered prototypes were followed by the first manned Ohka flight on 31 October 1944. On that day Lieutenant Kazutoshi Nagano straps into the prototype of the Ohka K-1 trainer, a version that had water tanks in the nose and tail instead of a bomb and rocket motors. For this test it has been equipped with two small solid propellant rocket boosters, one under each wing. Nagano is dropped by a Betty carrier plane at an altitude of 3.5 km (11,500 feet), pursues a stable glide for a few minutes and then fires the two motors. Immediately the machine begins to yaw, so Nagano quickly jettisons both rockets and manages to find a stable glide position once more. It is later found that the two rockets had not yielded the same amount of thrust and their positions on the wings caused the stronger rocket to attempt to turn the plane around its vertical axis (a similar problem had plagued the German Natter). Shortly prior to landing Nagano drains the water tanks in order to reduce the plane's weight and make it possible to fly and land at a relatively low speed on skids fitted beneath the fuselage and the wings (the operational Ohkas would of course not have an undercarriage, since they were not supposed to come back). Other than during its brief moment under rocket power the airplane handled well. It is decided not to use wing-mounted rocket motors anymore, and only equip the Type 11 with three in the tail, very close to the centerline so that any difference in thrust between the boosters will have little effect on the control of the vehicle. Subsequent test flights established how the Ohka should be operated. After being released from the bomber, the Ohka pilot would enter a shallow glide at a speed between 370 and 450 km per hour (230 and 280 miles per hour). Between 8 and 12 km (5 and 7 miles) from the target, and at an altitude of about 3.5 km (11,500 feet), he would ignite all three rocket motors and accelerate to a speed of 650 km per hour (400 miles per hour), then put his machine into a 50 degree dive to scream down at 930 km per hour (580 miles per hour), at the last moment pulling up the nose in order to hit the targeted ship at the waterline.

A navy flight unit was set up to operate the Ohka. Soon nicknamed the 'Thunder God Corps' it drew hundreds of volunteering pilots in spite of the nonexistent career prospects. After rejecting those who were married, were only sons, had too important family responsibilities, or were simply too old, some 600 remained. They trained first with a conventional Mitsubishi A6M 'Zero' fighter, practicing the attack profile with the engine off. Several pilots were given some flight instruction in one of the MXY7 prototypes of the K-1 two-seat trainer, but most were only able to rehearse using an Ohka while it sat on the ground.

It appears that some 751 Ohka Type 11 aircraft were built at two production sites but only a small number were deployed against the Allied fleet before the war ended, with poor results. The first time they are used is on 21 March 1945. Sixteen Betty bombers carrying Ohkas and two Bettys to provide navigation and observation are sent to attack US Navy Task Group 58, which includes the aircraft carriers *Hornet*, *Bennington*, *Wasp* and *Belleau Wood*. The bombers were to have been escorted by a force of 55 Zero fighters but owing to technical problems 25 of these either did not take off or had to turn back. Some 113 km (70 miles) from their target the planes are intercepted by 16 F6F Hellcat fighters. The Bettys immediately jettison their Ohkas, sending the pilots to watery graves without even a chance to impart damage on the enemy. All the Bettys are shot down and only 15 of the escorting Zeros make it back to base. On 1 April six Betty/Ohka combinations try again, attacking the US fleet off Okinawa. This time the bombers are able to approach close enough to their targets, and one Ohka completes a successful attack on the battleship *West Virginia*, causing moderate damage to one of its gun turrets. Three cargo ships are also hit by suicide aircraft but these may have been conventional kamikaze planes rather than Ohkas. The Bettys are all shot down. Eleven days later nine Bettys with Ohkas again attack the US fleet off Okinawa. One Ohka plunges onto the destroyer *Mannert L. Abele*, causing an explosion that rips the ship apart. The *Abele* becomes the first warship to be sunk by a kamikaze rocket plane, demonstrating the vulnerability of a ship to an accurately guided rocket missile flying at high speed. Another Ohka aiming for USS *Jeffers* is hit by anti-aircraft fire from the destroyer and blows up only 45 meters (148 feet) from the ship, imparting extensive damage. A pair of Ohkas target the destroyer *Stanly* and one hits it just above the waterline. However, the plane's explosive charge punches right through the ship and explodes only after emerging from the other side of the hull, causing little damage. The other Ohka narrowly misses the ship and falls into the sea. Only one Betty survives the mission.

Further attacks on 14 April (with seven Ohkas), 16 April (six Ohkas) and 28 April (a night attack with four Ohkas) fail to produce any Ohka hits and most of the Betty carriers are shot down. On 4 May the Japanese have better luck when one of seven Ohkas sent against the US fleet near Okinawa hits the bridge of the destroyer *Shea* and causes extensive damage and casualties. The minesweeper *Gayety* is damaged by a near-miss of another Ohka. Again the price for the Japanese is high, as all but one Betty is lost (in addition to all Ohkas and their pilots). On 11 May one of the four Ohkas sent out hits the destroyer *Hugh W. Hadley* and causes such extensive damage and flooding that the vessel is deemed to be beyond repair. But this proves to be the final Ohka success, as the attacks on 25 May (eleven Ohkas, with most returning to base owing to bad weather) and 22 June (with six Ohkas, two of the Bettys making it home) are ineffective. Initially presuming the Ohka to be just a new type of anti-ship bomb, the Allies learn of its true nature only after capturing some of the machines on Okinawa in June 1945.

Deployed in greater numbers, the Ohka might have played a more significant role in the war but the few successes do little to stop the mighty US fleet. These attacks do show how difficult it is to stop a rocket propelled missile, and that a single hit can severely damage or even sink a large warship. The post-war development of anti-ship missiles was rapid but the kamikaze pilots were superseded by electronic guidance and the rockets were launched from fast attack jets, not lumbering bombers like the Betty whose crews had also been effectively flying suicide missions.

In an artificial cave set behind the main buildings of Kenchoji Zen Temple in the ancient Japanese capital of Kamakure there is a monument to remember the Ohka missions. A steel plaque lists all of the Ohka pilots as well as the crews of the Betty bombers who died in the first ever (and hopefully final) kamikaze attacks using a rocket plane. Another plaque in the cave tells the Ohka story, albeit from a rather nationalistic Japanese perspective. The final part of the engraved text reads: "...Ohka attacks together with special attacks by Zero fighters carrying bombs were made repeatedly. That heroic battle tactic made American officers and men tremble with fear. This monument to the Jinrai warriors honors those pure and excellent young men who, without regard for their own sacrifice, courageously went to their place of death for their homeland and fellow countrymen."

Many Ohka Type 11 have survived and are on display at museums in the US, UK and Japan. The only remaining Type 22 jet-propelled version is at the National Air and Space Museum's Steven F. Udvar-Hazy Center, which also has a dual-cockpit trainer that seems to have been intended for preparing pilots for land-based launches, where Ohka's would have taken off with the help of a rail launch cart equipped with two solid propellant rocket boosters.

Japan also experimented with rocket boosters to assist aircraft take off. They were applied to the Nakajima 'Kitsuka', a Japanese version of the Me 262 jet fighter that suffered from underperforming jet engines. It was hoped that the additional power of jettisonable rocket boosters would limit the otherwise very long take-off run of this plane. On the second test flight of the first prototype, four solid rocket boosters with a thrust of 8,000 Newton each were installed. However, they had not been set at the correct angle. Rather than adjust them, which would take too much time, the thrust of each booster was halved in the expectation that at such low thrust the misalignments would not cause a significant disturbance to the balance of the plane. On 11 August 1945, after a delay of one day owing to the high level of activity of enemy aircraft in the vicinity, pilot Lieutenant-Commander Susumu Takaoka climbs into the cockpit, has the engines started and then taxies onto the runway where he stops, extends the wing flaps for additional lift at take-off and opens the throttles of the jet engines to build up thrust prior to releasing the brakes. Four seconds into the take-off run he ignites the four boosters and promptly finds himself in serious trouble. The sudden thrust of the four downrated boosters forces the nose of the plane up and makes the tail slam down onto the runway. He pushes the stick forward but it doesn't help. The boosters burn for 9 seconds with the plane essentially out of control. Just one second before the units burn out, the plane's elevators suddenly take effect and slam the nose down hard. Takaoka decides to abort the flight, but his brakes have little effect. He runs off the runway, the undercarriage collapses upon encountering a drainage ditch, and the plane slides to a standstill on its belly. The damage is severe, not only to the landing gear but more importantly to the engines slung under the wings. It is likely that the misalignment of the boosters was the cause of the problems. The Kitsuka never flew again because Japan surrendered several days after this aborted test flight and work on the project halted.

Rocket boosters were also envisaged to help the egg-shaped Kayaba 'Katsuodori' get up to take-off speed. This was to be a fighter plane powered by a ramjet engine. As explained earlier, a ramjet is a jet engine without a rotating compressor or turbine in which the air is instead compressed by the speed of the airplane and the shape of the air intake (in other words, the air is rammed into the engine as the plane flies through the atmosphere, increasing its pressure sufficiently for proper combustion and effective thrust). Owing to the lack of moving parts the design of a ramjet is relatively simple, is more reliable than a gas turbine engine, and ought to require less maintenance. By drawing oxygen from the air it can run for much longer than a rocket engine on the same amount of propellant. But for the ram effect to compress the air sufficiently for the engine to start to function, the plane must already be flying at a speed of at least Mach 0.3. Hence the engine could not be used for take-off. To get the Katsuodori up to speed, four solid propellant rocket boosters were to be mounted on the side of the fuselage beneath the wing roots, in pairs, one above the other. During the horizontal take-off run on an ejectable wheeled dolly the rockets would be fired in pairs, one on each side of the plane, with each lasting 5 seconds. After the second pair burned out the plane's speed would be sufficient for the ramjet to operate and all of the boosters would be jettisoned. But the project was shelved in 1943, probably in part due to the danger of using multiple solid propellant boosters: they cannot be stopped and if one misfires or has a lower thrust than its partners then the plane can quickly lurch out of control (recall that the German Luftwaffe used liquid propellant Walter rocket pods rather than solid propellant boosters to help their heavy bombers take off).

As in Germany, near the end of the war Japan also had a simple rocket aircraft for ramming purposes under development. Like the Zeppelin 'Rammjäger', its pilot was to ram his rocket powered plane into an enemy bomber, then glide back for a landing (followed by another hair-raising ramming mission in the same machine until either it or he were lost). But unlike the German machine it was not armed. It would have been accelerated by four of the same solid propellant boosters as used by the Ohka. While the Ohka had only three motors and was laden with a massive warhead, four rockets with a combined thrust of 11,000 Newton would probably have been able to drive the light ramming plane to a speed of over Mach 0.9. The concept remained a paper study but if this plane had taken to the air its test pilot would likely have found himself in serious trouble since the Japanese had virtually no knowledge of transonic aerodynamics. The plane had swept wings but it is not clear whether the designers understood the benefits of such wings at transonic speeds or whether they introduced this shape merely to provide an angled edge to cut more easily through the tail of an enemy plane. It is not known whether the Japanese rammer was to be towed into the air like its German counterpart or be launched from the ground.

The US had no appreciation of how far Japanese jet and rocket plane technology had advanced until after the surrender, when they got access to the aircraft factories, many of which were hidden in tunnels in the mountains, and discovered the fleet of advanced airplanes that Japan had been busy building in preparation for the defense of their home islands. Had Japan started its advanced projects a bit earlier, and had the US been forced to invade Japan itself by conventional military means rather than forcing a surrender by dropping atomic bombs, the US ships and planes would have been met by a variety of jet and rocket planes against which they would have had little defense.

RUSSIA'S TROUBLED START

The development of rocket planes in Russia before the war followed a very similar path as in Germany, with spaceflight enthusiasts developing rocket motors and later incorporating them into simple gliders. However, whereas in Germany rocket plane development was not deemed to be very urgent until the situation took a turn for the worse in 1943, in Russia it was the complete opposite. When German forces invaded the Soviet Union, Russia was ill prepared and its forces were initially easily overrun. The Luftwaffe pilots were able to accumulate individual scores of hundreds of Soviet planes shot down. Russia needed a 'wonder weapon' quickly. The development of an operational rocket propelled interceptor became an instant priority. As the situation improved, military support for rocket plane development actually lessened in Russia, while at the same time in Germany the need for such aircraft rapidly increased.

The history of rocket planes in Russia starts with a club of rocket enthusiasts led by rocket engine developer Fridrikh Tsander called GIRD, the Russian abbreviation for 'Group for Research of Reactive Propulsion'. Working without financial support, the members jokingly explain that the name of their organization actually means 'Group of Engineers Working without Money'. Inspired by the Russian theoretician Konstantin Tsiolkovsky and the German rocket pioneer Hermann Oberth, Tsander is a prominent advocate of spaceflight but he doesn't agree that other planets should be reached by expendable rockets. In an article 'Flights to Other Planets' published in the journal *Tekhnika I zhizn* in 1924 he explains that parachutes are less than optimal for landing a rocket on a planet which has an atmosphere, or indeed for returning a spacecraft to Earth, as parachutes do not offer the possibility of landing in a precise location. He continues: "For descending to a planet having sufficient atmosphere, using a rocket, as proposed by K.E. Tsiolkovsky, will also be less advantageous than using a glider or an aeroplane with an engine, because a rocket consumes much fuel during the descent, and its descent will cost, even if there is only one person in the rocket, tens of thousands of rubles, whereas descending with an aeroplane costs only several tens of rubles, and with a glider, nothing at all." In 1933 GIRD manages to launch the first Soviet liquid propellant rocket (called the GIRD 10) to an altitude of 400 meters (1,300 feet). The rocket and its engine were mostly designed by Tsander, but sadly he died shortly before the test.

Another prominent GIRD member was Sergei Korolev, who would go on to become famous as the chief designer of the early Soviet space rockets. During the Cold War Korolev was the Russian equivalent of Wernher von Braun, and was responsible for the launches of Sputnik in 1957 and cosmonaut Gagarin in 1961 that began the space race and caused President Kennedy to direct NASA to beat the Russians to a manned landing on the Moon.

Shortly after the creation of GIRD in 1931, Korolev approaches Tsander with an idea for a rocket plane. He proposes to base this RP-1 on the existing BICh 8 tailless flying-wing glider and to power it using Tsander's OR-2 rocket motor, which used gasoline and liquid oxygen and delivered a thrust of about 500 Newton. Later it is decided that the BICh 11 glider is better suited. In May 1932, Korolev becomes the director of GIRD and continues the design of the RP-1 and a successor called the RP-2. His work on the RP-1 and RP-2 designs evolves into the more ambitious RP-218, a two-seat rocket plane for high-altitude research equipped with a pressurized cabin. A fixed main landing gear is planned initially, but this is soon changed to a retractable undercarriage. The RP-218 was to be taken to altitude by a carrier

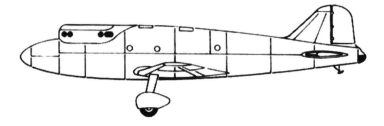

Concept for Korolev's RP-218.

airplane and released to boost itself to an altitude of 80 km (50 miles) or higher; to the edge of space, in fact. The plane is never built, partly because no rocket motor with the requisite thrust and weight is available. Interestingly, the envisaged mission scenario was very similar to that of the US X-15 rocket plane of the 1960s.

In 1935 Korolev designs the much simpler wooden SK-9 two-seater glider, which he intends to serve as a test bed for rocket motors. The RNII Rocket Scientific Research Institution, a new professional organization created by the merging of GIRD and the Gas Dynamics Laboratory in Leningrad in September 1933, sets out to transform the glider into a rocket plane by equipping it with an OR-2 engine. Being optimized to fly at relatively low speeds, modified gliders are useful for testing low-thrust engines such as this but they are not normally aerodynamically and structurally designed for the loads that come with powerful engines and high velocities. Otherwise gliders are well suited for conversion to rocket planes because they are fairly cheap and easy to modify, they do not have engines in the nose that need to be removed, they can be towed into the air by another airplane prior to starting the rocket engine, and they are able to glide to a safe landing once the propellants have been consumed. But great care has to be taken to ensure that the balance of the original design is not disturbed too much; hence most of the added weight has to be located near the original center of gravity. An added complication is that the weight of the propellant will decrease as the engine runs, whilst the fixed weight of the added structures, tanks and engine will remain unchanged. If this is not properly taken into account, an aircraft that is nicely balanced at take-off can become unstable as the flight progresses.

The rear seat of the SK-9 is replaced by a tank for 10 kg (22 pounds) of kerosene and two tanks for 20 kg (44 pounds) of fuming nitric acid. The rocket motor and its nitrogen pressurization system are installed in the aft fuselage, with the nozzle exit under the slightly modified rudder. The resulting RP-318 rocket plane has a take-off weight of 660 kg (1,460 pounds), a wingspan of 17 meters (56 feet) and a length of 7.4 meters (24 feet). The test phase begins with ground firing tests, initially with the OR-2 controlled from the cockpit but separated from the plane by an armor plate (in case it blows up) and later with the engine installed in the aircraft. More than 30 test firings are performed. In April 1938 the plane is deemed ready for flight but then the notorious 'purges' of the paranoid Soviet leader, Josef Stalin, take their toll. Already, in 1937, the director and the chief engineer of the RNII institute (now called

NII-3) were executed. But a war with Germany is looming and Stalin finally realizes that shooting Russian aircraft engineers is detrimental to the quality of his air defenses. By locking them up in development center prisons instead, it will be possible to keep an eye on them while they help to design planes for the war effort. Valentin Glushko, the chief engine designer, is sent to the Butyrka prison and works in a special design bureau for "subversive elements". When in June 1938 Korolev himself is declared an "enemy of the people" and sentenced to ten years' hard labor in the horrible Kolyma gold mines in which temperatures regularly drop to minus 38 degrees Celsius (minus 36 degrees Fahrenheit), the development of the RP-318 halts. Although Korolev is soon transferred to a prison design bureau on the request of famous aircraft pioneer Sergei Tupolev (himself in prison) he is not permitted to work on rocketry except at night in his own time.

Only near the end of 1938 is the RP-318 project resumed at NII-3, now under the leadership of Arvid V. Pallo and without the involvement of Korolev. It is decided to repair and modify the existing prototype into the RP-318-1, involving the rebuilding of the tail section that was damaged during the ground tests, the installation of a new landing ski, and the fitting of a shock absorber to the tail skid. The most important change is the replacement of the OR-2 engine by the more powerful ORM-65 rocket engine designed by Glushko and Leonid Dushkin. But concern soon arises about this choice. The ORM-65 has been ground tested many times and flown nine times on the winged experimental RP-212 cruise missile designed by Korolev, but it is not really suitable for a reusable piloted aircraft. The engine can only be ignited on the ground and, once operating, cannot be turned off. Also the heat of the combustion degrades the engine to such an extent that it will be dangerous to attempt to use it for several flights. The design is therefore modified to make it compatible with a manned rocket plane. The resulting RDA-1-150 engine is 2 kg (4 pounds) lighter, has an improved cooling system and redesigned propellant injectors, can be ignited in flight in a low thrust regime in which only about 10% of the normal amount of propellant flows into the combustion chamber, and its operating thrust can be regulated between 700 and 1,400 Newton. The new engine is ground tested over 100 times, including 16 times while installed in the actual aircraft.

The flight test program with Vladimir Pavlovich Fedorov as pilot begins in early February 1940 at a grassy airfield in Podlipki near Moscow. It starts with unpowered glide flights using a dummy engine for mass balance and with the plane being towed to altitude by an R-5 biplane. On the first flight the propellant tanks are empty, on the second they are half filled, and on the third the full tanks are slowly drained to mimic propellant consumption by the engine. In February 1940 three low-power flights are conducted during which the engine is run in the low-thrust ignition regime. Fedorov reports that the engine can be clearly heard in his open cockpit, which means that its proper functioning can be monitored by its sound as well as by the instrumentation in the cockpit.

Finally the aircraft is ready for a full powered flight. On 28 February 1940 the RP-318-1 is towed to an altitude of 2.8 km (1.7 miles) and released. It glides 200 meters (660 feet) down before Fedorov ignites the rocket engine. Gray smoke shows that the powder charge has fired to ignite the engine's liquid propellant, then brown smoke shows that the engine is operating in its starting regime. As Fedorov further opens the propellant flow, a bright, almost smokeless flame nearly 1.5 meters (5 feet) long streaks from the nozzle. It pushes the RP-318-1 from 80 to 140 km per hour (50 to 90 miles per hour) in 5 seconds. Next, the plane climbs up to an altitude of 2.9 km (1.8 miles). When the engine stops after 110 seconds of continuous operation, Fedorov glides the aircraft to a safe landing. The plane makes another two successful powered flights on 10 and 19 March, after which the arriving spring melts the snow and turns the airfield into an unusable mire. In the autumn the plane is returned to NII-3 and dissembled. There are plans for further tests using another engine and a jettisonable wheeled dolly but these never take place due to other priorities within the institute. In August 1941 the German Army approaches Moscow, and as the institute prepares for evacuation to the safety of the Ural mountains the RP-318-1 is burned to prevent the Germans from finding it.

Meanwhile, in 1940 NII-3 makes a study of a plane primarily powered by wing-mounted ramjet engines and an RD-1400 rocket engine in the tail to propel the plane up to the speed required for the ramjets to start working. In the spring of 1941 a draft concept called the Tikhonravov 302 is put together under the leadership of Mikhail Tikhonravov. This is later approved by the director of the institute, Andrei Kostikov, and Tikhonravov leads a team of engineers in developing the detailed design. He also assumes responsibility for the necessary aerodynamic calculations.

Kostikov had become director of the institute after the execution of his predecessor and the imprisonment of Korolev and Glushko; all of which he seemingly engineered to advance his own career. When war breaks out with Germany and Stalin takes an interest in the new project, Kostikov decides to claim the design in the expectation that its success will raise his standing with Stalin. In November 1942 Stalin names him chief designer, and Tikhonravov's revolutionary airplane concept becomes the Kostikov 302.

The Soviet State Defense Committee gives Kostikov only one year to get the 302 into the air. He duly promises that it will be able to fly for 20 minutes at an altitude of 8 km (26,000 feet) at a speed of 800 km per hour (500 miles per hour). A budget is allocated for the construction of two prototypes by NII-3's OKB-55 experimental production facility headed by Matus R. Bisnovat, and a series of flight tests. The 302 is mainly wood but the elevators are made of aluminum alloy. The straight, tapered, low-slung wings have a slight dihedral for enhanced roll stability. It has a pressurized cockpit, an undercarriage with retractable main wheels and a retractable tail wheel, and hydraulic actuators to help the pilot to handle the expected large forces on the control surfaces (normally fighter planes of that time were operated by muscle power alone, which could make steering a plane at high speed with high aerodynamic forces very 'heavy'). Because of the war, the test phase is to be minimized and the airplane turned into an operational fighter as soon as possible. Hence the 302 prototypes are equipped with armored glass in the canopy, an armor plate under the instrument panel and four ShVAK 20-mm cannon; two in the nose and two in the forward belly of the plane. In addition, the aircraft would be able to carry rockets on rails under its wings, or two 125 kg (276 pound) bombs for ground targets. The ramjets were to have been installed under the wings but development difficulties lead to the decision

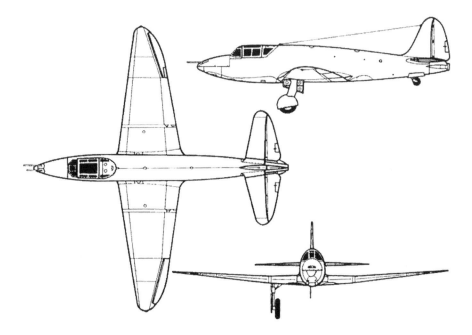

Three views on the Kostikov 302.

to cancel them altogether and operate the plane by rocket power alone, as the 302P ('Perekhvatchik', Russian for Interceptor).

The lack of ramjets greatly reduces the potential range of the aircraft but it is still valuable as a short-range interceptor for fast, short-duration attacks on enemy aircraft (essentially giving it the same role as the German Me 163). The new engine chosen for the plane is the RD-2M-3V developed by Dushkin and Glushko. Like the engine developed for the German Me 263 (and for the same reason) it has two combustion chambers: a large one with 11,000 Newton of thrust for take-off and ascent, and a smaller 4,000 Newton chamber for more economical cruise flight (these numbers are the thrust given at sea level; at higher altitudes the thrust is slightly greater because atmospheric pressure hinders the outflow of the exhaust). The propellant load for this engine consists of 505 kg (1,110 pounds) of kerosene and 1,230 kg (2,710 pounds) of 96% concentrated nitric acid; a nasty and dangerous substance. These propellants are not hypergolic, so an igniter is required to initiate combustion. The pumps providing the propellant to the combustion chambers operate using 80% concentrated hydrogen peroxide and provide gas at high pressure by the process of decomposition.

In the spring of 1943 the first of the two prototypes built by OKB-55 is sent for testing in the large T-104 wind tunnel of the TsAGI institute. Glide test flights begin in August 1943 with the 302P prototype being towed by a Tupolev SB bomber to its release altitude. V.N. Yelagin is the engineer responsible for these test flights, with pilots Sergey N. Anokhin, Mark L. Gallai and Boris N. Kudrin (the nation's oldest

One of the Kostikov 302 prototypes.

active test pilot). The twelve tow flights reveal serious stability problems at speeds over 200 km per hour (120 miles per hour), which is of course a major hurdle for a plane that is one day expected to fly at transonic speeds. Modifications are made and another series of glide flights are conducted in which the 302P is towed by Tupolev SB, Tupolev Tu-2, and North American B-25 bombers (the latter one of the planes donated by the US to help the Russians in their fight against Germany). This time the 302P proves to be exceptionally stable and rather easy to fly and land while gliding, so expectations for its handling under rocket power are high. However, development of the rocket engine is lagging far behind. In early 1944 it is still not able to reach the performance required to make the 302P the effective interceptor that Kostikov promised.

With the ram engines deleted, the plane is now expected to be able to fly at 8 km (26,000 feet) altitude for only 5.3 minutes instead of the specified 20 minutes, and at 725 km per hour (450 miles per hour); some 75 km per hour (50 miles per hour) less than originally specified. If the 302P were to fly at the original altitude and speed, it would run out of propellant in just 2.5 minutes. The absolute top speed also dropped from 900 km per hour (560 miles per hour) to 800 km per hour (500 miles per hour). The machine is still expected to be able to shoot up to 9 km (30,000 feet) in only 2.8 minutes but the military value of the plane as an interceptor is now highly doubtful. Moreover, whereas at the start of the project the Soviet Union was being overrun and overflown by the Germans, by early 1944 the need for an advanced interceptor plane is less urgent. In March 1944 it is decided to cancel the entire program, although one powered flight of the first 302P prototype remains in the planning. In the winter of 1944 it is fitted with skids and makes its first and final flight powered by an RD-2M3 rocket engine. Details are scarce but eyewitnesses say the flight was a success, even though one of the undercarriage ski legs failed at touchdown and caused the plane to slew into a snowdrift. However, the failure of the 302P as an effective weapon has dire consequences for Kostikov. Despite having the 'Hero of Socialist Labor' medal and the Stalin Prize, he is sent to prison on 15 March 1944 for obstructing the war effort and is not released until 1945.

During 1942 Aleksandr Yakovlev, the famous designer of Russian fighter aircraft, worked on the design of a concept very similar to the 302. It was to be based on his Yak-7, but instead of a piston engine and propeller it was to be driven by two ramjets mounted under the wings and a Dushkin D-1-A liquid propellant rocket motor in the rear fuselage. The result would be the Yak-7R (the 'R' standing for 'Reaktivnyy', meaning 'Reaction-propelled'). But the project never left the drawing board owing to the lack of reliable ramjets.

The most famous Soviet rocket plane of the Second World War is the Bereznyak-Isaev BI. This story begins in the spring of 1940, when the Zhukovsky Institute in Moscow (TsAGI) hosted a conference on ramjet and rocket propulsion. It is probable that the meeting was inspired, at least partially, by the advanced rocket and aircraft developments in Germany, which they knew of because Soviet intelligence had been able to recruit Willy Lehmann, a German Gestapo officer who kept them informed. Among the attendees were Viktor Fedorovich Bolkhovitinov, head of design bureau OKB-293, and Aleksandr Ya. Bereznyak and A.M. Isaev, two of his top engineers. Both Bereznyak and Isaev were very excited by the idea of rocket propelled aircraft and convinced Bolkhovitinov to let them start to design one. In the autumn of 1940 they showed fellow engineer Boris Chertok a preliminary design of what they called 'Project G'. It was a concept for a compact plane built of wood and duralumin (an aluminum alloy) with a take-off weight of 1,500 kg (3,300 pounds). To serve as an operational interceptor it was to have four machine guns; two 12.7-mm and two 7.6-mm caliber. The engineers intended to power their plane with a new 14,000 Newton rocket engine burning low-grade kerosene and red fuming nitric acid that was under development at NII-3 by a team led by Leonid Dushkin (who had also designed the RDA-1-150 that powered the RP-318-1). Since the thrust of the engine was close to the maximum weight of the plane, which would decrease as it burned propellant, it would be able to climb almost vertically once airborne. The top speed was estimated at 850 to 900 km per hour (530 to 560 miles per hour). According to the designers the most important selling points were the incredibly fast climbing rate (enabling it to reach enemy planes quickly and intercept them by surprise), its high speed (making it virtually invulnerable to enemy fighters), and the inherent simplicity of the rocket in terms of manufacturing and maintenance when compared to high-performance piston engines.

Bereznyak, Isaev and Chertok visited NII-3 in March 1941 to check on the status of the rocket engine, called the D-1-A-1100. This was state-of-the-art at the time. It weighed 48 kg (106 pounds) and consisted of several large forged-steel sections (the conical head with 60 propellant injectors, the cylindrical combustion chamber and the nozzle) joined using bolts and copper gaskets. Cooling was provided by pumping the kerosene fuel through the double-walled combustion chamber and the nitric acid oxidizer through the double-walled nozzle. The engine was ignited using a glowing plug of nichrome; later replaced by silicon carbide. But this marvelous technology was not working yet, mostly due to problems with Dushkin's innovative turbopump driven by hot gas and steam produced by a small combustion chamber that was fed a mixture of water and the same propellant as the main rocket combustion chamber; an efficient but complicated affair. Furthermore, the engine was not going to deliver the specified 14,000 Newton of thrust: the prediction was that it would deliver no more than 11,000 Newton. On 21 June Isaev proposed running the engine's turbopump on compressed air instead; it would result in a heavier rocket engine but would be much simpler.

The very next day Germany invaded the Soviet Union and the need for the rocket propelled interceptor instantly became very urgent. Bereznyak and Isaev set to work on a new, more detailed design. They finished it in only three weeks and on 9 July, together with Bolkhovitinov, met with the head of NII-3, the earlier mentioned Andrei Kostikov. Although Dushkin was not happy with the idea of altering his fuel pump, Kostikov agreed that the urgency of the situation meant that using compressed air made sense. A letter was sent to the Kremlin, and it was even shown to Stalin personally. The Project G team went to Moscow to report on their design and were ordered to build the plane in only 35 days. The engineers were given leave to visit their families, then literally lived in the factory to meet the extremely demanding deadline. The same order tasked NII-3 to finish the development of the D-1-A-1100 engine as soon as possible, and make it capable of multiple restarts in flight as well as thrust variation in the range 4,000 to 11,000 Newton.

The new design was called the BI for 'Blizhnii Istrebitel' (Close-range Fighter) and also the first letters of the two inventors Bereznyak and Isaev, although whether this was deliberate is unclear. It was now a sleek low-wing machine with a length of only 6.4 meters (21 ft) and a wingspan of 6.5 meters (21 ft). The take-off weight was 1,650 kg (3,640 pounds), of which 710 kg (1,570 pounds) was propellant. This made the BI a very diminutive fighter aircraft; smaller and lighter than the Me 163B. For comparison the conventional German Messerschmitt Bf 109G fighter had a length of 9.0 meters (29 ft), a wingspan of 9.9 meters (33 ft) and a maximum take-off weight of 3,400 kg (7,500 pounds), and was still regarded as being a relatively small fighter. The BI was kept as simple as possible so that it could be produced in short order and in sufficient numbers to overcome the German invasion of Soviet airspace. The four machine guns planned earlier were replaced by two more powerful 20-mm ShVAK cannon. The BI would have a wooden frame and a 2 mm (0.08 inch) plywood skin covered by a bonded fabric, and it would be easy to mass produce. The wings were to be relatively short to limit drag whilst still providing sufficient lift at the planned high flight speeds (lift is a function of both the area of the wings and the flight speed, so at higher speeds smaller wings are sufficient). However they were not particularly well designed for transonic flight phenomena (something that will result in a serious accident, as we shall see). The ailerons, elevators and rudder were covered by fabric but the flaps were duralumin. Ten tanks with compressed air (held at a pressure of 60 atmospheres), five in the forward section and five in the aft section, were required for the rocket engine's turbopump, to retract and deploy the landing gear, and to power the cannon. The forward section also housed two kerosene fuel tanks while the aft section had three tanks of nitric acid oxidizer. The air tanks were made from a high-strength steel (Chromansil) that was great for making light pressure vessels but not very resistant to corrosion. Their proximity to the extremely corrosive nitric acid was thus rather hazardous and required the acid tanks to be replaced periodically in order to ensure that no leaks could develop.

Working around the clock (with local furniture makers supplying the wood-and-fabric airframe) the team delivers the first prototype on 1 September 1941. A second prototype is also being assembled. However, Dushkin's engine is still not available. The first prototype, BI-1, is towed into the air by a Pe-2 bomber on 10 September with test pilot Boris Kudrin at the controls. Following release, he glides back to the airfield and makes a successful landing. Another fourteen unpowered flights follow and establish that the plane behaves well at low speeds. Interestingly, rival aircraft designer A.S. Yakovlev had the prototype towed to the TsAGI T-104 wind tunnel for testing. This alarmed the BI team because Bolkhovitinov had a rather rocky history with Yakovlev, but Yakovlev himself, and his aircraft designer, Ilya Florov, studied the results and gave the team good suggestions for improvements. In this way a yaw instability was corrected by enlarging of the rudder and adding two circular vertical plates at the tips of the horizontal stabilizer.

In addition to the problems with the engine, the BI project was delayed by the evacuation of both OKB-293 and NII-3 in October 1941 (in preparing for which, as mentioned above, the RP-318-1 rocket plane was burned). Most of Moscow's vital war industries were moved deep into the Ural mountains to ensure they would not be overrun by the rapidly advancing Germans. The BI team was stationed in Bilimbay but Dushkin's team ended up 60 km (37 miles) away in Sverdlovsk. Near their new accommodation the team built a test stand for their aircraft on the shore of the frozen Lake Bilimbay. It comprised a cradle that could hold the plane during engine tests, as well as measure the thrust. But there was still no engine to install. Dushkin was now increasingly absorbed by other projects (including NII-3's own Kostikov 302 rocket plane) but he assigned his engineer Arvid Pallo to oversee the installation and testing of the rocket engine. When the static test campaign finally begins in early 1942 it becomes immediately clear that the nasty nitric acid is trouble: it corrodes parts of the airplane as well team members, causing skin burns and respiratory irritation. Tanks of sodium carbonate solution have to be kept handy to neutralize the all-too-frequent acid spills. For these tests the new test pilot Grigory Yakovlevich Bakhchivandzhi (Kudrin was ill) operates the engine from the cockpit: a method fraught with risk, as we have seen with the He 112 which almost killed Erich Warsitz during a ground test. Indeed, on 20 February the BI-1's engine explodes, blasting the nozzle section into the lake. The forward assembly of the engine smashes through the airframe and strikes the rear of Bakhchivandzhi's seat, knocking him against the instrument panel. Fortunately his injuries are minor. Nitric acid spraying from a broken propellant line drenches Pallo. Luckily his eyes are saved by his protective glasses and the rest of his face is partly spared by alert mechanics who dunk him head-first into a tank of soda solution. His scars serve as a grim reminder of the dangers of rocket testing. A study of the engine debris reveals that the combustion chamber had succumbed to corrosion fatigue. In pushing on with the testing phase despite a shortage of engines, the one D-1-A-1100 available had been operated too many times and for too long. After the test bench is rebuilt and the engine's propellant supply system improved, static firing tests resume, with the undeterred Bakhchivandzhi performing three of them. A 5.5 mm (0.22 inch) steel plate is added to the rear of the pilot's seat as protection.

By April 1942 the BI-1 is deemed ready for flight testing at the nearby Koltsovo airfield (now the main airport of the city of Sverdlovsk). Interestingly, in spite of the obvious danger, the nitric acid for the engine is actually transported to the airfield in glass bottles and there poured into the containers from which it will subsequently be fed into the plane's tanks. All of this is done without any protective clothing.

On 2 May Bakhchivandzhi takes the controls of the BI for a short, low-thrust hop one meter above the ground. Then on 15 May he prepares for the first real test flight. Like the Me 163, the BI could explode if it were to make a hard landing before all the propellant was consumed, so it is loaded with only 240 kg (530 pounds) of nitric acid and 60 kg (130 pounds) of kerosene. He arrives wearing a new leather coat and shiny boots but when he climbs into the cockpit he is wearing his old flying gear: he explains that if he doesn't survive the flight the new coat and boots will be useful for his wife ... she could make some money by selling them! Taking off under rocket power, he leaps into the air and quickly reaches an altitude of 840 meters (2,800 feet) and a speed of 360 km per hour (220 miles per hour) with the thrust limited to 8,000 Newton. Another safety measure, a fairly standard one for the early flights of new airplane prototypes, is that the undercarriage is kept down. After one minute, an indicator lamp reports the rocket is overheating, so he shuts the engine off. Gliding in to land, the plane descends too rapidly owing to insufficient forward speed and the resulting lack of lift. The touchdown is hard and breaks the undercarriage but the plane is not significantly damaged. Bakhchivandzhi lives to wear his new coat and boots another day.

Compared to today's test flying, taking up a plane for the first time was in those days a truly scary business, especially if it was powered by a rocket engine. Modern airplane prototypes have already been exhaustively tested by sophisticated computer simulations long before metal is cut. The designs are based on an enormous database on subsonic, transonic and supersonic aerodynamics, control theory, materials and so on. Engines are tested hundreds of times before being deemed sufficiently reliable to be fitted into an airplane. All of this ensures that the question to be answered by the first flight is not whether it will fly, but rather how well it matches the performance predictions.

In contrast, pilots taking up revolutionary barely tested machines like the Me 163, Syūsui and BI were really pushing very experimental technology into the unknown. Rather than wondering how the planes would fly, their minds were probably more occupied with the question how these new beasts were going to try to kill them. And unlike modern pilots, these pioneers did not have teams of engineers monitoring the flight on the ground, ready to help in case of trouble; once in the air, they were truly on their own. Last but not least, they had no ejection seat to instantly boost them out of a hairy situation.

The BI's first flight lasted only 3 minutes and 9 seconds but was judged a success, with Bakhchivandzhi reporting, "the aircraft performed stable decelerations, gliding and handling like any ordinary aircraft". The State commission in charge of assessing the BI project delightedly noted: "The take-off and flight of the BI-1 aircraft with a rocket motor used for the first time as the aircraft's main engine has proved the practical feasibility of flight based on a new principle; this opens up a new direction in the development of aviation." This flight is sometimes hailed as that of the world's first operational rocket fighter plane, but this is somewhat of a stretch. Even though the BI was designed as an operational weapon, it was still very much experimental. It is certainly true that it flew before the first Me 163B in Germany, but the DFS 194 and Me 163A predecessors of that machine had taken to the air long before the first BI flight.

Encouraged, Stalin authorizes the production of a batch of 30 BI interceptors for operational military service. These are enhanced with racks for small bombs that can be dropped onto enemy bomber formations by flying above them, causing damage by a combination of shock waves and shrapnel. The decision to declare the BI ready for military service after only a single low-speed flight soon proves to be premature.

In July, Pallo is recalled to NII-3 by Dushkin to help him with the institute's own Kostikov 302 rocket plane. Isaev takes over management of the BI's rocket engine in OKB-293. To help get started, he goes to learn the tricks of the trade from Valentin Glushko in the Butyrka prison design bureau. Glushko shows Isaev how to improve the engine, and Isaev starts to work on what will become the RD-1 rocket engine that will later be incorporated in the BI aircraft.

After its first and only flight, the BI-1 is deemed too damaged by nitric acid spills for safe flight and it is retired. Testing continues using the second prototype. On 10 January 1943 Bakhchivandzhi flies the BI-2 for the first time. He still limits the D-1-A-1100's thrust to 8,000 Newton and the speed to 400 km per hour (250 miles per hour) and hence achieves an altitude of only 1,100 meters (3,600 feet). This time the undercarriage, with skids instead of wheels for taking off from the snowy airfield, is retracted, resulting in a smoother flight with aerodynamics more representative of a BI interceptor in operational use.

The second BI prototype.

A third flight (the second with the BI-2) is made two days later by another pilot, Konstantin Gruzdev. (Bakhchivandzhi was at that time visiting NII-3 to check upon progress with the Kostikov 302; there was clearly a great deal of interaction between the rocket plane teams of NII-3 and OKB-293.) This time the propellant flow is fully opened, allowing the engine to provide its full thrust of 11,000 Newton. At the peak altitude of 2,190 meters (7,190 feet) Gruzdev achieves a speed of 675 km per hour (420 miles per hour). When he extends the undercarriage one of the skids breaks off but he manages to land safely. Asked about the flight, Gruzdev comments, "It's fast, it's scary, and it really pushes you in the back. You feel like a devil riding a broom." Henceforth the 'devil's broomstick' nickname is used by everyone who works on the BI project. Film of this flight has survived, showing the BI accelerating quickly over the frozen lake, taking off and shooting up. The less than perfect landing was also captured, with the plane toppling forward because of the broken skid and performing a ground loop as it skidded across the ice. If such a thing had happened on a grassy airfield the plane would probably have burrowed its nose in the ground, with a much more severe outcome.

On his return Bakhchivandzhi takes over from Gruzdev and flies the refurbished ski-equipped BI-1 prototype on 11 and 14 March 1943, and on the 21st he takes off in the third BI prototype (also on skids) and climbs at a maximum rate of 83 meters (272 feet) per second; this is about half the rate of an Me 163B but still at least five times better than contemporary Russian piston-engine propeller fighters. On the 27th Bakhchivandzhi makes another low-altitude test flight. After 78 seconds, as he opens the throttle to push the plane beyond its speed record, the BI-3 suddenly goes into a 50 degree dive and smashes into a frozen lake instantly killing the pilot. The BI team has discovered the infamous transonic 'Mach tuck' phenomenon the hard way. This is confirmed by tests in TsAGI's new T-106 high-subsonic wind tunnel. The BI-3's onboard recording instruments were too badly mangled by the crash to give accurate data on the final speed, but estimates range from 800 up to an astonishing 990 km per hour (500 to 620 miles per hour). This disaster prompts the Air Force to cancel the pre-production order. For his achievements and ultimate sacrifice Bakhchivandzhi is posthumously awarded the 'Order of Lenin'. (In 1973 he gained the more prestigious 'Hero of the Soviet Union'.)

In May 1943, with OKB-293 relocated to Moscow following the German retreat, Bolkhovitinov writes a detailed report on the experiences with the BI prototypes. He emphasizes the need to study the shock effects that had caused the BI-3 to crash, and in order to investigate transonic and supersonic flight dynamics he recommends the development of a rocket plane capable of 2,000 km per hour (1,200 miles per hour).

No BI flights are made for over a year, probably in part because the urgent need for an operational rocket interceptor has lessened considerably, but also because time is required to build new airplanes. In 1944 five more prototypes are readied. BI-6 is given a pair of ramjet engines instead of a rocket engine. It is towed into the air on three occasions but the test pilot never manages to get the engines to work properly. The other prototypes are to be fitted with the new RD-1 rocket engine designed by Isaev, the first example of which is completed and tested in October 1944 (it should not be confused with the smaller RD-1 engine of Dushkin and Glushko).

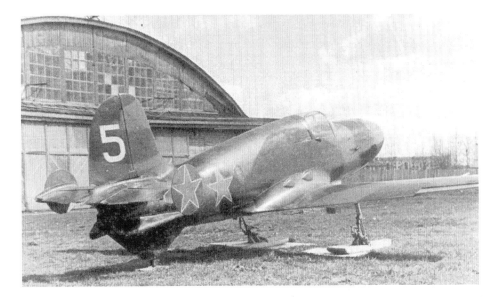

BI prototype number 5.

The general layout of the engine is similar to that of the D-1-A-1100 and also has a maximum thrust of 11,000 Newton, but Isaev has introduced numerous improvements, including a more reliable electric-arc igniter instead of the glow plug, and new injectors designed to improve the fuel and oxidizer mix in the combustion chamber. The BI-7 is fitted with the Isaev RD-1 and flown by test pilot Boris Kudrin on 24 January and 9 March 1945. The maximum speed attained is 587 km per hour (365 miles per hour), which is well short of the dangerous transonic flight regime. These flights reveal a problem of excessive vibration in the tail. Gliding tests using the BI-5 and BI-6 are made to investigate the problem but the pilots are unable to recreate the vibration. However, by this point the war is nearly over and there is no operational requirement for the BI interceptor, so work is halted. The 'devil's broomstick' is never flown in combat but its developers have gained valuable experience that will be put to good use after the war in developing new rocket propelled aircraft.

No examples of the BI survive but a reasonably accurate replica can be seen in the Russian Federation Air Force Museum in Monino. At the airport of Sverdlovsk, a monument of a BI replica shooting into the sky commemorates the rocket aircraft's first flight from that location.

Another rocket plane that however never progressed as far as the BI was the 'Malyutka' (Little One), the final aircraft designed by famous aviation pioneer N.N. Polikarpov. The construction of a prototype started in early 1944 and was nearly complete by the middle of the year. However, all work was suddenly stopped when Polikarpov died of a heart attack on 30 July of that year, and his facilities were absorbed into the rival Lavochkin design bureau. Work on the project was never resumed.

The Malyutka was initially planned to use a D-1-A-1100, the same rocket engine as powered the BI, but it was later decided to use a dual-chamber RD-2M-3V, which was the same engine as intended for the Kostikov 302P. The combined thrust of the two chambers was expected to be sufficient for a top speed of 845 km per hour (525 miles per hour). As the propellants were consumed, the changing center of gravity of the aircraft would cause stability problems. The solution was to equip the plane with a tank from which water could be discharged in order to compensate for the changing weight balance; simple but not very elegant. The fuselage would be made of wood, but unlike any previous Polikarpov fighter the wings and tail section would be lightweight aluminum alloy. The aerodynamic control surfaces would be operated by a pneumatic system, the cockpit would be pressurized for high-altitude flight, and the armament would comprise two powerful VYa-23 23-mm cannon. In contrast to the conventional undercarriage of the 302P and the BI, the Malyutka would employ an undercarriage and a retractable nose-wheel instead of a tail wheel. A plane with a tail wheel (a so-called 'tail-dragger') has its nose angled up whilst taxing, which makes such an undercarriage a good choice for a machine driven by a propeller that has to clear the ground. However, a jet or rocket aircraft has no need for a large ground clearance of its nose and so can be equipped with a nose wheel in order to align the plane more horizontally on the runway. This gives the pilot a much better view of where he is going during taxiing and the take-off run. Furthermore, with the plane in a horizontal position there is less danger of jet or rocket engine exhaust melting the runway. While most propeller fighter aircraft of the two world wars had tail wheels, nearly all jet fighters employ so-called tricycle undercarriages, with wheels under the wings and the nose.

The idea of adding rocket engines onto existing piston-engined airplanes (like the Germans with the Heinkel 112) was also explored as a stop-gap while waiting for the rocket and jet airplanes. The 3,000 Newton thrust Dushkin and Glushko RD-1 engine was fitted to various types of aircraft to boost their performance. On 1 October 1943 tests began with the engine in the aft fuselage of a Petlyakov Pe-2 dive bomber, with the turbopump driven by one of the two standard piston propeller engines. The tests with the Pe-2RD prototype revealed problems with the RD-1's electrical ignition, so it was replaced by a chemical ignition system. Both the RD-1 and the RD-1KhZ (the improved engine) were also flown on two Lavochkin La-7 fighters (the La-7R1 and La-7R2), an older La-7 prototype airframe (the La-120, converted into the La-120R), a Sukhoi Su-7 high-altitude interceptor (the original Su-7, not the jet fighter that was introduced in the 1950s with the same designation) and a Yakovlev Yak-3.

These tests proved the concept. For instance, on 11 May 1945 the rocket-equipped Yak-3, the Yak-3RD, reached a speed of 782 km per hour (486 miles per hour) at an altitude of 7.8 km (25,600 feet); some 130 km per hour (80 miles per hour) faster than the top speed of a conventional Yak-3. Indeed, it seems that a La-7R with an operating rocket motor successfully participated in the Moscow air displays of 1946 and 1947. However, on several flights the RD-1 exploded as kerosene fuel and nitric acid oxidizer came into contact outside the combustion chamber owing to leaks. The brave test pilots were usually able to land their damaged planes. One pilot of the sole rocket propelled Su-7 prototype was not so lucky, because while preparing the plane

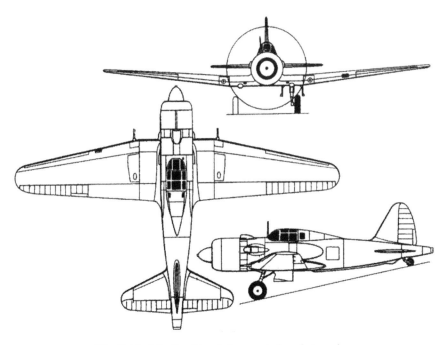

The Sukhoi Su-7 with a tail-mounted rocket engine.

for the first post-war air display over Moscow in 1945 the rocket motor exploded, destroying the aircraft and killing him. On 16 August 1945 test pilot V.L. Rastorguev died when his experimental Yak-3RD crashed for unknown reasons.

It became clear that apart from reliable and safe rocket engines, Russia also lacked vital knowledge on the dynamics of transonic flight. By 1935 Aleksandr Moskalyov had already drafted a concept for a rocket propelled aircraft that he thought should be able to exceed the speed of sound. The planform for this plane was based on that of his SAM-9 'Strela' (Arrow), a propeller aircraft that had a revolutionary ogival delta wing (also known as a Gothic delta because its shape resembles the arches in Gothic cathedrals). Moskalyov expected this type of wing to be well suited for transonic and supersonic flight, and this was later confirmed by the supersonic, ogival delta-winged British-French Concorde and Russian Tupolev 144 ('Konkordski') airliners. Piston engines and propellers were not going to show the full potential of his wing, but with a rocket engine he expected to be able to reach transonic speeds and beyond.

With Dushkin's help for the propulsion part, in 1944 Moskalyov came up with the design for the SAM-29, also known as the RM-1 (for 'Raketnyi Moskalyov', Russian for Moskalyov Rocket). Like the Strela, it had an ogival delta wing and big vertical stabilizer, but no horizontal tail. The planned engine was Dushkin's RD-2M-3V. To comply with the military's rather impractical demand that any new rocket plane must be armed to serve as an operational fighter, the experimental RM-1 would have two cannon in the nose. Unfortunately, at the end of the war the project

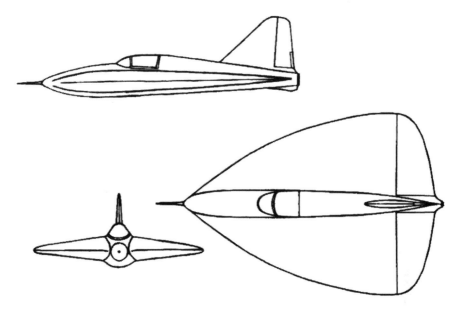

Concept for Moskalyov's RM-1.

was deemed too futuristic and in January 1946 Moskalyov's design bureau was closed. Had those in power understood the RM-1's potential and continued their support, then either it or a close descendant might well have become the first aircraft ever to fly faster than the speed of sound.

More or less in parallel with the RM-1, the development of another dedicated research rocket aircraft concept started in 1943. Designer Ilya Florentyevich Florov led the project for the Russian Air Force, and his Florov 4302/4303 was a relatively small rocket plane made entirely of light alloy. Its exterior had a very smooth finish to minimize aerodynamic drag, but unlike the Me 163 and various other high-speed German designs it had straight wings rather than swept back wings or delta wings. It is unlikely that the benefits of swept wings for transonic flight were fully understood in Russia at that time. (Nor indeed, as we shall see in the description of the post-war X-1, was the concept of the swept wing understood in the United States). The fully horizontal wings of Florov's design, which were set high on the fuselage, had down-ward angled 'flippers' at the tips. These effectively produced a negative dihedral in order to avoid 'Dutch roll', a stability problem common to high-winged aircraft that imparts an out-of-phase combination of 'tail-wagging' and rocking from side to side. (The German He 162 jet plane that flew near the end of the war also had a high wing and similar drooping wingtips, which the Germans called 'Lippisch Ears'.) The pilot was housed in a small pressurized cockpit. Three aircraft were built. The first had a fixed undercarriage with a tail wheel (using parts from a Lavochkin La-5FN fighter plane) and was intended for low-speed gliding flights only. The other two aircraft were for powered flight and (like the German Me 163B) were to take off employing an ejectable tricycle dolly, then land on a skid and a tail wheel. Aircraft 2 would be a

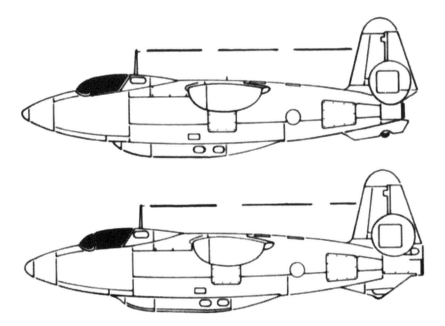

Florov 4302 (top) and 4303 (bottom).

Florov 4302 with a nitric acid/kerosene RD-1 rocket engine designed by Isaev with a maximum thrust of 11,000 Newton at sea level. Plane number 3 would be finished as a Florov 4303 with the RD-2M-3V two-chamber engine (the same engine that was planned for the Kostikov 302P and the Polikarpov Malyutka) delivering a combined maximum thrust of 15,000 Newton at sea level.

However, the flight test phase only begins in 1946, well after the end of the war. Pilots A.F. Pakhomov and I.F. Yakubov are appointed and that year make 46 towed glide flights with aircraft number 1. After some taxi tests and a short hop, Number 2 is flown for the first time under rocket power in August 1947, with Pakhomov at the controls. He is towed to an altitude of 5.0 km (16,400 feet) by a Tu-2 bomber. After release, he ignites the rocket engine and quickly accelerates to a speed of 826 km per hour (513 miles per hour), which is rather daring in a new experimental airplane that has not previously been tested under power at lower thrusts and speeds. Afterwards Pakhomov reports that the plane behaved well in all respects, and there were no vibrations. Several more flights with the 4302 then follow, and on one occasion a propellant feed line ruptures and noxious acid vapors slightly intoxicate the pilot. But by late summer 1947 the concept of the Florov 4302/4303 was obsolete thanks to the new information on high-speed flight and aerodynamics obtained from the defeated Germans. Moreover, jet fighters were now reaching similar speeds as those for which the 4302/4303 were designed. It was therefore decided that sufficient data had been gathered from the testing and that it would be better to concentrate effort and funding on the more advanced MiG I-270 rocket plane (described in the next chapter). When the 4302/4303 program was halted, aircraft number 3 was still awaiting its engine.

UK AND USA

While engineers in Germany, Russia and Japan were busy building experimental rocket propelled interceptors and designing concepts for even more advanced rocket aircraft, very little work was done on this subject in the United States and the United Kingdom. The success of Allied conventional fighter planes in gaining air superiority over European and Japanese territories gave no incentive to investigate the potential of rocket planes.

The first rocket research to receive financial assistance from the US government was the development of solid propellant RATO (Rocket Assisted Take-Off) units for aircraft by the Rocket Research Group of the Guggenheim Aeronautical Laboratory at the California Institute of Technology (GALCIT). On 12 August 1941 a tiny Ercoupe sports plane piloted by Captain Homer A. Boushey Jr., was fitted with such a solid propellant booster and launched from March Field in California. The booster burned for 12 seconds and gave a thrust of only about 130 Newton. In a later test the RATO unit exploded during level flight. The Ercoupe made a safe landing, but there was clearly something not entirely right with the propellant. On 16 August, Boushey nevertheless took off in the Ercoupe with six RATOs firing. The next step was to get the plane airborne by rocket thrust alone, and this was achieved on 23 August. The propeller was removed and 12 RATO units were fitted. The little plane was towed to a speed of about 40 km per hour (25 miles per hour) by a truck and then the rockets were ignited. Although only 11 units actually fired, the Ercoupe left the ground and reached an altitude of about 6 meters (20 feet).

After these flight tests the team discovered that RATO units tended to explode if they were stored for several days instead of being used immediately after production. The propellant grain shrank and caused cracks and openings to develop between the propellant and the casing, and these caused sudden destructive surges in the internal pressure. Once it became clear that this problem could only be solved by using a new type of propellant, the team developed an innovative solid propellant which was a paste created by mixing black gunpowder with common road asphalt. Its mechanical properties were much more stable under various storage conditions than those of the earlier powder propellant. Another major advantage of this composite propellant was that it could be cast into a predetermined shape, allowing propellant grains with pre-programmed thrust-over-time profiles (more or less equivalent to throttling a liquid propellant engine). This provided the basis for the solid propellants for all later large solid propellant rocket motors, including those on the Space Shuttle.

Based on the new propellant (called GALCIT 53) the team managed to deliver on a Navy contract for 100 RATO units capable of prolonged storage in hot deserts as well as arctic conditions and then delivering a thrust of 900 Newton for 8 seconds. Shortly thereafter, production of operational units for the Navy began at the Aerojet

Take-off of America's first RATO-equipped airplane on August 12, 1941 [NASA-JPL].

Engineering Corporation, a company set up by the GALCIT rocket research team for the commercial exploitation of their work.

Surprisingly, the only real American rocket aircraft under development during the Second World War was not an interceptor such as the German Me 163 or Japanese Syūsui, nor even an experimental research aircraft like the Russian Florov 4303, but more like the rather desperate German and Japanese back-against-the-wall, last-stand designs for ramming aircraft. In 1942 John K. Northrop, the famous aircraft designer with a fascination for 'flying wings', came up with the idea for a fighter which would be sufficiently sturdy that it could slice right through an enemy bomber. His 'Flying Ram' was a tailless flying wing built like a knife, with a reinforced leading edge over most of its span and no vulnerable air intakes. Northrop intended it to be powered by an Aerojet XCALR-2000A-1 'Rotojet' liquid propellant rocket engine (which did not yet exist) delivering a thrust of 9,000 Newton. Its take-off was to be assisted by a pair of 5,000 Newton thrust solid rocket boosters which would be dropped once airborne. The engine would run on mono-ethylaniline fuel and red fuming nitric acid oxidizer; a combination that we would not nowadays consider suitable for a manned military aircraft because it is rather toxic. Moreover this fuel would be especially corrosive to the innovative magnesium alloy structure which was to make the aircraft especially sturdy and resistant to damage. Northrop wanted the pilot to lie prone in the cockpit, in the expectation that this would enable him to survive the violent collision with an enemy plane. Moreover, this would make the flying wing much flatter so that the 8.5 meter (28 feet) wingspan would, seen from the front, present a minimum silhouette that would be difficult to hit by gunners in the targeted bomber and also enable it to more easily slice through the bomber.

Northrop managed to interest the US Army Air Forces (USAAF, forerunner of the US Air Force) in the project although at that time there was no obvious requirement for such a radical defense against enemy bombers. Perhaps intelligence forewarning of the secret long-range bomber developments underway in Germany was behind the support for this wild idea. In January 1943 the USAAF issued the Northrop company a contract for three rocket propelled prototypes under the designation XP-79, but in March it was decided to equip the third prototype with a pair of Westinghouse 19-B turbojets as the XP-79B. Similar to the development logic implemented in Germany, Japan and Russia, glider test vehicles were built to verify the aerodynamics of the aircraft design. Two prototypes (designated MX-334) were pure gliders but the third (designated MX-324) was fitted with an Aerojet XCAL-200 rocket engine that used mono-ethylaniline and red fuming nitric acid. Its thrust of 900 Newton was rather puny compared to the 17,000 Newton of the engine that was at that time propelling the Me 163B in Germany, and a clear indication of how far US rocketry was lagging behind the developments in Germany, Russia and Japan. The airframes of the gliders consisted of a center section of metal tubing covered with plywood, wooden wings, and a fixed tricycle undercarriage.

Flight tests of the MX-334 glider, towed into the air by a P-38 Lightning fighter, showed the flying wing to be rather dangerously unstable even after the addition of a simple vertical fin that was held in place using metal cables. On one flight, the glider got caught in the propeller wash of the tow plane just after release, which caused it to suddenly pitch up, stall and enter a spin. Test pilot Harry Crosby managed to recover from the spin but ended up flying inverted. Finding himself lying on the roof of the cockpit unable to reach the controls, he managed to open the hatch and parachute to safety. The MX-324 was tested at Harper Dry Lake in the Mojave Desert of southern California, where secret airplanes by both the Hughes and the Northrop companies were tested. On 20 June 1944 ground tests of the rocket motor were started, followed by taxiing tests with Crosby at the controls. On 5 July the engine and the plane were deemed flight-ready. He eased himself into a prone position in the cramped cockpit with his head in a sling to enable him to look ahead through the large windshield. In front of him he saw the P-38 that was going to tow him to release altitude. The two planes took off and climbed up to 2.5 km (8,000 feet), where Crosby triggered the towline release, then braced himself as he pressed the engine ignition button on the control stick. The thrust was very modest but still resulted in an acceleration of 0.08 G due to the plane's low total mass of about 1,130 kg (2,500 pounds). When the engine ran out of propellant after 4 minutes, Crosby glided the plane down to a gentle landing on the dry lake. Thus the MX-324 became the first true US rocket propelled aircraft to fly (fully five years after Germany's He 176). Several more flights were conducted, some equipped with transmitters that sent flight test data to ground-based recorders: an early use of telemetry that would

MX-324 at Harper Dry Lake.

eventually enable test engineers to monitor in real time how an airplane prototype was performing.

The actual XP-79 rocket plane was never built. Delays in the development of the complicated Rotojet rocket engine, which could never be made to work at full scale, eventually led the USAAF to cancel the XP-79. The development of the jet-powered XP-79B continued. On 12 September 1945 Crosby finally took the XP-79B up from Mojave's Rogers Dry Lake (later to become famous as Edwards Air Force Base). All was well for about 15 minutes, then the plane suddenly entered a spin from which he could not recover. Crosby failed to bail out and was killed when the plane struck the desert floor. The magnesium alloy structure was almost completely consumed by the resulting fire. After this disaster the USAAF decided to abandon the project. Flying wings like the MX-334 and the XP-79 were signature designs of Northrop and had many advantages over more conventional aircraft shapes, but they proved difficult to fly. Only when sophisticated control electronics became available did a flying wing finally become operational as the Northrop Grumman B-2 Spirit stealth bomber in 1997.

In the United Kingdom rocket aircraft developments were even more modest. A launch sled propelled by solid rocket motors was developed to 'catapult' especially modified Hurricane fighters dubbed 'Hurricats' off the bows of merchant ships as a means of protecting convoys from marauding German Condor long range bombers. Steam catapults, as used to launch small float planes from Navy ships, were too weak to get the heavy fighter airborne. The Hurricat was a stop-gap measure, as the plane could be used only once: after its mission it had nowhere to land and the pilot had to bail out or ditch his aircraft as near as possible to the convoy that he was defending, and hope that one of the ships would pick him up.

5

The rise and fall of the rocket interceptor

"The only time you have too much fuel is when you're on fire." – *Anonymous*

The world had entered the Second World War with piston-engine propeller aircraft technology, and these were sufficient for the duration of the war. However, shortly after Germany's surrender the former Allies suddenly found themselves in the jet age, as well as kicking off a conflict between the US and Western Europe on the one hand and the Soviet Union and Eastern Europe on the other. It was clear that any new air war would be fought by planes equipped with turbojets and/or rocket engines: no piston-engine aircraft could ever hope to keep up with the new jet fighters that were already flying at the end of the war, particularly the German Me 262 and Dornier 234 and the British Gloster Meteor turbojet.

Because of this paradigm shift in aircraft propulsion, advanced but conventional propeller fighters developed at the end of the war became obsolete almost overnight, just as a new generation of aircraft became necessary in order to maintain the uneven military balance of the Cold War and to fight in the limited-scope conflicts that this spawned (such as in Korea and the Middle East). It was soon recognized that major development efforts were required for the new generation of high-speed fighters, but whether these would primarily be propelled by rocket or jet engines remained to be determined.

The need for rapid interceptors quickly became very urgent in Europe because of the proximity of the countries of the Warsaw Pact and NATO: their bombers could attack one another's cities and military facilities within minutes of crossing the Iron Curtain. Moreover, an incoming bomber could spend a lot of time getting up to high altitude and speed in friendly airspace prior to crossing the border, but a defending interceptor would have little time to react. In fact, the situation was very similar to that which had faced the German Me 163 pilots in 1944 and 1945. Interceptors which could achieve high speeds and high altitudes in little time were therefore a priority in post-war Europe.

As we have already seen, the push for fast developments in aeronautics during the 1940s meant that new aircraft designs followed one another in rapid succession. This situation would continue into the 1950s and 1960s, with concepts sometimes already being obsolete prior to their first flight. Amidst this design fury, rocket aircraft had to compete with turbojet aircraft for development funding.

By 1945 several rocket planes had been successfully flown, starting with simple gliders fitted with solid rocket boosters, through modified piston-powered planes to dedicated rocket propelled aircraft. However, only the Me 163B had seen operational service. Overall, this experience did not bode well for the rocket powered interceptor concept. There were many accidents due to the poorly understood aerodynamics of transonic flight. Also the propellants tended to be difficult to handle and downright dangerous (notably the corrosive hydrogen peroxide of the German Walter engines and the almost equally nasty nitric acid used by the Russian engines), especially in combination with the rather low reliability of the rocket engines used. Moreover, the very short range of the rocket aircraft restricted it to quick attacks on enemy aircraft flying in the vicinity of the interceptor's base. Protecting an entire country such as Russia would require vast numbers of rocket interceptors and airfields. Furthermore, Germany's investments in often rather fanciful 'wonder aircraft' and rockets were of little help during the war: if the Luftwaffe and the Wehrmacht (Army) had bought conventional weapons for the money they invested in highly novel technology, they might have been able to extend the war considerably. In other words, the value of rocket aircraft remained to be proven. However, near the end of the war the problem of how to efficiently operate rocket airplanes did not seem to be so much intrinsic to the type of technology than to its immaturity.

The aerodynamic problems associated with transonic speeds, which were an issue for new jet aircraft as well, could only be solved using the proper wind tunnel tests, theoretical modeling and innovative wing design. Of course, the route to supersonic flight had already been partly explored by Lippisch and other aerodynamicists. The German, Japanese and Russian high-speed interceptors of that time looked distinctly futuristic, but their swept-back or delta wings, tail exhaust nozzles, cockpits placed in front of the wings and lack of propellers would soon become the norm for fighter aircraft. Some of the early jets and rocket aircraft still look rather modern today! As regards the unpleasant rocket propellants, fortunately less vicious alternatives were able to be produced.

The only true disadvantage inherent in rocket power was short range, limiting the role of rocket-propelled fighter planes to point defense: an interceptor used to defend a specific target by taking off and climbing to altitude as rapidly as possible in order to counter an approaching threat, and then land and prepare for another mission. But it had already become apparent during the war that the range issue could be partly resolved by a combination of rocket and jet engines, with the relatively simple rocket providing the thrust for a rapid ascent and the more complicated jet engine enabling the plane to cruise for relatively long periods.

Because of the rapid developments during the war, jet engines had become more competitive, in terms of thrust per engine weight, in comparison to rocket engines. The early Heinkel HeS 3 jet engine that powered the world's first jet aircraft in 1939, the He 178, weighed 420 kg (930 pounds) and produced a maximum thrust of 4,400 Newton (equivalent to about 440 kg). This meager thrust to weight ratio meant the engine could vertically lift just 1.05 times its own weight. In 1941 the 'cold' Walter RII-203 rocket of the Me 163A was far better. It had a thrust of 7,400 Newton and weighed only 76 kg (170 pounds), which is a thrust to weight ratio of around 10. The early rocket engine could thus produce some ten times more thrust per unit of engine weight than the early jet engine and was simpler to manufacture and maintain. The jet engine was however more fuel-efficient, requiring less propellant than the rocket engine for the same amount of thrust. Near the end of the war the Junkers Jumo 004D turbojet engine, weighing 745 kg (1,640 pounds), could already deliver a thrust of 10,300 Newton at low altitudes and had a thrust to weight ratio of about 1.4. The Me 163B's Walter HWK 109-509A2 of 1944 still had a higher maximum thrust of 17,000 Newton, weighed only 160 kg (350 pounds) and so had a thrust to weight ratio of about 11. Then again, at least in terms of thrust, the difference was shrinking. Not only were jets more economical in propellant consumption than aircraft rocket engines (translating into longer flights and greater ranges), they were now producing similar amounts of thrust while their thrust to weight ratio was rapidly improving.

A disadvantage early jet engines had with respect to rockets was that they required very careful throttle movements: changing power rapidly would often extinguish the combustion in the turbojets (so-called flameouts). The start procedures for early jet engines were also very sensitive, as there was no automatic sequencing of the various events built in. An error could cause the engine to quit or possibly overheat. Even worse for the pilots was that everyone on the airfield would know about his mistake because the engine would produce a loud rumble and shoot a huge flame out of the tailpipe. Rocket engines were in general much less fidgety. However, the sensitivity issues of turbojets were soon solved by more advanced fuel controls.

In other words many advantages rocket engines still had over jet engines for use in fighter planes were being eliminated one after another. One strong inherent plus was the lack of the need for an inflow of air and associated air intakes, which gave rocket aircraft an advantage in terms of aerodynamic drag. Air intakes also cause technical complexity because the air going into a conventional jet turbine engine needs to flow at subsonic speed even when the plane itself is flying supersonically. Hence using rocket engines also meant that the propulsion was independent of the vehicle's flight speed. A more important remaining advantage of rocket engines was that they were much more effective than jet engines at high altitudes. Rockets have their own oxidizer whereas jets need to scoop up air to use its oxygen. The density of the atmosphere drops with altitude, so for each jet engine there is an altitude at which it can no longer be fed sufficient air to work properly. Rocket engines can even work in vacuum, and in principle actually work better there since their exhaust flow is no longer hindered by air blocking the nozzle's exit. As a result, the thrust of a jet engine decreases with altitude whilst that of a rocket engine increases (by up to 25%, at least for a rocket engine with a nozzle that is optimized for use at high altitudes).

THE LAST ROCKET FIGHTER

At the end of the war Russia captured from Germany three Me 163B interceptors and seven Me 163S glider trainers. These were tested in the large TsAGI wind tunnels as well as in gliding flight but a lack of the necessary propellants and unfamiliarity with the Walter engine prevented powered flights. The TsAGI analyses confirmed that the aerodynamics of the Me 163 were very suitable for high-speed flight. The Russians were of course even more interested in the Komet's advanced successor, the Me 263, having captured some equipment and documentation along with technical staff who had worked on the project. The brewing Cold War and the threat of Western bombers renewed Soviet interest in fast, high-altitude rocket interceptors. Already in February 1946 the Lavochkin OKB was ordered to develop such an interceptor to detect and destroy enemy bombers. It had to be able to reach an altitude of 18 km (59,000 feet), achieve a top speed of 1,100 km per hour (680 miles per hour) in level flight at 5 km (16,000 feet), and be able fly at maximum thrust for six minutes. The plane was also to be capable of attacking at any time of the day and in all weather conditions, which meant that it had to be equipped with radar. Its armament was to be six 83-mm TRS-82 rockets. And to make the challenge even more difficult, the military wanted flight testing to start on 1 May 1947.

By the end of 1946 Lavochkin had finished the design for 'Aircraft 162': an all-metal plane powered by the familiar Dushkin RD-2M-3V, with a pressurized cockpit and air-to-air rocket launching tubes integrated in the fuselage. It was calculated that the rocket thrust would enable the plane to reach an altitude of 18 km (59,000 feet) in 2.5 minutes. Inspired by captured German research, the wings of the Lavochkin 162 were extremely innovative: not only were they swept, they were swept forward. Such wings provide most of the favorable high-speed characteristics of swept-back wings and offer the benefit of making the aircraft less sensitive to stall. When a plane with swept-back wings stalls (i.e. when the air can no longer correctly follow the contours of the wing because the angle of attack is too high) the air usually starts to let go at the root of the wing, and this effect quickly travels along the length of the wing to its tip causing the whole wing to rapidly lose lift and making the plane uncontrollable. When a forward-swept wing stalls, the disruption at the root cannot easily progress to the wingtips and ailerons. Giving a pilot much better control at high angles of attack makes his plane more maneuverable, which is very welcome on a fighter aircraft. But forward-swept wings were extremely difficult to build using standard 1940s aircraft materials since they require to be extraordinarily stiff. Otherwise, when the wingtips bend up or down at high speeds the airflow can rip the wings off (on a swept-back wing the airflow tends to push a twisting wingtip back to the horizontal). Because of this, along with the many unknown aerodynamic characteristics of swept wings, the La-162 was soon redesigned with well-understood straight wings.

A full-scale mockup was built but the chief designer, Semyon Lavochkin himself, soon began to express doubts about pursuing the concept. The plane would have a phenomenal rate of climb but its flying time and range would be extremely short. He believed further development of jet engines would result in better interceptor fighter designs. He was also concerned about the nitric acid propellant, which had proved to be very corrosive and unsafe in experiments during the war with RD-1 engines fitted in La-7 and La-120 aircraft. And there was the issue of the high operating pressures in the various pipes in the engine, which made it prone to leaks that resulted in a low reliability and heavy maintenance. Another issue was that the turbopump of the RD-2M-3V would have not only to feed propellants to the combustion chamber but also drive an electric generator, which was calculated to be too weak to provide sufficient electrical power for the onboard radar. Furthermore, it was soon found that the high aerodynamic drag of the new straight wings would prevent the rocket engine from driving the plane to the required transonic flight speeds. All this, in combination with the troublesome development of the designated radar system, led Lavochkin to call a halt to work on the 162.

However, there was another Russian design bureau still working on a pure rocket-powered interceptor. The Mikoyan and Gurevich (MiG) design bureau had set out to develop a local version of the German Me 263 rocket fighter called the MiG I-270. The Soviet military pushed their aircraft designers to copy as many as possible of the advanced German design features to get a head start in the new jet age. But Russian rocket engines were based on very different propellants than the Walter engine, and designers strongly resisted a radical switch from their own known and proven rocket technology. In addition, the MiG engineers and TsAGI aerodynamicists were not yet completely familiar with the characteristics and peculiarities of swept and delta-shaped wings and therefore preferred to employ conventional straight ones (the same was done in the US for the X-1 experimental rocket plane, for the same reason).

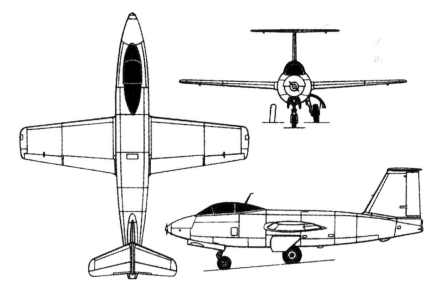

Planform of the MiG I-270.

MiG I-270.

As a result the I-270 essentially became an Me 263 with a Russian engine, a more conventional wing and tail design and a longer fuselage. The tail was different in that it had a horizontal stabilizer high on the vertical fin in a T-arrangement to diminish aerodynamical interference from the main wings. But the new plane had an ejection seat, a first for Soviet fighter aircraft and essential for bailing out at transonic speeds. In this respect it was better than its German predecessors. The rocket engine was the RD-2M-3V (described earlier) with two combustion chambers. It gave the I-270 less thrust than the original HWK 109-509C of the Me 263. While the main combustion chamber of the German engine provided 20,000 Newton its counterpart on the RD-2M-3V could yield only 11,000 Newton. The smaller 'cruise flight' chambers of the two engines were comparable at about 4,000 Newton. Even so, the combined thrust of both RD-2M-3V chambers was still expected to give the plane a maximum speed close to supersonic and make it possible to reach an altitude of 17 km (56,000 feet) in 3.2 minutes. For ease of maintenance and replacement of the engine, the fuselage could be split into two sections. Power came from a generator connected to a small propeller on the nose (like in the Me 163B) in addition to a generator cleverly connected to the turbopump of the rocket engine. The cockpit was pressurized, and armor plate in the forward structure and an armored windscreen would protect the pilot from enemy gun fire. Like the Me 263, the MiG I-270 had a tricycle undercarriage. The two main wheels retracted into the fuselage since there was no room for them in the thin wings. As a result these wheels had a very narrow separation, making the plane wobbly on the ground and difficult to land on a rough field or in the presence of a cross wind. The maximum take-off weight was 4,100 kg (9,100 pounds), of which 2,100 kg (4,700 pounds) was propellant. It was armed with two 23-mm NS-23 cannon and eight RS-82 solid propellant air-to-air rockets, and it was meant to defend large industrial installations as well as military bases. The threat was anticipated to be American B-29/B-50 Superfortress bombers and the new B-36 Peacemaker that could fly at altitudes up to 15 km (48,000 feet).

Two prototypes were built, Zh-1 and Zh-2. The first was intended for gliding tests and was fitted with a ballast mass instead of an engine. Pilots trained for the gliding flights on a Yak-9 fighter used as a glider, loaded with lead weights to reproduce the weight and balance of the I-270. Both the Yak-9 trainer and the glider version of the I-270 were towed into the air by a Tu-2 bomber. The first glide test of the I-270 Zh-1 occurred on 3 February 1947 with test pilot V.N Yuganov at the controls. Until July of that year the Zh-1 remained connected to the tow airplane for the entire flight, but for the second phase of the unpowered tests it was released at altitudes of 5 to 7 km (3 to 4.5 miles) to glide home and land. It reached a maximum speed of 600 km per hour (370 miles per hour) during these unpowered tests. In early 1947 the Air Force put pressure on MiG's chief designer, Artem Mikoyan, to get the I-270 ready for a powered flight demonstration during the annual air display in Moscow on 18 August. The gliding flight test phase was curtailed and an RD-2M-3V engine installed in the second prototype, Zh-2, in May. Unfortunately the smaller combustion chamber blew up in ground tests on 16 July and damaged the tail section. The aircraft was repaired but could not be readied in time for the air display. An aircraft exploding during that international showcase would not make a good impression on the Soviet leadership and the rest of the word!

On 26 August, well after the show, the Zh-2 made its first two powered taxi runs and a short hop with A.K. Pakhomov at the controls. Its first (and unfortunately also last) flight was on 2 September and it lasted just 7 minutes. Pakhomov successfully climbed to an altitude of 3 km (10,000 feet), then initiated a gliding descent back but widely overshot the planned landing point and crashed beyond the airfield perimeter. He was unhurt but the aircraft was damaged beyond repair. Fortunately by then the first prototype had been upgraded by replacing the dummy engine with a real one, and powered tests resumed using the Zh-1 with V.N. Yuganov at the controls. A taxi run was made on 29 September. On the first flight on 4 October it took off under the roaring combined thrust of the two combustion chambers. The main chamber was shut down 130.5 seconds into the flight at an altitude of 4,450 meters (14,600 feet). The plane reached a maximum speed of 615 km per hour (382 miles per hour) at an altitude of 2,900 meters (9,510 feet) operating on the smaller combustion chamber only. As he glided back Yuganov found that the undercarriage would not extend so he made a belly landing that was so soft that the plane was barely damaged. The poor reliability of the RD-2M-3V rocket engine however continued to slow down the test flight phase when its largest combustion chamber exploded during a ground test and blew off a large part of the tail. The Zh-1 was repaired but was not ready for its third powered flight until January 1948. By then, however, a new problem had shown up: after each flight the engine had to be rinsed with water to remove the dangerous and corrosive nitric acid propellant, but during the freezing Russian winter this was impossible without also turning the motor into a block of ice. There was no alternative to postponing further testing until March.

In the meantime, the Aviation Technology Committee of the Air Force carried out a reassessment of the I-270's potential as an operational interceptor, with shatteringly negative conclusions. A major complaint was that the engine couldn't be restarted in flight without the high risk of an explosion caused by nitric acid accumulating in the combustion chamber between extinction and reignition. The corrosive and dangerous nature of the propellant was another major issue: the I-270's parts and particularly its oxidizer tank, as well as the technicians working on the plane, were constantly under attack by corrosive nitric acid vapors. Personnel dealing with the engine had to wear bulky protective suits that made work on the plane arduous. The acid tank required labor-intensive removal, checks and replacement every 2 months. Special materials and acid-resistant coatings (as many as nine layers for the most critical areas) were used but corrosion occurred anyway. The difficulty in landing an unpowered I-270 (as shown during its first powered flight) and the impossibility of rinsing the engine with water during the winter were also listed as severe operational limitations. It is clear that the Air Force had lost its enthusiasm for the design.

By then the idea of an interceptor powered only by a rocket engine was rapidly becoming outdated. The turbojet-propelled MiG 15 had already flown in December 1947, would soon enter service, and would satisfy most of the requirements initially set for the I-270. The MiG 15bis version had a nearly supersonic maximum speed of 1,075 km per hour (668 miles per hour) and a flight ceiling of 15.5 km (50,900 feet). Furthermore, it had a maximum range of 1,200 km (750 miles), which could even be extended to 1,980 km (1,230 miles) using externally carried drop tanks. That made it much more useful than the I-270, which was basically only capable of a brief sortie to attack enemy aircraft that came within several tens of kilometers of its base. The much longer powered-flight endurance of the jet also meant it could attack bombers multiple times. It would even have time to dogfight with enemy fighters, something that was becoming an important requirement. Whereas in powered flight the earlier Me 163B could easily leave behind the enemy propeller fighters that it encountered, the I-270 would have had to engage new Western jet fighters which could match it in horizontal flight. The only advantage the MiG I-270 still had over the MiG 15 and other early jets was its rate of climb. The MiG 15 could reach an altitude of 15 km (49,000 feet) in about 5 minutes, but the I-270 could get there in 3 minutes (with its more powerful engine the Me 263 would have done it in 2 minutes). But this single advantage didn't outweigh all the weak points of the I-270 and the strong points of the MiG 15.

Moreover, the development of surface-to-air guided missiles was advancing at a great rate. The S-25 (NATO codename SA-1 Guild) and the infamous S-75 (SA-2 Guideline) entered service during the 1950s and could reach and destroy high-flying intruders even more rapidly than a rocket propelled interceptor because they had no pilot for whose survival acceleration levels had to be kept to a reasonable maximum. Between them the MiG 15 turbojet and the S-25 and S-75 missiles rendered the I-270 obsolete, with the inevitable abandonment of the idea of a manned rocket interceptor. The MiG design bureau focused its efforts on jet interceptors and fighters as the new generation of military aircraft.

The pure rocket fighter was briefly revived in the late 1940s at the OKB-2 design bureau, where a team led by A. Ya. Bereznyak (designer of the wartime BI rocket plane) and the German Siegfried Günther jointly worked on a multi-chamber rocket interceptor. During the war Günther had been a leading designer at the Heinkel company, and after failing to find a job in the US or the UK had offered his services to the Soviets. Günther and Bereznyak worked on different but similar versions of their supersonic rocket aircraft indicated by project numbers 466, 468 and 470. These were all delta-winged designs powered by a rocket engine with four combustion chambers installed in the tail. The maximum total thrust of this engine was to be 82,000 Newton at sea level increasing to 96,000 Newton at an altitude of 20 km (66,000 feet). The plane had no horizontal stabilizers (typical of delta-winged aircraft) but had an enormous central vertical fin as well as two smaller fins under the wings. It would take off with a jettisonable dolly or a sled driven by solid propellant boosters, and land on skids. The pilot would have a pressurized, armored cockpit equipped with an ejection seat. A radar in the nose would help him to find his target, which he would engage using either four 23-mm cannon or two canisters with six unguided missiles each. A strike version of the aircraft carrying four high-explosive bombs was also considered as an option. The 470 interceptor was envisaged to fly at a top speed of 1,910 km per hour (1,190 miles per hour) above 11.5 km (38,000 feet) altitude, equivalent to Mach 1.8, and to climb to 20 km (66,000 feet) in 2 minutes and 14 seconds. But none of these aircraft designs made it off the drawing board. Inevitable competition with surface-to-air missiles and jet fighter designs promising much larger range meant that the entire project was judged to be obsolete by June 1951. OKB-2 was disbanded and its personnel transferred to other activities. Günter returned to West Germany in 1954 and joined Heinkel AG when that company reopened in 1955. He never spoke about his work in the USSR, and neither did the Russians, but it seems that he had worked on various other advanced Soviet aircraft apart from the 466/468/470. In 1950 Ernst Heinkel said of his former employee, "I am convinced that Günter worked on those Soviet designs that today have become a problem for the Western World."

The British also initiated a military rocket aircraft project just after the Second World War. The Fairey company planned a new delta-winged research aircraft called the Delta 1, which they initially envisaged as taking off vertically from a very short and steep ramp. It was to lead to an operational interceptor that would use a similar mode of take-off. The company was granted a contract in July 1946 to develop an unmanned rocket aircraft to test the concept. Simply called the Fairey VTO (Vertical Take-Off) it had an Armstrong Siddeley Beta rocket engine which was derived from the Walter HWK 109-509 'hot' engine and used the same propellants. However, the Beta had two combustion chambers and nozzles each developing a thrust of 4,000 Newton, and these could be independently swiveled, one side-to-side and the other vertically so that, together, they could steer the aircraft (by what would now be called thrust vector control) at low velocities. This was important, because when the VTO left the ramp it would not be flying fast enough for its aerodynamic control surfaces to have sufficient 'grip'. Two 3,000 Newton solid propellant boosters were added for take-off. The starting method and propulsion were so similar to the German wartime Natter that the Fairey VTO soon gained the nickname 'Son of Natter'.

Tests started in 1949 from a ship moored in Cardigan Bay in Wales, then in 1951 resumed at the vast Woomera Rocket Range in Australia. There were stabilization

The unmanned Fairey VTO [Fairey Aviation Company].

problems early on because the autopilot, which was derived from that of the German A4/V2 rocket, had to be carefully adjusted to accommodate the completely different aerodynamics. When this was finished the take-off launches and ensuing flights were satisfactory. About 40 of these expendable test vehicles were built and launched, but it was eventually decided that the manned Fairey Delta 1 should be a jet-powered research vehicle that would take off in a conventional manner from a runway. As in the USSR, unmanned surface-to-air missiles had already taken over the role intended for the VTO rocket interceptor.

MIXED UP; COMBINED JET/ROCKET INTERCEPTORS

Of course, as some German and Russian projects had explored during the war, there was a way to combine the fantastic rate of climb and high-altitude performance of a rocket and the endurance and range of a jet in a single aircraft.

After the failed attempts to create mixed-power ramjet/rocket aircraft during the war, the Russians did not resume their interest until the mid-1950s. The MiG 19SU (also known as the SM-50) was a MiG 19S equipped with a U-19 rocket boost unit based on a 29,000 Newton S3-20M engine running on nitric acid and kerosene. This was specifically intended for intercepting the elusive Lockheed U-2 reconnaissance aircraft. This CIA spy plane with its extremely long wings could fly at an altitude of 21 km (70,000 feet), beyond the reach of normal fighter aircraft. The rocket booster was a self-contained unit that required only electrical and mechanical attachment to the aircraft. It could be carried on the belly of a MiG 19S which, apart from some structural reinforcements and the interfaces, was otherwise a standard version of this jet fighter. Ground clearance was narrow with the booster under the fuselage, so the aircraft needed a smooth runway and the rotation immediately prior to taking off had to be shallow. From November 1957 to February 1958 five prototypes made a total of 44 rocket propelled flights that demonstrated the aircraft could reach a maximum altitude of 24 km (79,000 feet); an impressive improvement of the normal ceiling at 17.5 km (57,400 feet). This altitude was reached using a 'standard intercept' profile in which the MiG used its jet engine to get to 14 km (46,000 feet) and then ignited its rocket engine to make a quick jump further into the thin air. However, the total time to reach its peak altitude was 12 minutes, which was not particularly spectacular, and it could not remain there for any length of time. The wings and control surfaces were never designed to operate in rarefied air, and above 20 km (66,000 feet) it was about as maneuverable as a brick and difficult to point in the right direction for firing its guns and missiles (which in those days had to be aimed much more accurately than their modern, highly autonomous descendants). In addition, the aircraft was prone to stalling at high altitude and the lack of oxygen tended to make its turbojet flame out. This meant that even under the control of an exceptionally skilled pilot, the aircraft could only make a running jump up at a U-2 in the hope of firing a few shots in the general direction of the target at the top of its parabolic trajectory, before it tumbled down again. In short, the chance of a successful intercept was remote. A number of MiG 19P fighters which had the 'Gorizont-1' (Horizon) radio datalink for automatic guidance to a target by ground control were also modified for the U-19 booster but these MiG 19PUs were not very effective either. By then surface-to-air missiles were capable of shooting down a U-2; as was shown on 1 May 1960 with the downing of Gary Powers by an S-75 (SA-2) missile. (A second missile struck a high-flying MiG which was also attempting to intercept this particular U-2.) After this missile success, the MiG 19SU and MiG 19PU aircraft were retired.

In 1954 MiG also initiated the development of a mixed-power, short-range point defense interceptor variant of the Ye-2 swept-wing jet prototype fighter they were in the process of creating (the delta-winged version was the MiG 21, a famous aircraft type that became a huge success after its introduction in 1959). The new aircraft was powered by a Tumansky AM-9Ye turbojet that delivered a thrust of 37,000 Newton with afterburner and a Dushkin S-155 rocket motor whose maximum thrust was also 37,000 Newton. The S-155 ran on (non-hypergolic) nitric acid and kerosene, which meant it could draw fuel from the same kerosene supply as the turbojet engine. This plane was named the Ye-50 and three prototypes were built. The standard Ye-2 was already capable of flying at Mach 2 but the rocket engine was to increase its rate of climb as well as its maximum altitude and maximum speed. The S-155 engine and the tanks for the hydrogen peroxide running the propellant pumps were fitted in the tail, with the rocket nozzle located just above the jet engine exhaust.

The first prototype, the Ye-50/1, flew on jet power on 9 January 1956 under the control of test pilot V.G. Mukhin, and first used its rocket engine on 8 June that same year. It made a total of 18 successful flights, three of which used combined jet and rocket power. On 14 July the Ye-50/1 made a routine checkout flight after its jet had been replaced but the new engine failed and the plane crashed. The pilot was unhurt but the plane was badly damaged and detailed inspection showed that it was beyond repair. Test flights could not resume until a second prototype of a slightly improved configuration was built. This Ye-50/2 aircraft was under construction from 18 July to 7 December 1956, and on 3 January 1957 made its maiden flight with test pilot V.P. Vasin. No fewer than 25 flights were carried out during the first half of the year, and the plane attained a maximum altitude of 25.6 km (84,000 feet) on 17 June with the help of rocket thrust; considerably higher than the normal ceiling of the standard Ye-2/MiG 21 of about 19 km (62,000 feet). On that same flight the Ye-50/2 also reached a maximum speed of 2,460 km per hour (1,530 miles per hour), or Mach 2.33, which was about one-third of a Mach number faster than when flown under jet power alone. The Ye-50/2 was joined by the Ye-50/3, which featured a longer forward fuselage in order to increase the internal fuel load. This made its first flight in April 1957. After nine successful flights disaster struck on 8 August when the rocket engine exploded and the plane fell out of the sky. Unfortunately the ejection seat failed and test pilot, N.A. Korovin, was killed when the plane hit the ground. This curtailed the Ye-50 test flights.

In the meantime MiG's Gorky factory had been ordered to build a batch of 20 pre-production aircraft in preparation for operational deployment as the MiG 23U (not to be confused with the trainer version of the MiG 23 'Flogger' jet which had the same designation but was a totally different aircraft developed a decade later). None were ever built, however, because the Dushkin OKB which built the S-155 rocket motors had in the meanwhile closed down. In any case, the Air Force had seemingly lost its enthusiasm for mixed-power interceptors, at least in part because of the difficulties associated with operating and maintaining the rocket engine. Furthermore, the sheer vastness of the Soviet Union meant that a great number of aircraft would have been needed even to defend a limited number of high-value targets.

In parallel with the MiG Ye-50, a mixed-power version of the Yakovlev Yak-27 twin-engine fighter was under development. This Yak-27V first flew in May 1957 and operated the standard two RD-9Ye turbojets, each with an afterburning thrust of 38,000 Newton, in combination with an S-155 engine mounted in the tail. With this rocket boost the prototype managed to zoom to 25 km (16,000 feet) during its 2-year test program before closure of the Dushkin OKB and diminishing support from the Air Force curtailed further development.

It appears that in addition to Dushkin, the Kosberg rocket design bureau had been working on rocket engines based on oxygen and kerosene propellants for both the MiG Ye-50 and the Yak-27V but it is not clear whether these were ever flown.

These two planes proved to be the Soviet Union's final attempts at a mixed-power fighter interceptor, because the concept was soon rendered obsolete by improvements in jet engine technology.

The US also experimented with mixed propulsion to satisfy a requirement issued in 1945 for a short-range interceptor. Operating in the immediate vicinity of its base this new aircraft was to provide the last line of defense against incoming high-speed bombers, and the specifications emphasized high speed, high flight ceiling and rapid climb. Bell proposed a weaponized, delta-wing version of its X-1 supersonic research rocket plane. Republic offered its AP-31 mixed-power aircraft with inversely tapered wings (wider at the tip than at the root in order to avoid wingtip stalls at low speeds) and the capability to vary the angle of attack of the wing with respect to the fuselage in flight (for a high angle of attack and thus high lift during take-off and landing, and a low angle of attack and low drag for high-speed flight). Northrop proposed the XP-79Z, which would be similar to the aforementioned jet-powered XP-79B except that it would be based on a rocket motor (something other than the complex Rotojet that had been planned for the original XP-79). Convair offered a swept-wing aircraft with a very unusual 'ducted rocket' powerplant. This was basically a ramjet with several small rocket motors inside the combustion chamber. As well as delivering their own thrust, the rockets would act as igniters for the ramjet. Another four rocket engines, each with a thrust of 5,300 Newton, would be fitted outside the rear fuselage to get the plane into the air and to sufficient speed to enable the ramjet to function. A small Westinghouse J-30 jet engine would provide electrical and hydraulic power, and fly the aircraft back to base once it ran out of rocket and ramjet propellant (it was later proposed that this jet be replaced by a midget racing car engine for supplying power only). As it was to take off using a jettisonable dolly, the plane would be fitted with a lightweight undercarriage for landing only.

In April 1946 Convair was announced to have won the contest by Headquarters Air Material Command of the Air Force and the interceptor project was designated XP-92. Wind tunnel tests soon showed that instead of swept wings the aircraft would behave better with delta wings similar to those developed by Alexander Lippisch for the Me 163 (after Germany's defeat Lippisch had been relocated to the US as part of Operation Paperclip). The pilot would operate the controls in the prone position in a cockpit that was mounted on the nose of the aircraft like a spike projecting from the circular ramjet air intake, and which could be completely ejected in an emergency. The four external rockets would now each have a thrust of 27,000 Newton and there would be a total of sixteen 220 Newton rocket engines inside the ramjet.

The design emerging was not particularly appealing to the eye: a tubular vehicle with a conical nose and a large triangular fin and delta-wings that made it look like a small dart. There is a saying in aircraft design which asserts "if it looks good, it flies good". The XP-92 Dart did not look particularly trustworthy. The cockpit position in particular would severely limit the pilot's view, making it impossible to check his 'six o'clock' (i.e. directly aft). In addition the whole propulsion concept proved to be overly complicated, with its large numbers and types of engines. Just one prototype was built with a conventional cockpit and only a jet engine. It was used primarily for research into delta-wing aerodynamics. In that role it did become the first US delta-wing aircraft (and the first of many delta-winged Convair fighter designs). When the test flights ended in late 1953 the mixed-propulsion interceptor concept was already considered outdated by the USAF.

Although Republic did not win the contest it was awarded a contract to build an experimental mixed-power interceptor based on its AP-31 design. This became the XF-91 Thunderceptor, a heavily modified version of the F-84F Thunderstreak with a liquid propellant rocket engine added for fast climb and high-speed interception. The original idea was to use a Curtiss-Wright XLR27 four-chamber rocket motor with a total thrust of 60,000 Newton but due to development problems a Reaction Motors XLR11-RM-9 with about half the power was mounted instead (the first XF-91 flew with empty XLR27 housing pods and holes for two nozzles above and two below its jet exhaust). The XLR11 was the first US rocket engine developed specifically for use in aircraft, and was also used in the X-1 and D-558-2 high-speed research aircraft described in the next chapter. The compact rocket engine was housed in a pod under the jet tail pipe and had four individual thrust chambers that burned ethyl alcohol and liquid oxygen. Each chamber had a fixed, non-throttleable thrust of 7,000 Newton but the total thrust of the cluster could be regulated by turning individual chambers on or off. It was fairly reliable.

The first XF-91 Thunderceptor [US Air Force].

The XF-91's four-chamber XLR-11 rocket engine was placed directly under its jet engine's exhaust pipe [US Air Force].

The two XF-91 prototypes retained the originally proposed and unusual inversely tapered wing, with the angle of attack relative to the fuselage being variable in flight between minus 2 and plus 6 degrees. The innovative wing provided excellent control at supersonic as well as at very low speeds, reducing the tendency of the tips to stall before the rest of the wing at low speeds (which can lead to a loss of control). The problem with limited space to house the main undercarriage was solved with a highly original arrangement of narrow twin wheels in a bicycle arrangement which retracted outwards into the broad wingtips. Otherwise the design was fairly conventional, with a cross-shaped tail and a regular cockpit with an ejection seat. The armament planned for its eventual operational service consisted of various types of cannon and air-to-air rockets. The prototypes were built and tested at Edwards Air Force Base,

California. The first flight was on 9 May 1949 with Republic's chief test pilot Carl Bellinger at the controls, and it was powered only by the turbojet engine. It was 30 months before the rocket engine was delivered and fitted. On its maiden flight using dual-power on 9 December 1952, the Thunderceptor flown by Russel 'Rusty' Roth became the first US fighter (as it was intended to become an operational military combat aircraft) to fly faster than the speed of sound in level flight. It achieved Mach 1.07 powered by the jet engine with afterburner and two of the four rocket chambers. The aircraft later reached a maximum speed of 1,812 km per hour (1,126 miles per hour, Mach 1.71) on the combined thrust of the 24,000 Newton jet engine with afterburner and 27,000 Newton from the four rockets, which was very good for the time. With the XLR27 rocket engine it would probably have been able to reach Mach 2. It could climb to an altitude of 14.5 km (47,500 feet) in just 2.5 minutes and its maximum altitude was 16.8 km (55,000 feet). But the rocket equipment and propellant took so much space that the Thunderceptor could carry fuel to run the jet engine without afterburner for a mere 25 minutes, and even for that it had to be equipped with two large drop tanks with extra jet and rocket propellant. This would have restricted the plane to a point defense role.

The second XF-91 prototype was similar to its predecessor, although it was later retrofitted with a V-shaped 'butterfly' tail. It suffered a serious engine failure during take-off in the summer of 1951 with Bellinger at the controls, while famous test pilot Chuck Yeager flew chase in an F-86 Sabre (as described later in this book, in 1947 Yeager became the first man to break the sound barrier in the X-1 rocket plane). In *Yeager, an Autobiography*, Carl Bellinger describes what happened: "Chuck radioed, 'Man, you won't believe what's coming out of your engine.' A moment later I got a fire warning light. 'Christ,' I said, 'I think I'm on fire.' He replied, 'Old buddy, I hate to tell you, but a piece of molten engine just shot out your exhaust, and you'd better do something quick.' " Bellinger was still too low to use his ejection seat and black smoke was filling the cockpit, blinding him. Assisted by Yeager's calm instructions Bellinger managed to land the Thunderceptor with its rear fuselage burning violently. He scrambled out and ran for his life. Only 90 seconds had elapsed between Yeager's warning and the emergency landing. In spite of the serious damage, the aircraft was repaired and equipped with a V-shaped tail for further experiments.

Together the two XF-91s completed 192 test flights over the course of five years, but no more were built. As with the Soviet combined-propulsion interceptors, the XF-91 suffered from its short range and very limited total flight time, with the result that it was soon rendered obsolete by a new generation of powerful, more versatile turbojet-only fighters. The second prototype was scrapped but the first one (minus engines) is on display at the National Museum of the US Air Force near Dayton, Ohio.

Around the same time as the new mixed-propulsion aircraft, there were also plans to equip existing jet fighters with auxiliary rocket engines for increased performance for short periods. The Aerojet LR63-AJ-1 that burned white fuming nitric acid and JP-4 jet fuel (a 50-50 blend of kerosene and gasoline) and could produce a thrust of 23,000 Newton was under development for the F-84 Thunderstreak but during the Korean War the Air Force decided to use it in the F-86 Sabre because the early-type

Sabres were being outperformed by the Soviet MiG 15s in North Korean service and it was hoped that the rocket engine would enable the Sabre to counter the MiG threat. Versions of the engine were tried on an F-84G in 1952 and on an F-84F in 1954, but by the time the LR63 had been modified for the F-86 and flight tests on a prototype started in 1956 the Korean War was long over. Neither the mixed-propulsion version of the F-86 Sabre nor that of the F-84 Thunderstreak became operational.

The US Navy still saw some use for carrier-based mixed-power interceptors. In 1955 they ordered six North American FJ-4 Fury fighters with a Rocketdyne AR-1 rocket engine mounted above the standard jet engine exhaust pipe. The AR-1 ran on hydrogen peroxide and standard JP-4 jet fuel, taking the latter from the same tanks as fed the jet engine. The rocket gave the Fury an additional 22,000 Newton of thrust for short periods, sufficient to reach a speed of Mach 1.41 and a maximum altitude of 21.6 km (71,000 feet); considerably better than the normal limits of the standard FJ-4 at just over Mach 1 and 14.3 km (46,800 feet). Development of the rocket-boosted Fury, called the FJ-4F, became a top priority for a while after the US got into a bit of a panic concerning intelligence estimates for the high-altitude capabilities of the new Soviet Tupolev Tu-16 'Badger' and the Myasishchev Mya-4 'Bison' bombers. They urgently needed a naval high-altitude interceptor to thwart the new threat. The first two FJ-4Fs entered flight testing in April 1957 at the Naval Air Test Center, Patuxent River, Maryland, and achieved the aforementioned performance. Between them they made 103 flights, accumulating a total of 3.5 hours of rocket operation. However, the other four designated FJ-4s were never converted into mixed-power interceptors and, unsurprisingly, the FJ-4F concept was soon abandoned. Apart from the obsolescence of the rocket propelled interceptor idea, the risks involved with storing and handling hydrogen peroxide on warships put an end to the development.

Around the same time, in 1957 the Vought company was thinking about boosting their F8U Crusader naval fighter with a rocket engine. The idea was to fit a Reaction Motors XLF-40 rocket engine above the standard jet tail pipe in order to increase the rate of climb for high-altitude intercepts. The XLF-40 burned jet fuel and hydrogen peroxide and delivered 35,600 Newton of thrust. Unfortunately an engine exploded during a ground test and killed two Reaction Motors workers, prompting the engine company to withdraw from the project. Vought continued development for a while with the intention of switching to a Rocketdyne XLR54 rocket engine with a thrust of 26,000 Newton. Dummy engines were installed in two Crusaders to verify that the rocket engines would fit, but no real engines were ever installed. Vought had also planned to equip its new XF8U-3 Crusader III with a Rocketdyne AR-1 engine (as on the FJ-4F) but this was canceled together with the entire Crusader III project.

In the early 1960s the Navy introduced the North American Vigilante supersonic carrier-based bomber. The company planned to fit this aircraft with an auxiliary jet fuel/hydrogen peroxide rocket engine but this never happened. Neither was a mixed-propulsion interceptor version equipped with a Rocketdyne XLR46-NA-2 engine that they tried to sell to the Air Force as the Retaliator.

Back in Europe, France had entered the Second World War with obsolete aircraft that were quickly defeated by the Luftwaffe, and during the war there were no further developments. Afterwards, however, the country was determined to become a strong military power again and worked indigenously on advanced military aircraft. A large number of innovative test planes were built and flown during the late 1940s and the 1950s. In 1944 the government had created SEPR (Société d'Etude de la Propulsion par Reaction) in order to develop its own rocket engine capabilities. The French do not appear to have had any interest in a pure rocket fighter like the Me 163 with its inherently limited range and endurance. They were more interested in aircraft with combined rocket and turbojet propulsion. Studies for a rocket propelled interceptor with auxiliary jet engines began in 1948 at the aircraft manufacturer SNCASO (Société Nationale des Constructions Aéronautiques du Sud-Ouest), commonly referred to as Sud-Ouest. The general idea was to use both jet and rocket engines for a rapid climb and high-altitude intercept, and jet engines alone for the return to base. The rocket engine was to be based on the SEPR design successfully employed by the Matra M04 missile and its successors. However, an engine on a missile has to work only once and for a short time; a rocket motor for a manned aircraft must be more reliable, reusable and maintainable. For operational readiness of the aircraft and engine, the rocket propellant would have to be readily storable (exit liquid oxygen with its extreme cooling requirements), fairly insensitive to changes in temperature and to impurities in the tanks (exit hydrogen peroxide) and easy to produce and handle (again exit hydrogen peroxide). Somewhat surprisingly, it was decided to use nitric acid as oxidizer even although the corrosion of the engine and aircraft structure must have been apparent. (As we have seen, this problem made the Russians eventually discard nitric acid as a propellant for rocket interceptor aircraft.) The fuel was 'furaline', a mixture of 41% furfuryl-alcohol, 41% xylidene (dimethylaminobenzene) and 18% methanol; also a rather surprising choice since it was fairly difficult to produce, especially in comparison with the low-grade kerosene used in the Russian rocket engines for aircraft. An advantage over kerosene was that furaline is hypergolic with nitric acid, eliminating the need for an ignition system.

As the concept of rocket thrust (let alone mixed-propulsion) on a manned aircraft was totally new to France, it was decided first to build a relatively simple proof-of-concept design before embarking on the expensive and risky development of a high-performance aircraft. Therefore one of the existing SNCASO SO-6020 'Espadon' (Swordfish) jet prototypes was fitted with a SEPR 25 engine that ran on nitric acid and furaline fed from tanks on the wingtips, and renamed the SO-6025 Espadon. The light alloy engine was slung under the rear fuselage and produced a modest thrust of 1,500 Newton. It was regeneratively cooled by a flow of nitric acid running around the chamber and nozzle before being injected into and burned in the combustion chamber. The pumps supplying the propellants were powered by the aircraft's jet engine, the method pioneered by BMW during the war in their 003R combined jet/rocket engine. The modified Espadon was intended to demonstrate the safety and reliability of the rocket engine and the capabilities of the mixed-propulsion concept. The first ground test of the plane fitted with the rocket engine was in May 1951 but

The Espadon under rocket power.

the first flight did not occur until 10 June 1952. A total of 76 rocket propelled flights had been made by July 1955, with the rocket-equipped Espadon becoming the first European aircraft to reach Mach 1 in level flight on 15 December 1953. An improved engine, the SEPR 251, was installed in another Espadon and flown 13 times between March 1953 and October 1954.

Based on the successful trials with the two Espadons, the French Air Force invited proposals from French aircraft manufacturers to design a lightweight interceptor with jet engines, rocket engines, or both, which would be able to achieve Mach 1.3, have a high climb rate and be able to operate from rough airfields.

Nord Aviation proposed the Nord 5000 'Harpon' (Harpoon), which was initially conceived as a rocket propelled interceptor with sharp swept-back canard stabilizers and double-delta wings (in which the sweep angle of the leading edge changes at a certain point, as on the Space Shuttle). Morane-Saulnier proposed the MS-1000, of which not much is known. Neither was selected for further development. The Nord 5000 never got further than a wooden scale model.

SNCASO won a contract with designer Lucien Servanty's SO-9000 Trident. This fast looking, bullet-shaped supersonic interceptor had a rocket engine fitted in its tail, augmented by a pair of small Turboméca Marboré II jet engines with 4,000 Newton of thrust each for a fuel-economic return flight after the rapid climb and intercept on combined propulsion. Its wings were straight but relatively thin to minimize drag at high speeds, and had an overall span of 7.57 meters (24.8 feet). The light jet engines were on the wing tips. This kept the fuselage aerodynamically clean but required the wings to be especially strong and stiff, and hence rather heavy. The SEPR 481 rocket engine had three combustion chambers and these could be ignited and extinguished individually to give it a throttling capability in three steps of 15,000

SO-9000 Trident [SNCASO].

Newton and an impressive maximum thrust at sea level of 45,000 Newton. At an altitude of 11 km (36,000 feet) the engine even produced 52,000 Newton owing to the lower external air pressure trying to push the rocket exhaust back in. The gas generator driving the turbopumps ran on the same propellants as the combustion chambers, meaning that the rocket engine did not rely on power from the jet engines; it was an independent unit. One complication was that a mixture of water and methanol had to be injected into the gas generator to reduce the gas temperatures and prevent the turbine blades of the turbopump from melting.

To enable the pilot to escape from a stricken plane at high speed, the entire nose section of the Trident could be jettisoned; a system which offered the pilot a lot of protection but was a much heavier solution than an ejection seat (a similar system had already been used in the Heinkel 176 rocket plane and was also incorporated in the Douglas D-558-1 Skystreak, D-558-2 Skyrocket and Bell X-2 research aircraft, and similar concepts would later be suggested as possible methods for astronauts to escape from malfunctioning shuttles and spaceplanes). The control system was well suited to transonic and supersonic flight speeds, as all three tail surfaces (the vertical and two horizontal stabilizers) were all-moving, rather than small separate rudders and elevators. This prevented the controls from locking up due to Mach shock waves forming at the control hinge lines (the aforementioned dangerous 'shock stall'). At low speeds normal ailerons on the wings were used, but at high speeds these would be locked in order to prevent shocks forming and in this configuration roll would be controlled by differential use of the horizontal tail surfaces (which would thus act as elevons). The tricycle undercarriage was retracted into the fuselage. The fully-loaded weight was 5,055 kg (11,140 pounds); more than the original 4,000 kg requirement but still relatively low for an interceptor.

SNCASO built two Trident prototypes. The rocket engine was not fitted for the initial test flights, as the engineers first wanted to determine how the plane handled at low speeds under jet power alone. The first jet-only prototype took off from Melun-Villaroche on its maiden flight on 2 March 1953 with test pilot Jacques Guignard at the controls. After completing a number of other successful flights the prototype was revealed to the public at the famous international air show at Le Bourget near Paris in June. Guignard also took the second Trident (still lacking a rocket engine) for its first flight on 1 September 1953 but owing to trouble with the relatively weak jet engines he was unable to gain sufficient height after take-off and hit an electricity pylon. The plane struck the ground so violently that the escape module broke off. Guignard was badly injured but survived. To preclude further take-off accidents the one remaining Trident was refitted with two licence-built early versions of the British Armstrong Siddeley Viper jet engine that had almost twice the thrust of the previous Marboré II engine. Test pilot Charles Goujon resumed the test flights on 16 January 1954. Later in the autumn the plane and the rocket engine were deemed to be ready and the first rocket propelled flight using one chamber of the SEPR 481 rocket engine was made on 9 September. Over the next 2 years the Trident made a total of 24 rocket propelled flights to explore its performance. It was capable of exceeding Mach 1 on jet power alone when pushed into a shallow dive but with the SEPR 481 it could achieve a top speed of Mach 1.63 and an altitude of nearly 15 km (52,000 feet). After the flight test program was terminated in April 1956 the first and only surviving Trident prototype was retired to the Le Bourget Musée de l'Air et de l'Espace museum. It is still there, suspended from the ceiling flying at Mach 0 but nevertheless still looking fast and beautiful in its bare metal color scheme with French national markings. The wingtip-mounted jet engines do give it a decidedly 1950s' look, though.

The success of the Trident test program prompted the French Air Force to place an order for an improved version to serve as an operational mixed-power

SO-9000 Trident in the museum at Le Bourget [Mikaël Restoux].

SO-9050 Trident II [SNCASO].

interceptor. Designated the SO-9050 Trident II, this plane was equipped with the new SEPR 631 two-chamber rocket engine with each chamber delivering 15,000 Newton of thrust (although less than the three-chamber engine this was a more propellant-economical compromise between climb rate and endurance). The fully-laden take-off weight was about 5,900 kg (13,000 pounds) and the wingspan was 6.95 meters (22.8 feet). The Trident II was still fitted with the Viper engines of British origin but the detachable cockpit concept of the SO-9000 was abandoned in favor of an ejection seat. In order to be able to mount air-to-air missiles under the fuselage the undercarriage legs were lengthened.

Flight testing of the first Trident II began on 19 July 1955 powered by the jet engines only. The first rocket powered flight was on 21 December. The second plane was damaged beyond repair shortly thereafter when both jet engines flamed out soon after take-off and test pilot Guignard had to make an emergency crash landing. This time he was only slightly injured. A third prototype, equipped to be flown without a pilot, controlled from the ground by radio link, was used as a replacement for the lost aircraft (the remote control feature was never actually used). The first prototype was used to explore the highest performance of the design while the third was for tests at low and medium speeds. The first Trident II disintegrated in mid-air during a dive on 21 May 1957 killing pilot Charles Goujon. The real cause for this disaster was never discovered but an explosion due to propellant leakages and structural failure induced by the abnormal aerodynamic loads were high on the list. However, confidence remained high that the Trident II would enter operational service. The magazine *Flight* reported that at the Le Bourget air show, which opened only 3 days after the crash, "a fascinating large-scale model shows a hypothetical Trident II base with camouflaged dispersals, refueling and servicing points, missile protection and other modern conveniences". The surviving prototype was quickly brought up to the standard of its predecessors so that high-performance test flights could continue. It eventually reached a top speed of Mach 1.95 and a record altitude of 24 km (79,000 feet), both very impressive for the time. Jacques Guignard set two time-to-altitude records using this aircraft: 2 minutes and 36 seconds to 15 km (49,000 feet) altitude, and also 3 minutes and 17 seconds to 18 km (59,000 feet).

Despite losing two aircraft during the test campaign (not unusual for experimental planes in those days), in June 1956 the government placed a pre-production order for the operational Trident III. This had a Turboméca Gabizo jet engine at the tip of each wing, each developing 11,000 Newton of thrust on its afterburner. The rocket main engine was the SEPR 632, which was very similar to the 631 except for using 'tonka' fuel (an approximately 50-50 mix of triethylamine and xylidine which is hypergolic with the nitric acid oxidizer) rather than furaline, which was difficult to produce. The new aircraft was to be equipped with a fire control radar and Matra R 511 air-to-air missiles. However, the first three production aircraft had the 631 engine and they did not carry armament. These early planes were tested until October 1958, with Captain Jean Pierre Rozier reaching an altitude of 26 km (85,000 feet), although this was not registered as a record because the flight was not supervised by the FAI (Fédération Aéronautique Internationale), the international society responsible for logging world aviation records. The top speed set with a Trident III was Mach 1.95 at an altitude of 22.1 km (72,500 feet). But the whole program had been canceled in April 1958 with only four aircraft completed, another two almost finished, and four more in various stages of construction. Unfortunately all of the Trident IIs and IIIs were scrapped, leaving us not a single example of this impressive product of the rebirth of French aeronautical prowess. This is all the more lamentable because the Trident III set an official FAI altitude record for a completely self-propelled rocket plane of 24,217 meters (79,452 feet) on 2 May 1958, achieved by pilot Roger Carpentier in response to the project's cancellation a week earlier. The Trident III actually had the potential to push to 30 km (100,000 feet). Other experimental rocket aircraft had already gone higher than that but, unlike the Trident III, they were air-launched by carrier planes and therefore could not claim official FAI records for self-propelled flight.

As the Russians and Americans had found out, mixed-propulsion planes based on small jet engines and big rocket engines had much too little range and endurance for practical military use. At full rocket power the Trident III could stay in the air only for about 4 minutes! Like other air forces, by the end of the 1950s the French shifted their interest away from short-range interceptors to multi-purpose long-range fighter aircraft. In the future, interceptors would be powered by *large* jet engines and exploit rocket power only for extra boost. Following this philosophy, aircraft manufacturer SNCASE (Société Nationale des Constructions Aéronautiques du Sud-Est, not to be confused with SNCASO, although in 1957 they merged to form Sud Aviation) won a contract to develop the SE-212 Durandal, a delta-wing mixed-propulsion interceptor named for a mythical medieval French sword. Developed by a team headed by Pierre Satre, it was primarily powered by a SNECMA Atar 101F jet engine which delivered a thrust of 43,000 Newton with afterburner, and had a single-chamber 7,500 Newton SEPR 65 auxiliary rocket engine. With a total weight of 6,700 kg (15,000 pounds) it

The second Durandal at the 1957 Paris Air Show.

was a bit heavier than the Trident II and III. The Durandal was intended to be armed with a single AA-20 air-to-air missile carried under the centerline or two 30-mm cannon. As with the Espadon, the pumps of the rocket engine were driven by the jet engine through an auxiliary gearbox. The rocket engine ran on the hypergolic combination of nitric acid and triethylamine-xylidine (TX).

Development of the SEPR 65 started in December 1953 and its first ground test was performed on 4 November 1954. The initial Durandal prototype took to the sky on 20 April 1956 and its first rocket propelled flight was on 19 December that year. The second prototype first flew on 30 March 1957 and it employed rocket power for the first time in April 1957. During its test program the Durandal reached a speed of 1,444 km per hour (898 miles per hour, Mach 1.36) on jet power alone and 1,667 km per hour (1,036 miles per hour, Mach 1.57) with rocket thrust, both at an altitude of about 12 km (39,000 feet). The two aircraft made a total of 45 rocket flights before the program was terminated in May 1957. The main reason for the cancellation was the fact that carrying a single missile it would have only one chance to hit its target and would then be unable to defend itself. This gave it little advantage over a guided surface-to-air missile, and so rendered it of very limited use to the French Air Force. Parts of the first Durandal are in storage at the Musée de l'Air et de l'Espace at Le Bourget, near Paris. It was intended to fit a derivative of the SEPR 65 engine, the SEPR 651, to the small Nord 1405 Gerfaut II delta-wing experimental aircraft but this was never done.

Aircraft company Dassault also developed a small mixed-propulsion interceptor, the delta-winged MD-550 – Mystère Delta 550. It had no horizontal stabilizers but an exceptionally large vertical stabilizer. It was to have two Turboméca Gabizo engines side-by-side for jet propulsion and a single-chamber SEPR 66 rocket engine located between and beneath them. The rocket had a gas generator driving the turbopumps and burned nitric acid and furaline to deliver a thrust of 15,000 Newton. After the development started in September 1953 the first SEPR 66 ground tests took place on 10 January 1955. Over a year later, on 19 January 1956, the engine made its first test flight on the Mystère IV B05, a 'flying test stand' based on the Dassault Mystère IV fighter-bomber. On the tenth rocket flight an explosion of the combustion chamber due to an ignition delay severely damaged the tail of that aircraft. Nevertheless, the pilot managed to land. The problem was quickly traced to the very low temperature at the high altitude at which the plane was flying, and rapidly resolved. At the end of this testing the Mystère IV B05 had made 55 rocket flights and proven that the concept worked. A prototype of the MD-550 (lacking a rocket engine) was built and flown for the first time in June 1955 on jet power provided by two temporary Viper engines (which were, however, never replaced with the intended Gabizo turbojets). This aircraft was then modified, equipped with a smaller tail fin, afterburners, and a SEPR 66, and named the Mirage I. Pilot Gerard Muselli ignited the SEPR 66 for the first time in flight on 17 December 1956, pushing the Mirage I up to an altitude of 12 km (39,000 feet) and a speed of Mach 1.6. Previously, the maximum speed attained on jet power alone had been Mach 1.2 so the rocket thrust added a considerable 0.4 Mach. But the project was terminated in late January 1957 with the aircraft having made only five rocket flights. Like the SNCASE Durandal (canceled 4 months later) the Mirage I had very limited military use because it could only carry a single air-to-air missile.

In reaction to the Air Force's shift in interest to multi-purpose fighters with a long range and multiple armament, Dassault designed the Mirage II. This was a bigger aircraft powered by two Gabizo engines and a SEPR 661, which was a dual-chamber version of the SEPR 66 with the same 15,000 Newton maximum thrust but the option of a 7,500 Newton intermediate thrust by using just one chamber. The SEPR 661 was only ever fired on a ground test bench and the single Mirage II prototype only flew on jet power, but this aircraft design formed the basis for the subsequent, much more successful mixed-power Mirage III.

The Mirage III was a very slender delta-wing interceptor based on the innovative 'area ruling' concept which dictates that to limit aerodynamic drag, changes to the frontal cross section of an aircraft should be made as gradual as possible. This in turn means that near the wings, the fuselage must become thinner in order to maintain a constant cross section with respect to the nose and tail. This gave rise to the famous 'wasp waist' (or 'Cola Cola' or 'Marilyn Monroe') configuration that can be seen on many supersonic fighters. With its maximum take-off weight of 13,500 kg (29,700 pounds) it was much larger and heavier than the Tridents and Durandal. The Mirage III was optimized as a high-altitude interceptor, emphasizing rate of climb and speed rather than maneuverability and low-speed handling. Its primary propulsion was a single SNECMA Atar 101G.1 turbojet which delivered a thrust of 44,000 Newton with afterburner. The pumps of its SEPR 84 rocket engine were powered by the jet engine, feeding the motor with nitric acid and TX. The combustion chamber had a double wall through which nitric acid flowed en route to the chamber, thereby cooling the engine. Another smart feature was that the high-pressure kerosene also served as a hydraulic fluid for operating various valves. Even although it had only one chamber, the SEPR 84 could provide a maximum of 15,000 Newton as well as an intermediate level of 7,500 Newton of thrust. The development of the new rocket engine started at the end of 1956, and it was first ignited in the air on 12 July 1957 on the Mirage III 001 prototype piloted by Roland Glavany. This plane flew a total of 30 missions, during which Glavany managed to reach Mach 1.8 in level flight at an altitude of 12 km (39,000 feet).

The success of the prototype led to a pre-production order for ten Mirage IIIAs, which were intended to be multipurpose fighters rather than merely interceptors. The Mirage IIIA was longer than the prototype, had a larger wing, and was fitted with the powerful Atar 9B turbojet which could deliver a thrust of 58,900 Newton with afterburner. The Mirage IIIA 02 prototype was initially fitted with the first improved pre-production SEPR 840 engine and had sufficient propellant for an 80 second burst of thrust. The aircraft made its first rocket propelled flight on 18 February 1958, and it performed a total of 15 rocket flights with the SEPR 840. It went on to make an impressive 444 rocket propelled test flights with the operational SEPR 841 version of the engine. On 24 October 1958 Glavany nudged the Mirage IIIA 01 prototype past Mach 2 in level flight with the help of a SEPR 840, becoming the first European pilot to do so. On 25 January 1960 a rocket boost enabled the aircraft to reach an altitude of 25.5 km (83,700 feet). The Mirage IIIA design suffered from corrosion of the nitric acid tank and damage to the underside of the fuselage caused by divergence of the rocket exhaust at high altitude (where the lower air pressure permits the exhaust to expand into a wide plume). However, these problems were solved by changing the material of the acid tank and beefing up the fuselage thermal protection.

The first major production version, the Mirage IIIC, had its maiden flight in 1960 and entered operational service in December the following year (the IIIB was a two-seat trainer and had no provision for a rocket engine). It was armed with two 30-mm cannon and had five attachment points for missiles, bombs or extra fuel tanks. The SEPR 841 with its propellant tanks could be added as a separate package to the aft fuselage. If the rocket motor was not needed for the mission, it could be left vacant or replaced by an additional jet fuel tank. The Mirage IIIC remained in use with the French Air Force until 1970, making 1,505 rocket propelled flights involving 2,064 ignitions of the SEPR 841. The rocket engine could give a very powerful kick which allowed the aircraft to accelerate from Mach 1 to Mach 2 in only 90 seconds; on jet power alone this would take about 4 minutes. In terms of added engine weight, this boost came fairly cheaply. The Atar 9B-3 turbojet had a weight of 1,460 kg (3,210 pounds) and on afterburner could deliver 58,900 Newton (equivalent to 5,890 kg) of thrust; a thrust to weight ratio of 4. The SEPR 841 had a weight of only 158 kg (348 pounds) and delivered 15,000 Newton (equivalent to 1,500 kg) of thrust: a thrust to weight ratio of 9.5; more than double that of the jet engine. Operating the rocket engine from the cockpit was also very simple: the instrumentation was limited to an ignition switch, a handle for thrust level selection, a control light to indicate that the engine was operating and another light to indicate a malfunction.

The ensuing Mirage IIIE, which had various improvements to better suit the role of ground attack aircraft, could be fitted with a SEPR 844 engine in which the TX

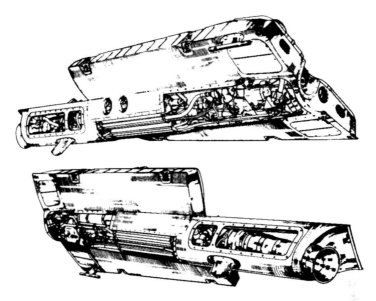

The SEPR 844 rocket pack for the Mirage IIIE [SEP].

Ground testing of an SEPR 844 on a Swiss Mirage IIIS [F + W Emmen].

fuel of the 841 was replaced by kerosene (the IIID was the two-seat trainer version of the IIIE). The main benefit of this was that the rocket engine could get its fuel from the same supply as the jet engine. But because this new propellant combination was not hypergolic, a small tank of TX still had to be carried to ignite the engine (the tank was sufficient for two in-flight starts). The rocket engine could operate for a

A SEPR 844 on a Swiss Mirage IIIS operating in flight [F + W Emmen].

total of 2 minutes and could be stopped and reignited in flight. The SEPR 844 could deliver a thrust of 12,600 Newton at sea level and 15,600 Newton at an altitude of 20 km (66,000 feet) for a specific impulse of 220 seconds. The Mirage IIIE was also built under licence in Switzerland by F + W Emmen as the Mirage IIIS for the Swiss Air Force, which opted to equip its aircraft with SEPR 844 rocket engines.

The Mirage III went on to become the most successful operational mixed-power fighter aircraft ever by logging over 20,000 rocket propelled flights. Despite the use of nitric acid, a spent rocket pack could be made ready for a new mission in less than 15 minutes. This demonstrated that rocket engines could be used by an operational fast-reaction interceptor aircraft. Acid corrosion problems such as experienced by the Russians were avoided by adding a corrosion inhibiter to the nitric acid propellant. A total of 164 SEPR 841 and 111 SEPR 844 engines were built and each could be used for an average of 35 flights. SEPR rocket packs were used on Mirage III fighters by the air forces of no fewer than six countries: France, Libya, Pakistan, South Africa, Spain and Switzerland. This made the SEPR 84 family the most built, most reused, and most operational man-rated (i.e. intended for use on piloted aircraft) series of rocket engines ever. Nevertheless, improved jet engines, accurate long-distance air-to-air missiles, and ground radar networks capable of providing advanced warning of airspace intruders meant that the auxiliary rocket engine was no longer needed, and subsequent French fighter aircraft did not have an add-on rocket capability (although some could be fitted with RATO boosters for take-off). France ended the operational service of the SEPR 844 in 1984 but the Swiss continued to use them until 1996.

Hence the French interceptor philosophy had moved from the concept of a rocket plane with auxiliary jet engines, to a turbojet aircraft with auxiliary rocket power, to a pure jet fighter without any provision for in-flight rocket propulsion. It meant the end for rocket propelled military aircraft in France.

The United Kingdom, although at the forefront of turbojet engine technology, had not done much on rocket and rocket plane development during the war. Based on the knowledge and equipment they managed to capture in Germany in 1945, and assisted by several German rocket scientists, Britain rapidly caught up on rocketry. The initial focus was on the development of jettisonable RATO units, most notably the Lizzie liquid oxygen/petrol engine for the Vickers Wellington bomber, a modified version of the German RI-202 for the Avro Lancastrian (a passenger aircraft based on the Lancaster bomber), and the 'cold', compressed-air-fed hydrogen peroxide/calcium permanganate Sprite engine for the de Havilland Comet jet airliner.

The Sprite, developed in-house by the aircraft company, was meant to help the airliner take off from the many 'hot and high' airports in the British Empire where the air density, and therefore the aircraft's lift at take-off, were diminished by high temperatures and high elevations. The two Sprites on the Comet could each provide a 22,000 Newton boost for 16 seconds. Flight tests on the aircraft prototype started in May 1951 but the system was never used for real operations; injecting water into the combustion chambers of the jet engines proved to be a simpler and cheaper method of gaining the necessary extra boost.

An important innovation for an upgraded Sprite was the use of permanent silver plated nickel gauze packs as the catalyst, replacing the expendable liquid potassium permanganate solution used in the Walter engines as well as the early version of the Sprite. This type of catalyst, which was not consumed and hence did not need to be replenished for each flight, would be incorporated in all following British hydrogen peroxide rocket engines. The Sprite later became the Super Sprite, in which kerosene was injected into the combustion chamber and then spontaneously ignited due to the high temperature of the steam and oxygen gas developed by the decomposition of hydrogen peroxide (which the British named High Test Peroxide, HTP). At 19,000 Newton the engine's thrust was actually less than that of its predecessor, but it had an increased burn time of 40 seconds. The Super Sprite was used operationally by the Vickers Valiant bombers of the Royal Air Force in the early 1950s.

In early 1951 the estimated capabilities of near-future Soviet long-range bombers led the British to believe they would need to be able to intercept bombers flying at extreme altitudes as well as at speeds up to Mach 2. The limitations of the available early warning radar system, in combination with the expected large scale of an air attack, meant that the United Kingdom required a large force of high-performance interceptor aircraft with very fast climb rates and high service ceilings. However, the existing British fighters were nowhere near capable of thwarting the perceived Soviet threat. Based on their experience with RATO rocket engines, the British made plans to use rocket engines for new high-speed, high-altitude manned interceptors.

Initially there was also some interest in pure rocket propelled interceptors, with a government specification leading to the Short P.D.7, Bristol 178 and Hawker P.1089 concepts. The Short P.D.7 design had large delta wings that provided room for four large internal kerosene tanks, while the fuselage housed the additional liquid oxygen needed for the Screamer rocket engine (more on this engine later). The high-subsonic airplane was to have been able to reach an altitude of 9 km (30,000 feet) in less then 3 minutes from the moment that its wheels started to roll over the runway. But once at the altitude of enemy bombers the aircraft would have hardly any propellant left, making it difficult to engage anything. This very limited range meant that the rocket interceptors would have to be stationed as close as possible to their potential targets and in very large numbers, which translated into a large number of individual bases along the coast. Like the French, the British quickly found range and endurance of a pure rocket fighter too limited to be of much use, and hence regarded rocket power on aircraft as being practical only if combined with turbojet thrust. They therefore asked their industries to propose concepts for interceptors using both jet and rocket engines. In addition, they wanted competitive industries to integrate rocket engines based on different propellants in their designs, in order to preclude becoming stuck in dead-end developments.

For the Royal Navy, the Hawker company proposed a new version of their Sea Hawk carrier-based jet fighter augmented with an Armstrong Siddeley Snarler rocket engine running on liquid oxygen and water-methanol (i.e. methyl-alcohol-water) and producing a thrust of 9,000 Newton for an increased rate of climb. The engine was installed in the P.1040 prototype of the Sea Hawk, which was fairly straightforward because the double exhausts of the single Rolls Royce Nene turbojet engine were positioned at the sides of the fuselage, thus leaving the tail conveniently free for the rocket. The fuselage was reinforced to deal with the extra rocket thrust, the Snarler's turbopumps were connected to the 23,000 Newton jet engine via a gearbox, and the jet fuel tank capacity was decreased in order to make room for the new liquid oxygen and water-methanol tanks. There was enough propellant to run the rocket engine for 2 minutes and 45 seconds at full power. The aircraft was subsequently redesignated P.1072.

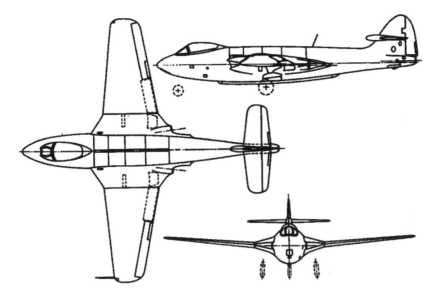

Design of the Hawker P.1072.

The plane's Snarler engine was first ignited during a flight on 20 November 1950 by Hawker's chief test pilot Trevor Wade. The August 1954 edition of the magazine *Flight* described it as follows: "The day was dull and overcast, and the P.1072 could be heard only distantly through dense cloud. Those on Bitteswell airfield were quite ignorant of whether Wade had fired the rocket or not, and were still waiting expectantly when the fighter appeared low down at full power. Passing overhead, Wade did a smart roll; all had gone well." Six rocket propelled flights were made in total, three piloted by Wade and three by Neville Duke. The rocket engine was only fired during a steep climb in order to preclude the plane developing so much speed as to enter the transonic regime, for which its aerodynamics were inappropriate. With rocket thrust the plane could (theoretically) reach an altitude of 15 km (50,000 feet) within 3.5 minutes of its wheels starting to roll over the runway. This translated into an impressive rate of climb of over 73 meters per second (240 feet per second). But this performance could never be verified because the absence of a pressurized cabin meant that the pilot could not fly higher than 12 km (40,000 feet). Nevertheless the experimental prototype did establish that the concept of an auxiliary rocket engine worked. The test campaign also established that despite the cryogenic liquid oxygen the aircraft could be kept fully fueled for extended times without the engine's valves freezing up (although the oxygen would boil off and so require constant topping up). Interestingly, the aforementioned *Flight* article does not comment on the fact that the Snarler engine exploded when Duke attempted a restart during the sixth flight on 19 January 1951; presumably that was still secret information at the time. This incident is believed to have been caused by a small quantity of fuel left in the system after the previous firing of the engine, which then instantly exploded upon reignition. Luckily the damage was relatively small, and the pilot landed the plane safely. The P.1072's tail section was repaired but by then the Navy had decided that using afterburners on their jet engines was an effective and less complicated means of increasing thrust on fighter aircraft. The rocket propelled Sea Hawk was thus canceled. The P.1072 never flew again, and unfortunately it does not exist anymore.

The Avro company responded to the government's request for a mixed-power interceptor with a proposal for an aircraft based on the Armstrong Siddeley Viper 2 turbojet which produced a sea level thrust of 7,000 Newton, in combination with an Armstrong Siddeley successor to the Snarler rocket engine with the equally inspired name Screamer. The Screamer burned liquid oxygen and standard aviation kerosene, produced a maximum thrust of 35,000 Newton, and could be restarted in flight. The Snarler had used water-methanol as fuel because of its good cooling characteristics. The Screamer could run on the same kerosene fuel as an aircraft's jet engine, which meant it did not require separate fuel tanks, but the disadvantage was that kerosene is a much less effective coolant than water-methanol. Hence the Screamer depended on a water-cooled combustion chamber and nozzle. In comparison to an engine based on hydrogen peroxide the Snarler's and especially the Screamer's propellant was less costly and safer, but the liquid oxygen had the disadvantage of requiring very cold storage. Unlike the Snarler, the Screamer engine was required to run independently of the jet engines, which could therefore be turned off. It had its own gas generator in which a small amount of liquid oxygen and kerosene were burned to provide gas to drive the turbopumps to feed propellant to the rocket engine's combustion chamber (water was injected into the gas generator's exhaust to cool the very hot gas in order to prevent it damaging the turbopump assembly).

The resulting Avro 720 aircraft was a tailless delta-wing design somewhat similar to the later French Mirage. It was to be primarily constructed of metal honeycomb sandwich, a light but strong material (although difficult to repair) now widely used for spacecraft. It was intended to arm it with a pair of de Havilland Blue Jay (later renamed Firestreak) passive infrared air-to-air missiles carried on underwing pylons. The Type 720 was meant for the Royal Air Force but a derivative called the Type 728 would satisfy the specific requirements of the Royal Navy (it would be equipped with the much more powerful Gyron Junior jet engine instead of the Viper, and have a strengthened undercarriage for safe operation from aircraft carriers). Thanks to the low drag, exceptionally low weight and good altitude performance of the Screamer, both versions were to be able to reach Mach 2, be able to climb to 12 km (40,000 feet) in only 1 minute 50 seconds and have a maximum altitude of 18 km (60,000 feet). Armed with two heat seeking missiles under the wings the 720/728 was expected to be fully capable of intercepting high-performance bombers and blowing them out of the sky. A contract was issued for the construction of two prototypes and a structural model for ground tests. In 1956 the first aircraft was almost finished when the entire project, including the development of the Screamer, was canceled. The military had somewhat belatedly come to the conclusion that using liquid oxygen in an interceptor aircraft was not such a good idea after all. Oxygen has to be kept at the very low temperature of minus 183 degrees Celsius (minus 279 degrees Fahrenheit) in order to keep it liquid. Even with extensive insulation around the aircraft's internal tanks it was expected that onboard oxygen would boil off at a rate of 4% per hour. So either it must be stored in special tanks and pumped into the aircraft shortly before flight, or if it is stored in tanks in the plane it must be topped off. As already determined by the Germans in 1939 when they reviewed von Braun's early design for a vertical take-off rocket interceptor, this is a major problem for a quick-reaction interceptor that must always be ready for a rapid take-off.

Despite the cancellation of the Avro 720/728, a scheduled test of the Screamer on a Gloster F8 Meteor jet fighter continued. The rocket engine was to be placed on the underside of the aircraft, along with a liquid oxygen tank that would be jettisoned if the pilot ever had to make an emergency belly landing. The rocket's fuel and water tanks were inside the aircraft's fuselage. Ground firing tests were performed, but the full modification of the Meteor was not finished and its Screamer was never used in flight.

With Avro also out of the competition, Saunders Roe was the only company left working on a true, newly-designed mixed-power interceptor. Their SR.53 developed under the leadership of chief designer Maurice Brennan was a single-seat delta-wing aircraft which combined the de Havilland Spectre rocket with an auxiliary turbojet. The rocket engine would be used for rapid ascent and attack, while the jet engine mounted above it would enable the aircraft to make a low-power return to base. Like the Avro Type 720, the SR.53 would only be armed with two de Havilland Blue Jay

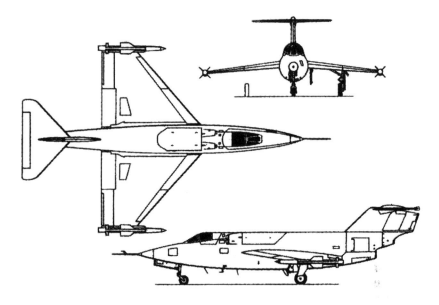

Design of the Saunders Roe SR.53.

(Firestreak) missiles, in this case mounted on the wingtips. The SR.53 was equipped with a T-tail so that the horizontal stabilizers were kept out of the wake of the wings. These stabilizers were of the all-moving type, thus avoiding the shock-stall problems of conventional elevators. Both the wings and the stabilizers were kept thin in order to limit high-speed drag and had a thickness ratio of only 6%. The take-off weight of 8,400 kg (18,400 pounds) included 5,000 kg (11,000 pounds) of propellant.

The de Havilland Spectre rocket engine was specifically developed for a mixed-propulsion interceptor and was heavily based on wartime German research by the Walter company. Hydrogen peroxide was decomposed by being passed through the silver plated nickel gauze catalyst and the resulting hot oxygen used to burn standard aviation kerosene. Cooling of the combustion chamber was achieved by running the hydrogen peroxide through a double wall. Pumps driven by the SR.53's small jet engine fed the propellants to the Spectre's combustion chamber from tanks that were pressurized with air. This pressurization served to suppress cavitation (the forming of bubbles due to local low pressure) in the pumps. Static firings of the rocket engine began in 1952 and demonstrated that it could be throttled from 8.9 to a maximum of 36,000 Newton. The Spectre was flown on a modified Canberra jet bomber test bed (described later) in 1957 before it was deemed ready for incorporation in the SR.53. The auxiliary jet engine was the Armstrong Siddeley Viper 8, which could produce a maximum thrust of 7,300 Newton. This also powered the onboard systems. It was not equipped with an afterburner, and in fact did not provide enough power to enable the plane to fly around after an overshoot on landing (as with the German wartime Komet, the pilot was committed to his approach). The rocket and jet engine received their kerosene from separate tanks but if necessary the turbojet could draw upon the rocket engine's kerosene wing tanks. The combined power of both engines gave the SR.53 a top speed of Mach 2.2 and a maximum altitude of 20.4 km (67,000 feet). From standing on the runway, it could climb to an altitude of 15 km (50,000 feet) in just 2 minutes and 12 seconds with a very impressive climb rate of 270 meters per second (890 feet per second). However, it had sufficient propellant on board only for 7.5 minutes of full rocket thrust.

The US backed the development financially with $1.5 million (equivalent to $12 million in 2010) and the UK Ministry of Defence signed a contract with the company in May 1953 to build three SR.53 prototypes, although the third one was canceled in January 1954 to cut costs. The Ministry of Defence had expected the first SR.53 to fly in mid-1955 but problems with the Spectre and its propellant supply system (one engine even exploded in a ground test) delayed delivery of the first SR.53 prototype to January 1957. With registration number XD 145, this machine was sent by water (the factory was on the Isle of Wight) and by road to the Ministry's aircraft testing site at Boscombe Down for ground test firings of its Spectre engine. The aircraft took to the skies under rocket power for the first time on 16 May 1957 with test pilot John Booth at the controls. It was found to behave as expected with good flight characteristics in general, although the somewhat lower than advertised performance of the Viper engine in combination with an increase in aircraft weight meant that its performance fell short of the specification.

A potential problem with the aircraft's recovery from a so-called 'deep stall' was discovered when a Gloster Javelin jet fighter, which had a similar configuration with

The SR.53 at the Farnborough air show in 1957.

delta wings and T-shaped tail, crashed during a test flight because at extreme angles of attack the turbulent wake of the stalled wings blanketed the horizontal stabilizer, rendering its elevators ineffective and thereby making a recovery impossible (as this involved using the elevators to push the nose down). To prevent this from happening to the SR.53 a set of four small solid propellant rockets were installed in the rocket engine's cowling in the tail: if the pilot ever discovered himself in a deep stall, he could ignite these motors to pitch the aircraft back down (the motors were units that the Royal Marines used to shoot hooks and lines up for cliff assaults). There was also an issue with the Viper flaming out when the SR.53 was pushed over after a steep climb, resulting in a short duration of zero G (the principle on which aircraft for 0-G weightlessness-simulating research are based) that disrupted the flow of propellants. More important, however, was a serious design flaw that was discovered during the test flight campaign: the rear airframe structure suffered badly from the high level of acoustic energy (i.e. sound vibrations) generated by the powerful Spectre engine. The loads were not sufficient to immediately damage the aircraft but prolonged firing of the engine rapidly weakened the structure. This is called 'fatigue' and is a well known issue in aircraft design and maintenance (the problem is similar to how a spoon weakens when you bend it several times until it eventually breaks due to the growth of micro-fractures in the metal). Fatigue on the rear structure of the aircraft delayed the development program even further.

Meanwhile the Royal Air Force had come to the conclusion that it really required something better than the SR.53, which was hampered by its weak turbojet and very limited range and endurance. The small Viper could only be used to fly back to base, as it did not even provide enough power for the SR.53 to take off without additional rocket thrust. Saunders Roe responded with the SR.177, an improved design in which the 7,300 Newton Viper jet engine was replaced by the more powerful de Havilland Gyron Junior that could deliver a thrust of 32,000 Newton in normal use and almost twice as much with afterburner. This engine was the first British turbojet specifically designed for sustained running at supersonic speeds, enabling the SR.177 to fly faster than Mach 1 for relatively long periods if assisted by rocket thrust. In addition, a new version of the Spectre engine was to be installed. The Spectre 5 delivered the same maximum thrust as its predecessor in the SR.53 but could be throttled over a wider range down to just 10% of its maximum power, enabling it to be run in combination with the jet engine for prolonged low supersonic flight. The airframe was enlarged to accept the big Gyron Junior, and the positions of the jet and rocket engines were reversed with respect to the SR.53 (i.e. the turbojet was now located below the rocket engine). It also had a larger nose to accommodate the intended new radar system that would assist the pilot to find his target. Like the SR.53, the SR.177 had a pair of Blue Jay (Firestreak) missiles at its wingtips but these could be replaced with pods, each containing 24 unguided air-to-air rockets. An auxiliary power unit with a hydrogen-peroxide steam turbine was installed to power the onboard systems in the event that the turbojet flamed out at high altitude. All in all, the take-off weight of the SR.177 was one-and-a-half times that of the SR.53. Although the general layout was similar to its predecessor (including the 6% thickness ratio of the wings and tailplane) the SR.177 was virtually a new design. The large Gyron Junior required a lot more air than the Viper, so the sleek lines of the SR.53 with its modest air inlets just behind the cockpit canopy were sacrificed for a huge chin-mounted intake. Of course, the benefits outweighed aesthetics: the new jet engine would considerably improve endurance, enabling the SR.177 to cruise for a long time on jet power alone. This time could be extended by fitting jet propellant drop tanks under the wings and also by aerial refueling of kerosene from tanker aircraft. The limited supply of rocket propellant could thus be conserved for making a dash once a target was identified. The combined rocket and jet thrust were expected to give the SR.177 a top speed of Mach 2.35 at 12 km (40,000 feet) and Mach 2.75 at 21 km (70,000 feet). For a short-warning interception mission it would be able to achieve Mach 1.6 at 18 km (60,000 feet) in just 4 minutes, but this would consume its entire propellant load and require a gliding return. The powerful jet engine was also expected to enable the SR.177 to be effective in light ground attack and photographic reconnaissance roles, for which the rocket engine would be removed.

In July 1956 funding was secured for the construction of 27 SR.177s, with the first prototype expected to fly by April 1958. It was a large project employing some 400 Saunders Roe and subcontractor engineers on design work alone. Flight testing with the SR.53 was planned to continue, but only to assist in the development of its bigger sister.

However, as elsewhere, changes in military requirements and the evolution of the interceptor concept from a pure rocket plane, to a mixed-power aircraft with a small auxiliary jet engine (the SR.53) and to a mixed-power design with a large jet engine were inevitably leading to the next step of discarding the rocket engine entirely. The death knell for the SR.177 as a Royal Air Force interceptor was the infamous 1957 Defence White Paper in which Minister of Defence, Duncan Sandys, expressed the belief that all manned fighter aircraft would soon be rendered obsolete by guided missiles. Only the development and production of the English Electric Lightning jet fighter would be continued. For the short term this pure jet interceptor was deemed sufficient, because it had become clear that the new Soviet bombers were nowhere near as fast and high-flying as had been feared earlier. No more-capable jet or rocket interceptors would be required, because in the longer term the defense of the country could be safely turned over to surface-to-air missiles. The SR.177 fell victim to what would prove to have been a disastrously erroneous policy that all but killed the UK's independent military aviation industry. The fact that new, manned fighter aircraft are still being developed and produced today is testimony to how overoptimistic Sandys was about the capabilities of missile systems.

Saunders Roe still tried to sell the SR.177 design to the Royal Navy, as well as to likely foreign customers. The Germans were very interested in its rapid interception capabilities, since their border with the Warsaw Pact meant the Luftwaffe had little time to react to intruding aircraft. When in August 1957 the British decided against the Royal Navy version of the SR.177 the writing was on the wall. Development did continue for a while longer with the British Ministry of Supply agreeing to fund the construction of five of the planned six prototypes in anticipation of sales to Germany. However, the Germans decided not to take financial risks and ordered the

Artistic impression of an SR.177 in Luftwaffe service.

Lockheed F-104G Starfighter instead, as did many other European NATO countries. This was a pure jet which combined the roles of ground-attack fighter, photo-reconnaissance and interceptor into a single all-weather aircraft. It was later discovered that Lockheed had paid out millions of dollars in 'Sales Incentives' (i.e. bribes) to several European politicians to buy their aircraft and secure what was called the 'Deal of the Century'. Whatever the reason for the Germans' change of mind, in hindsight their decision to buy a multipurpose jet fighter instead of a mixed-power interceptor and its derived ground-attack and reconnaissance variants was sensible. The Starfighter was a highly successful, albeit rather dangerous, fighter that remained in operational service with the Luftwaffe until 1987.

With no support left the SR.177 project was canceled just before Christmas 1957, with the design being 90% complete and the production of various equipment for the first six aircraft already well under way. No complete prototype had yet been made other than a full-scale wooden mockup (unfortunately no longer in existence) which was used for assessing the cockpit layout and the proper fitting of the various pipes, cables and control systems (nowadays this is done by using 3-dimensional computer models, which can be updated and improved much faster than a mockup).

The SR.53 test flight campaign continued, with the first prototype being joined by the second aircraft with registration number XD 151. This made its first flight on 8 December 1957. Unfortunately it crashed on 5 June 1958 shortly after taking off for its 12th flight. The exact cause was never established but it seems the aircraft lacked the power to climb because its rocket engine had shut down just as the plane left the ground. It struck a pylon at the end of the runway and violently exploded, killing test pilot John Booth. The test program resumed using the first prototype. Saunders Roe proposed turning XD 145 into a true test bed for new rocket engines, including air-launching it (from the back of a Valiant bomber) rather like the experimental X-15 rocket aircraft in the US, but the Ministry of Defence felt the aircraft was not suitable for such a program and halted funding. The decision was not unreasonable, because the SR.53's aluminum structure would limit its maximum speed to around Mach 2 to avoid structural failure by overheating. Also its air intakes and its still relatively thick wings would probably have caused problems at speeds far over Mach 2. Moreover the aircraft lacked reaction control thrusters for attitude control at extreme altitudes (although these could probably have been retrofitted, as was done to the Starfighter-derived NF-104A described in the next chapter). In October 1958 the flight campaign ended and the only remaining SR.53 was transferred to the Royal Air Force Museum storage facility at Henlow. It remained there until 1978 when it was donated to the Brize Norton Aviation Society, which restored the aircraft for static display. It can now be seen in the Royal Air Force Museum at Cosford. The two SR.53 prototypes together accumulated a total flight time of 22 hours and 20 minutes over 56 flights by three pilots.

During the late 1950s the Spectre rocket engine was also intended to be used in the Italian 'Leone' (Lion), which was an attempt by aircraft manufacturer Aerfer to develop their Sagittario 2 supersonic (in a dive) jet into a competitive interceptor, but the project was abandoned before a single prototype could be built.

Saunders Roe had also investigated a third mixed-power interceptor, the rather large P.187. This would have been powered by two Gyron turbojets with afterburners and four Spectre 5 rocket motors, sufficient to enable it to attack targets at Mach 2 within a range of 420 km (260 miles) and up to an altitude of 18 km (60,000 feet). It would have taken 3 minutes and 50 seconds to climb to 12 km (40,000 feet), which was a relatively slow rate of 52 meters per second (171 feet per second). It was to be equipped with an advanced radar and four air-to-air missiles. The propellant required for this performance would have made its take-off weight almost four times that of the SR.177! The aircraft would have been crewed by a pilot and a radar operator. It featured a sliding nose for improved viewing during landing, an idea that was later adopted by the Concorde supersonic airliner. The P.187 project never left the drawing board.

The cancellation of the SR.53 and SR.177 meant the end of the development of true mixed-power interceptors in the UK but there were also several projects focused on fitting auxiliary rocket engines into existing military jets.

One idea was to put a rocket on the English Electric Lightning jet interceptor, which was still under development when this option was considered. Like on the Mirage III the rocket system would be designed as a self-contained pack that could easily be fitted on the belly of the aircraft if needed. The rocket would considerably increase the maximum altitude of the Lightning, enabling the plane to attack high-flying intruders. The Napier Double Scorpion was developed for this purpose. It had two thrust chambers which burned kerosene in hot oxygen gas that was made by the decomposition of hydrogen peroxide with catalyst packs of metal gauzes and pellets. The engine had a turbopump powered by hydrogen peroxide that was decomposed by a separate, smaller catalyst pack. The total thrust was almost 36,000 Newton, and each combustion chamber could be individually throttled in steps of 0 to 50 to 100% thrust. To test the engine, it was installed on an English Electric Canberra jet bomber. The rocket system, complete with engines, tanks and plumbing, formed an integrated unit that could be carried in the standard bomb bay of the Canberra, thus minimizing the required modifications. To prevent the Canberra from starting to pitch due to the additional rocket thrust from its belly, the nozzles were angled slightly upward so as to aim the thrust through the aircraft's center of mass. Test flights in early 1956 with Canberra number WK163 started off with a single Scorpion thrust chamber, in which configuration it was even flown at the Paris air show that year. Flights with several versions of the Double Scorpion followed. The system was found to be very reliable and to require relatively little maintenance. A complete overhaul was required after a total firing time of 1 hour (similar to the SEPR 84 series of the Mirage III) but since the Scorpion could only be operated for several minutes per flight, due to the limited propellant supply on board, this was not needed very often. The aircraft successfully demonstrated the Double Scorpion at the Farnborough air shows of 1956, 1957 and 1958, as well as taking part in the 1958 Battle of Britain memorial show.

The Canberra was a subsonic aircraft, and the rocket engine could not be used to increase its maximum speed because the aerodynamics were unsuitable for transonic flight. But it could boost the aircraft's maximum flight altitude, as it was intended to do for the Lightning interceptor. On 28 August 1957 the rocket enabled WK163 to capture the world altitude record with a flight up to 21.4 km (70,310 feet) under the control of pilot Michael Randrup and observer Walter Shirley. It was the third record attempt: on the first flight the plane had not flown high enough, while on the second the official barograph with which they were to measure the altitude malfunctioned. Experimental rocket planes had flown higher in the US but they were dropped from carrier aircraft rather than taking off on their own power and thus their achievements were not recognized by the FAI as official records. A red scorpion and details of the achievement were painted on WK163's nose. Now retired after a long career as a flying test bed for all manner of new equipment, this aircraft has been preserved by Classic Aircraft Projects at Coventry Airport and still flies (without rocket power) at air shows in the UK.

Apart from testing engines, there was actually an operational job for which rocket-enhanced Canberra's proved to be ideal. In the 1950s the British performed atomic bomb tests as part of the development of their own nuclear deterrent. The explosions shot radioactive particles high into the atmosphere and the bomb developers needed samples for study. Because the tests with WK 163 had shown that the rocket engine improved the standard service ceiling of 15 km (49,000 feet) by about 30%, a pair of Royal Air Force B6 Canberras were fitted with Double Scorpions to fly high-altitude missions over Australia and the Pacific and collect samples from test explosions.

One of these aircraft, registration number WT207, is loaned to Napiers to conduct further flight testing of their Double Scorpion propulsion unit. On 9 April 1958 pilot John de Salis and navigator Patrick Lowe are flying a test, climbing from 13.7 km to 18.3 km (45,000 to 60,000 feet) in just under 4 minutes by combined jet and rocket power (a very modest climb rate, but the mixed-power Canberra was meant simply to reach high altitudes, not rapid rates of climb and high speeds). After shutting down the Double Scorpion, the aircraft descends to 17 km (56,000 feet). At that point de Salis ignites the rocket engine once again in order to burn off the residual hydrogen peroxide to ensure that none of this dangerous liquid remains in the tank for landing. However, upon ignition the crew immediately notices a rumbling noise and buffeting of the aircraft, and a warning light for the rear bomb bay flips on: the rocket engine is on fire! Before they can start the carbon dioxide fire extinguisher in the bomb bay an explosion blows apart the central section of the plane, ripping both wings and the tail off. Mist appears in the crew cabin due to rapid depressurization. Lowe jettisons the hatch above his seat and ejects first, then de Salis shoots out through the clear canopy above the cockpit. They set a record for the highest ejection seat escape ever, which is only broken in 1966 when the crew of an SR-71 Blackbird escapes at an altitude of 24 km (80,000 feet). The likely cause of the loss of WT207 was a leak in the engine which allowed kerosene and HTP (hydrogen peroxide) to mix and ignite outside the combustion chamber. After modifications to the Scorpion the flight testing resumed, but not for long because the British decided that there was after all little need for an auxiliary rocket pack for the Lightning interceptor (which was already demonstrating a rather phenomenal rate of climb and speed for its day). The Scorpion engine was canceled, and with it all rocket development work at Napier ceased. It also meant the end of the Canberra flights using that company's rocket engine after a total of 500 rocket propelled flights.

Another engine that was mated with the Canberra was the 36,000 Newton thrust de Havilland Spectre, which also used hydrogen peroxide and kerosene. It was fitted in a Canberra and first ignited in the air in January 1957. The test flights showed that the propellant delivery system required further development work. The Spectre was removed from the aircraft and subsequently fitted on another Canberra. This aircraft was never used for nuclear sampling missions but the test flights cleared the engine for use on the aforementioned SR.53 mixed-power interceptor.

In the mid-1960s the South African Air Force was interested in purchasing the subsonic Blackburn Buccaneer, which was in service in the British Royal Navy as an attack aircraft. The Buccaneer was meant to fly very low in order to avoid detection by radar. However, due to their 'hot and high' conditions, the South Africans wanted the aircraft to be equipped with rocket engines for extra thrust during take-off. This resulted in the Buccaneer S.50 version with two retractable Bristol Siddeley BS.605 rocket engines fitted just behind the bomb bay. Together these could give 30 seconds of thrust at 36,000 Newton to supplement the 180,000 Newton from the twin jet engines running on afterburner. As with the Super Sprite and Scorpion, the BS.605 used kerosene and hydrogen peroxide as propellants. The modifications required to the standard Buccaneer design were modest. South Africa bought 16 of the planes, which flew with 24 Squadron from 1965 until 1991. Ironically, the rockets turned out not to be necessary and were eventually removed from all operational Buccaneers. The roar of the twin BS.605 rocket engines was only unleashed four times, during air show displays.

The number of mixed-power interceptors developed in the 1950s and early 1960s is astonishing, especially considering that the results in terms of operationally useful aircraft were fairly disappointing. In those years, Cold War tensions produced large aeronautical development budgets. Remaining at the forefront of technology meant a parallel investigation of every possible means of gaining military advantage. In the end the mixed-propulsion interceptor proved to be a dead end, since improvements in turbojet technology, advanced radars for early detection of airspace intruders, and the growing military desire for multi-purpose fighter aircraft, combined to render placing rocket engines on aircraft redundant. The added performance simply was not worth the disadvantages of shorter range and endurance, greater aircraft weight, and added complexity. And the role of the interceptor itself became much less important during the 1950s as it became possible to use surface-to-air missiles to shoot down high-flying bombers. Moreover, the Soviet Union was developing ballistic missiles to deliver nuclear warheads, and these could not be intercepted by any manned aircraft. Fast, low-flying bombers carrying nuclear weapons remained a serious threat, but rocket propelled interceptors were ill-equipped to do anything about these.

There was still a role for high-altitude, high-speed reconnaissance aircraft ('spy planes') but they required endurance to be able to penetrate far into enemy territory and cover an adequate area. Advanced jet aircraft, specifically the Lockheed SR-71 Blackbird, soon filled that role. The SR-71 could not fly quite as fast and as high as some of the latest experimental rocket planes, but it could undertake missions lasting hours rather than minutes.

The mixed-power high altitude aircraft did come back one more time in 1963 in the form of the NF-104A AeroSpace Trainer, a heavily modified F-104A Starfighter with a rocket engine on its tail that was developed to train pilots in flying at extreme altitudes (of which more later).

Had mixed-powered interceptors remained of interest to the military, then their evolution would probably have eventually led to advanced systems combining rocket and jet functionalities into a single engine. Such engines, which operate as turbojets at low speeds and altitudes, ramjets at high speeds, and rockets at high altitudes, are currently seen as the key to making fully reusable single-stage spaceplanes possible. But the demise of the mixed-power interceptor sent jet engines and rocket engines their separate ways, with jet engines powering aircraft and rocket engines powering expendable missiles, space launchers and a few specialized high-speed, high-altitude experimental aircraft.

ZERO-LENGTH LAUNCH

The use of Rocket Assisted Take-Off (RATO) boosters became very common shortly after the Second World War because the early jet engines delivered relatively low thrust when the aircraft was moving slowly. Heavy jet aircraft in particular needed a bit of help to get going. But soon the take-off thrust of jet engines had increased to the point that RATO was only required for heavy cargo aircraft using short runways or airfields in hot places and at high elevations (they are still in use today, but only in very limited circumstances).

A different idea for mixing rocket and jet propulsion that was investigated was to use a powerful rocket booster to shoot an airplane straight into the sky, dispensing with the need for a runway. With this ultimate stretch of the RATO concept, fighter aircraft could be launched anywhere and anytime, even from a truck trailer!

For Cold War military planners the vulnerability of airfields and their concentration of aircraft was always a major issue, especially for the early, underpowered jets that could not take off from rough fields and required especially long runways. This concern led to the need for Vertical Take-Off and Landing (VTOL) aircraft: jet fighters that (as with helicopters) would not need much more than a clearing in a forest to operate from. Most of the VTOL fighter concepts proved to be impractical but the developments ultimately led to aircraft like the swivel-nozzle Harrier, which can take off and land virtually anywhere whilst operating as a conventional jet fighter when in the air.

During the 1950s, blasting jet fighters into the air using rockets was a simpler way of liberating aircraft from their dependence on runways for take-off (although the returning aircraft would normally still need a prepared airfield for landing). The idea had already been pioneered by the British ship-launched, rocket-catapulted Hurricat fighter of the Second World War, but the new jet fighters were much heavier than the old Hurricane propeller fighter and required much larger boosters to get airborne: so large, in fact, that they were impossible to fit into the airframe and (like RATO units) had to be attached externally and jettisoned immediately after use. This had several benefits though. One was that the heavy rocket equipment did not need to be taken with the aircraft during its entire mission. Another was that not much modification to existing aircraft would be needed to accommodate the external rocket boosters. Also of benefit was that because the rockets would be jettisoned soon after take-off, they would not need to be especially light and efficient: relatively simple solid propellant boosters similar to those used in surface-to-air missiles would suffice. This concept became known as Zero-Length Launch (ZEL).

In the early 1950s the US Air Force began a program called 'Zero Length Launch, Mat Landing' (ZELMAL) in which a Republic F-84G Thunderjet fighter was to be shot into the air using a large solid propellant rocket booster. (The F-84 was selected because it was sufficiently light that it could be launched by already available rocket boosters.) To solve the problem of the need for a landing strip upon return, the idea was for the fighter to be equipped with a hook to snag an arresting cable suspended close to the ground in order to come to a quick stop, rather like on an aircraft carrier, except that instead of rolling to a halt the aircraft (without lowering its undercarriage) would smack down onto an inflatable mattress measuring 25×245 meters (80×800 feet) and 1 meter (3 feet) thick. An additional perceived benefit of this technique was that ZELMAL aircraft would not need an undercarriage and so would be lighter than comparable conventional fighter aircraft.

The Glenn L. Martin Company was selected to manage the development of the system, with the Goodyear Tire & Rubber Company making the air-filled mat. Tests started at Edwards Air Force Base, California, on 15 December 1953 with a pilotless F-84G being launched from a trailer normally used to fire Matador cruise missiles (it seems that the same type of rocket booster was employed). As planned, the aircraft was lobbed into the air and then crashed onto the hard desert floor. The next test less than a month later, on 5 January 1954, was equally promising. It was manned by test pilot Robert Turner and the G-levels that he experienced during the launch were no worse than during a conventional catapult launch from an aircraft carrier although he accidentally jerked one hand and throttled the engine back, almost stalling it. Turner made another flight on 28 January. Both flights went surprisingly well, and surviving movie footage (which is on the Internet) shows a very smooth operation with a fluent acceleration and a clean separation of the booster. This indicated that rocket-boosted take-offs were a feasible operational military possibility. In both tests Turner landed conventionally. Landing on the mat would prove much more problematic. The first time the rubber mat was inflated after being transported on a couple of trailer trucks, it was found to leak so badly that parts of it had to be sent back to the manufacturer. The first mat landing on 2 June 1954 became a fiasco when the aircraft's arresting hook tore the mat wide open and caused a very hard landing. The plane was damaged beyond repair and Turner was rendered inactive for months due to back injuries. Two more mat landings were conducted but the sudden impact on the mat remained too hazardous. Test pilot George Rodney suffered a neck injury from his mat landing: "We tied ourselves into the seat real well, so we wouldn't pitch forward into the control column and the instrument panel, but unfortunately your head, it goes into a big arc and comes down on your chest." After 28 rocket launches ZELMAL was terminated.

The sudden lift-off of a ZEL launch must have been rather strange for the pilots: they were sitting in the cockpit of a familiar aircraft but rather than seeing a runway in front of them they were looking up into the sky at a steep angle. And instead of the reassuring, slowly growing push of the jet engine there was a sudden explosion of power hurtling them into the air. It is a bit like sitting in your own car but with an additional dragster racing car's engine in the boot.

Despite the termination of ZELMAL the idea of launching a fighter with a rocket booster was still believed valid. The Air Force initiated a new program in 1957 that dispensed with the mat landing and so was simply named ZEL. It was to involve the launch of a nuclear-armed strike aircraft from a truck trailer which, since it could be hidden anywhere, would be hard for the enemy to destroy during a first strike. After dropping his bomb, the pilot of the undercarriage-less aircraft would simply bail out over friendly territory. To be able to carry a heavy atomic bomb into Soviet territory the aircraft would have to be much larger than the F-84. The selected F-100 Super Sabre was about twice as heavy as the F-84, so a new rocket booster was developed by Rocketdyne. This solid propellant rocket could deliver a thrust of almost 578,000 Newton for 4 seconds and accelerate the F-100 at about four G. It was affixed under the aircraft's rear fuselage, at a slight angle so that its thrust was aimed through the center of gravity and would thus not cause any rotation. At burnout, the plane would be flying at 450 km per hour (280 miles per hour) at an altitude of 120 meters (400 feet). Preliminary tests were started with a so-called 'iron bird', a structure of steel and concrete that simulated the weight and the mass distribution of the F-100. These tests showed that if the booster were not precisely aligned with respect to the center of gravity of the entire contraption it could perform some very impressive backward summersaults. But this was soon fixed, and the earlier problem of the pilots' hand on the throttle moving backwards due to the acceleration was remedied by introducing a fold-out handle which the pilot could slip his hand into.

The first manned launch of an F-100 at Edwards Air Force Base on 26 March 1958 went perfectly. Test pilot Al Blackburn said he found the flight "better than any ride you can find at Disneyland". On his second launch, however, the rocket did not separate, even when he tried to shake it free using wild maneuvers. He had to use his ejection seat and let the plane crash in the desert because it was impossible to land it with the big booster attached. The investigation showed the attachment bolts had not sheared off as they were intended to. Thereafter explosive charges were provided that could blow the boosters off on command. Another 14 successful flights were made by October. A sign on the trailer claimed it to be the 'World's Shortest Runway'. The tests were not kept secret: footage of one of these launches was used in the *Steve Canyon* television series. There was even a public demonstration of the system, with the pilot showing off by performing a slow roll immediately after booster separation. The technical feasibility of the ZEL concept was proven, but its operational role was not so clear.

First of all there were the practical as well as safety and security issues relating to driving a fighter with a nuclear weapon on a truck through dense forests, over narrow roads and through tunnels. Critics said it would be better to launch ZEL aircraft from fixed, protected positions. To test this idea, a few launches were performed out of a hardened shelter at Holloman Air Force Base in New Mexico, the last of which took place on 26 August 1959. But launching from fixed positions denied the flexibility and elusiveness of the mobile system. A more serious threat to the project was that the idea of sending nuclear-armed fighter aircraft into the Soviet

An F-100 ZEL with its impressive rocket booster [US Air Force].

Launch of an F-100 ZEL [US Air Force].

Union was rapidly being made obsolete by the increasing reliability and accuracy of unmanned ballistic missiles. Although 148 F-100s were modified to enable ZEL launches, the program went nowhere.

The concept of a ZEL nuclear-armed strike fighter was picked up once more, this time by the German Luftwaffe in 1963. Working with Lockheed, they organized rocket-launch experiments using an F-104G Starfighter at Edwards Air Force Base. The Rocketdyne booster could push the Starfighter to a speed of 500 km per hour (310 miles per hour) in just 8 seconds. Lockheed test pilot Ed Brown, who performed the flights, was very impressed: "All I did was push the rocket booster button and sit back. The plane was on its own for the first few seconds and then I took over. I was surprised at the smoothness, even smoother than a steam catapult launch from an aircraft carrier." The successful experiments were followed up by further tests at the German Air Force base at Lechfield. But this project was also canceled for the same reasons as the F-100 ZEL and because the tests using the expendable rocket boosters were quite expensive. In addition, the much more practical VTOL Hawker Siddeley GR.1 Harrier was by then under development. This made its first flight in December 1967. It could not only take off vertically without rocket booster assistance, it could also land vertically. A German F-104G equipped with a rocket booster and a dummy nuclear missile is on display at the Luftwaffe museum in Berlin-Gatow in Germany.

In France there was a proposal for a ZEL version of the mixed-power Durandal in which the aircraft would be launched from a mobile trailer using a cluster of solid propellant rockets attached at an angle on the tail, but this was not developed into a real system.

Soon after the US started to experiment with rocket launched F-84Gs, the Soviets initiated a very similar project using their MiG 19 (which, as described earlier, was also converted into a high-altitude mixed-power interceptor around this same time). Rather than launching nuclear-armed strike aircraft, the Soviets intended their rocket-launched interceptor to play a role similar to that they had envisaged for their earlier rocket propelled aircraft, namely a fast-reaction point-defense interceptor. Launching these from truck trailers with large rocket boosters would make it possible to station them at remote locations or in battle areas where there were no suitable runways. The MiG design bureau prepared a proposal for a trailer-launch system for the MiG-19. This was given the go-ahead in 1955. MiG came up with a modified version of the MiG-19S, designated the SM-30, which had a reinforced structure to handle the high rocket thrust, the ventral fin was replaced by two new fins straddling the rocket booster, and a special headrest to protect the pilot from whiplash at the onset of the sudden acceleration.

Like the US Air Force and the Luftwaffe, the Soviets used a large, jettisonable

A standard MiG 19 at the Letecké museum near Prague, Czech Republic [Michel van Pelt].

solid propellant rocket booster which was mounted on the rear fuselage and pointed slightly downwards. The PRD-22 booster had sufficient thrust to shoot the 8,000 kg (17,000 pound) plane into the sky. The SM-30 could be transported on a large trailer-truck combination but had to be placed on another type of trailer for the launch. The plane was connected to this trailer by bolts that would shear and release the aircraft upon ignition of the booster. When launched from soft surfaces, such as from within a forest, soldiers would be required to dig a trench behind the booster to prevent the powerful exhaust from creating a fountain of loose dirt that would be visible to the enemy from far away. The first test launch was performed using a remote-controlled unmanned airplane in the autumn of 1956. The launch went smoothly (just like the Americans had experienced) but the trailer was wrecked, thus proving that it needed to be fitted with a blast shield for protection. On 13 April 1957 test pilot Georgi M. Shiyanov made the first manned flight, which was a big success. He had trained with a special launch simulator catapult (even successfully enduring an excessive 18 G acceleration during one test when a technician made an error arming the catapult). Several more SM-30 launches were performed, all of which were successful (film of the launches is on the Internet). However, landing the heavy fighter on rough and small landing strips like those envisioned to be available near forward battle areas was not so easy; getting the plane to stop before it ran out of runway was very tricky using the standard MiG-19 drag parachute and brakes. An arresting-cable system (like on an aircraft carrier) was therefore tried out.

In the end, however, the SM-30 project was terminated because of the difficulty of driving the large, heavy aircraft across the countryside and because an airplane which does not require a runway for take-off but does require one for landing is not all that useful. Just as the F-100 and F-104G ZEL were rendered obsolete by the introduction of long-range nuclear missiles, the SM-30 turned out to be less effective for forward air defense than the new mobile battlefield surface-to-air missiles.

6

Breaking the barrier

"You've never been lost until you've been lost at Mach 3." – Paul F. Crickmore (test pilot)

Until about the start of the Second World War the strange phenomena which develop at transonic speeds were academic, since the propeller aircraft of the day did not fly that fast. But starting in 1937 mysterious accidents began to occur at high speeds. An experimental early version of what was to become Germany's most potent fighter of the war, the Messerschmitt Bf 109, disintegrated as its pilot lost control in a fast dive. Pretty soon other new, high performance military airplanes were running into similar difficulties. For these fast, propeller-driven fighters the airflow over the wings could achieve Mach 1 in a dive, making air compressibility a real rather than a theoretical issue.

In the US, the aeronautics community was rudely awakened to the realities of this unknown flight regime in November 1941 when Lockheed test pilot Ralph Virden was unable to pull his new P-38 fighter out of a high-speed dive and crashed (due to the problem of 'Mach tuck' described in a previous chapter). It became clear that any propeller fighter pilot who inadvertently pushed his fast plane into a steep dive was risking his life. Aggravating this problem was that bombers were flying ever higher, which meant that in order to reach their prey interceptors had to venture into the thin, cold air of the stratosphere in which the speed of sound was lower, and thus issues of air compressibility occurred at slower flight speeds than they did when flying nearer the ground.

A really thorough understanding of high-speed aerodynamics was initially not necessary, because measures to prevent control problems focused on limiting the dive speed and temporarily disrupting the airflow to prevent shock waves from forming on the wings and controls. Due to the limitations of propellers and piston engines, it was accepted that conventional aircraft would never be able to fly faster than sound.

Then jet and rocket aircraft appeared. These were quickly realized to require the potential to fly at Mach 1 or even faster for extended durations, so a real understanding of transonic and supersonic aerodynamics rapidly became a 'hot' issue that promised real military advantages. The name 'sound barrier' had been coined by a journalist in 1935 when the British aerodynamicist W.F. Hilton explained to him the high-speed experiments he was conducting. In the course of the conversation Hilton showed the newsman a plot of airfoil drag, explaining: "See how the resistance of a wing shoots up like a barrier against higher speed as we approach the speed of sound." The next morning, it was incorrectly referred to in the newspaper as "the sound barrier". The name caught on because the issues which conventional high-speed aircraft invariably encountered on reaching transonic speeds gave the impression that the magic Mach number was indeed a barrier that would need to be overcome.

It not only represented a physical barrier, but also a psychological one: there were many skeptics who said supersonic flight was impossible because aerodynamic drag increased exponentially until a veritable wall of air emerged. They pointed to the loss of the second prototype of the de Havilland DH 108 Swallow on 27 September 1946. This high-speed jet disintegrated while diving at Mach 0.9, killing pilot Geoffrey de Havilland Jr., and crashing into the Thames Estuary. Shock stall had pitched the nose downwards, and the resulting extreme aerodynamic loads on the aircraft cracked the main spar and rapidly folded the wings backwards. Others, however, noted that rifle bullets could fly at supersonic speeds, so the sound barrier was not an impenetrable wall. Indeed, during the war it had been realized that streamlined bullet shapes were ideal for supersonic speeds. This is how rockets such as the German A4/ V2 got their familiar shape. The V2 achieved Mach 4 as it fell from the sky towards London. In fact, because it fell faster than the speed of sound there was no audible warning of the imminent danger until the impact reduced whole blocks of houses to rubble; only afterwards did the sound arrive. But aircraft need wings, rudders, ailerons and other devices to develop lift and facilitate control, making their aerodynamics much more complicated than those of bullets and rockets. The traditional tool for gathering aerodynamic data and developing new aircraft and wing shapes was the wind tunnel. However, the technology available at that time did not permit accurate and reliable measurement of airflow conditions at transonic speeds: the aircraft models placed in the wind tunnels would generate shock waves in the high-speed air flowing around them, and these in turn would reverberate and reflect across the test section of the tunnel. As a result, there was a lot of interference and the measurements of the model did not correlate to the real world in which aircraft flew in the open air rather than in an enclosed tunnel. Also, you can scale an aircraft but you cannot scale the air, so air flowing around a small-scale model does not necessarily behave in the same way as air flowing around a real airplane.

The least understood area was from about Mach 0.75 to 1.25, the transonic regime where the airflow would be unstable and evolve quickly, and for which no accurate aerodynamic drag measurements and theoretical models were available. It was called the 'transonic gap'; the aerodynamicists nightmare equivalent to the 'sound barrier' so dreaded by pilots. The aerodynamic drag is especially high in this range of speeds, peaking at just below Mach 1. However, it actually diminishes considerably at higher supersonic speeds (which is why modern aircraft either fly well below or well above Mach 1, spending as little time and fuel as possible at transonic speeds). Drag occurs at transonic speeds for two reasons: firstly as a result of the build-up of shock waves where the airflow reaches Mach 1 (typically over the wings), and also because the air behind the shock waves often separates from the wing and creates a high-drag wake. At even higher speeds the shock waves move to the trailing edge of the wing and the drag-inducing air-separation diminishes and finally vanishes, leaving only the shock waves.

All major military powers realized that if their aircraft were to remain state-of-the-art and competitive, then transonic aerodynamics was an area that really needed to be explored and mastered. Specialized and heavily instrumented research aircraft would be needed, speeding through the real atmosphere rather than a wind tunnel. In effect these airplanes were to be flying laboratories. In the US, work was started on the Bell X-1, the first of the famous X-plane series and the first aircraft to break the dreaded sound barrier. The Russians initiated their transonic research using the captured German DFS 346, but soon moved on to designing their own aircraft.

The UK started development of the Miles M.52 research aircraft in 1943. It was to be powered by an advanced turbojet (because the British had considerable experience on such engines, and little on rockets). The jet's fuel economy meant it would be able to take off using its own power. The M.52 might have become the first plane ever to exceed Mach 1 if the secret project hadn't been canceled by the new government in early 1946, weeks before completion of the first prototype for subsonic testing. Apart from dramatic government budget cutbacks, one reason for the cancellation was that, based on captured German research, it was feared that the M.52's razor sharp but straight wings were unfit for high-speeds and that swept-back wings were a must for supersonic flight. The Miles engineers had thought about a delta wing for the M.52 during the war, but discarded it as being too experimental for their short-term project. However, the Bell X-1 did not have swept wings either, and both aircraft used all-moving horizontal stabilizers to preclude shock-stall problems (interestingly, the Bell engineers got the idea for the special tailplane from the M.52 team during a visit to Miles Aircraft in 1944). Had the British continued their ambitious project, they could have beaten the US in breaking the sound barrier: in 1970 a review by jet engine manufacturer Rolls Royce concluded that the M.52 would probably have been able to fly at supersonic speeds in level flight.

A 30%-scale radio-controlled model of the original M.52 design powered by an Armstrong Siddeley Beta rocket engine and launched from a de Havilland Mosquito did reach a speed of Mach 1.38 on 10 October 1948. This was quite an achievement, but by then the manned X-1 had stolen the show with its record-braking Mach 1-plus flight over the desert of California about a year earlier. In spite of the UK's prowess in aeronautical design and records, there never would be a British counterpart to the X-plane series of the Unites States.

The development of specialized rocket aircraft purely to reach extreme speeds and altitudes went in parallel with that of the rocket/mixed-power operational interceptor. However, where the interceptors needed only to go as high as the maximum altitude that enemy bombers could achieve, there were no limits for the experimental aircraft: they were meant to provide information on entirely new areas of aerodynamics and aircraft design, and their designers and pilots kept on coaxing ever more impressive performance from them. Rocket engines proved to be very appropriate for propelling research aircraft up to extreme speeds and altitudes, since endurance was not of great importance. Rocket engines were light and relatively simple compared to jet engines of similar thrust, and because they did not need air intakes this made it much easier to design airframes suitable for supersonic flight. From the 1940s through to the late 1960s the rocket propelled X-planes achieved velocities and altitudes unrivaled by contemporary jet aircraft, with some of their pilots gaining 'astronaut wings'. The rocket powered interceptor turned out to be a dead end but the early rocket research aircraft led to the Space Shuttle and the current designs for future spaceplanes. New experimental rocket planes are still being developed, although when they fly they are often unmanned.

If flying mixed-propulsion interceptor prototypes was a risky business, then the pilots of early experimental research rocket aircraft had a truly dangerous job. These aircraft were, by definition, going beyond the known boundaries of velocity, altitude and aerodynamics; what pilots refer to as "pushing the envelope". Such planes had to incorporate new, often hardly tested, technology such as experimental wing designs, powerful rocket engines and innovative control systems. Unsurprisingly, whilst being tested several of these experimental aircraft crashed, blew up, or were ripped apart by aerodynamic forces. There were no accurate computer simulations and knowledge databases to warn of design errors, incorrect assumptions and unexpected situations that are nowadays resolved long before a new airplane makes its first test flight. In fact, the research aircraft of the 1940s, 1950s and 1960s were providing the data required to set up such models, and they had to obtain it the hard way. Modern aerodynamic design tools still depend on the experience gained in those years.

In addition, the means of escaping from a plane heading for disaster were much more limited than for today's test pilots, who have sophisticated avionics on board to tell them what is happening to their aircraft, and reliable ejection seats which permit a bail-out at any speed and altitude. In a recent interview for *NOVA Online*, Chuck Yeager, the first man to break the sound barrier in the X-1, summarized the test pilot philosophy of time as follows: "Duty above all else. See, if you have no control over the outcome of something, forget it. I learned that in combat, you know ... you know somebody's going to get killed, you just hope it isn't you. But you've got a mission to fly and you fly. And the same way with the X-1. When I was assigned to the X-1 and was flying it I gave no thought to the outcome of whether the airplane would blow up or something would happen to me. It wasn't my job to think about that. It was my job to do the flying." The urgency of Cold War developments, as well as an acceptance of loss of life ingrained into pilots and aircraft developers during the Second World War, meant high risks were taken and many test pilots perished as their new aircraft succumbed to some overlooked detail in the design.

CHASING THE DEMON IN THE AIR

The US Army Air Force and NACA (National Advisory Committee for Aeronautics; forerunner to NASA) in 1944 initiated a program called X-1 (originally it was XS-1 for 'Experimental Sonic One' but the 'S' was dropped early on). Its purpose was the development and use of a rocket research aircraft specifically in order to investigate the mysterious transonic region of speed, determine whether there was such a thing as the sound barrier and, if there were not, pass beyond Mach 1. Initially NACA had expected to use an advanced turbojet-powered aircraft which would take off under its own power (just like the British M.52) and, in a very scientific way over a series of flights, study transonic phenomena at different subsonic speeds just short of Mach 1 (because the initial design was not expected to be capable of exceeding the speed of sound). But the Army Air Force was in a hurry to find out whether the sound barrier was a myth, and they pushed for a simpler design based on existing technology that would soon be able to reach and hopefully even surpass Mach 1. Based on previous experience with the Northrop XP-79, as well as early information about the Me 163 Komet, they were confident that a rocket propelled and air-launched, but otherwise fairly conventional aircraft would suffice. As the military was paying for the project, their views prevailed.

The Bell Aircraft Company was awarded a contract for three prototype aircraft in March 1945, just before the war in Europe ended. Consequently, when the X-1 was designed the important German wartime discoveries about transonic flight were not yet available. As a result, the X-1 had conventional wings rather than the swept-back wings of the revolutionary German type. But the wings were relatively thin, with a maximum thickness of only 10% of the chord (the width of the wing at any point). In comparison, the wing of the Me 163 varied in thickness between 14% at the root and 8% at the tip; for the DFS 346 the maximum thickness was 12%. But because their wings where swept the effective thickness with respect to the air flow was actually less (as explained in the description of the Me 163). Conventional straight wings for subsonic propeller fighter aircraft were generally thicker, with a typical ratio of 15%. The wings of the X-1 were made especially strong to be able to handle the powerful shock waves that were expected in spite of their narrow width.

To compensate for the huge amount of aerodynamic drag, a powerful engine was needed. But at that time the US was not as advanced as the UK and Germany in turbojet technology and it did not yet have a jet engine that could provide sufficient thrust to push an aircraft beyond Mach 1. Also, problems were foreseen in ensuring a proper airflow into a jet engine during transonic flight. So as not to delay the project, the X-1 designers opted to install a relatively simple, home-grown liquid propellant rocket engine. A liquid propellant rocket engine would also be much smaller than the giant jet engine of the M.52 and would not need air intakes, making integration with the aircraft (both in design and construction) less complicated, which would in turn enable the development to progress faster and with fewer surprises. For instance, not requiring an enormous air duct to pass right through the length of the fuselage meant the wings could be connected by a single spar, resulting a simple, sturdy design with a relatively low weight.

An important requirement was that the propellants be relatively safe and easy to handle, as well as available in large quantities. This excluded the nasty and difficult-to-produce hydrogen peroxide used in Germany, as well as the dangerous nitric acid favored by Russian rocket plane designers. The engine selected for the X-1 was the

Reaction Motors XLR11-RM-3, which burned a fuel that consisted of a mixture of five parts ethyl alcohol to one part water, in combination with liquid oxygen. These propellants were non-toxic, did not spontaneously ignite on coming into contact, and gave reasonable performance. Moreover, unlike (for instance) gasoline, the alcohol-water fuel mixture could be used to cool the engine: the water improved the cooling capabilities for only a modest decrease in specific impulse. This early version of the XLR11 did not have turbopumps but relied on pressure from a nitrogen tank to drive the propellants into the combustion chambers. It was a pure American design and not based on any German technology, since that was not available when the engine was developed. The four combustion chambers of the XLR11 each produced a thrust of 6,700 Newton, and the engine could be throttled simply by varying the number of chambers ignited at any time. At full power the engine would consume the onboard supply of propellants in less than 3 minutes but this was expected to be sufficient for a short leap beyond Mach 1 if the aircraft were dropped from a carrier plane at high altitude (in contrast, the M.52 would have been able to fly under power for about 20 minutes). After its powered run, the X-1 would glide back for landing.

The airframe was constructed from high-strength aluminum, with propellant tanks welded from steel (the patch of frost you can see on many of the rocket X-planes is caused by water vapor in the air freezing on the fuselage at the location of the frigid liquid oxygen tank). For the shape of the fuselage, the designers decided to model it after a 0.50 caliber gun bullet; a piece of hardware which was known to be able to fly faster than Mach 1 and whose shape was based on extensive earlier research on the aerodynamics of munitions. The X-1 was basically a bullet with wings. It looks very stubby to us today, and also in comparison to the previously described German DFS 346 that was otherwise very similar in purpose and concept. In order to adhere to the bullet shape there was an unconventional cockpit with its window streamlined flush with the fuselage. Bailing out would have been terribly difficult, because the pilot did not have an ejection seat (a novel technology at that time) or an escape capsule (like the DFS 346 or M.52); he would have had to exit through a small hatch on the starboard side of the nose. It would have been quite a feat in a rapidly tumbling, disintegrating airplane that might be on fire. And even if the pilot were to make it through the hatch, he would have almost certainly struck either the sharp wing or the tail. Health and Safety did not really exist in those days.

In addition to these rather blunt aerodynamic design solutions, the X-1 employed one sophisticated idea: an all-moveable horizontal tail plane (inspired by that of the British M.52 concept) set high on the vertical tail fin to avoid the turbulence from the wings. It was known that the elevon controls on conventional stabilizers generated strong shock waves at high speeds, making the airplane impossible to control in the all-important pitch direction and ultimately producing the infamous 'Mach tuck' that caused it to nose over into a terminal dive. But if the entire stabilizer is moved, not just a part of it, no shock wave forms on its surface and there is no elevon to become blocked; in other words, it allowed control of an aircraft at transonic and supersonic speeds. This was such a revolutionary discovery that the US hid it from

Bell X-1 number 1 in flight [US Air Force].

the Soviets for as long as possible. During the Korean War the all-moving tail gave the US F-86 Sabre jet fighter a real advantage over the agile Soviet MiG-15, whose conventional tail had elevons which made it difficult to control at speeds approaching Mach 1. The all-moving horizontal stabilizer promptly became a standard feature on all supersonic aircraft, including the Russian successors to the MiG 15.

The X-1 had good flight characteristics at transonic as well as lower speeds, both under rocket power and while gliding. Pilots found it a delight to fly, very agile with the handling characteristics of a fighter. It had a length of 9.5 meters (31 feet) and a wingspan of 7.0 meters (23 feet). Fully loaded with propellant it weighed 6,690 kg (14,750 pounds). Any propellant left after a powered flight was jettisoned in order to avoid landing with the hazardous liquids on board, and its dry weight was 3,107 kg (6,850 pounds)

Although originally designed for a conventional ground take-off, the X-1 was air-launched from a high-altitude B-29 Superfortress bomber to maximize the use of its own propellant to accelerate to supersonic speed in the higher atmosphere, where both the aerodynamic drag and the speed of sound were significantly lower. At sea level a plane must exceed 1,225 km per hour (761 miles per hour) to surpass the speed of sound but at an altitude of 12 km (39,000 feet) Mach 1 is 'only' 1,062 km per hour (660 miles per hour). This meant the transonic and phenomena which the researchers were interested in would occur at slower, easier to attain speeds.

X-1 number 3 being mated with its B-50 Superfortress carrier [US Air Force].

The X-1 flight tests were to be undertaken at Edwards Air Force Base, at that time named Muroc Army Airfield, the famous test flight airfield out in the Mojave Desert of California. The base is next to Rogers Dry Lake, a large expanse of flat, hard salt that offers a natural runway. The desert also offers excellent year-round weather, as well as a vast, virtually uninhabited area with plenty of free airspace. All this made the base perfect for testing new high-speed and potentially dangerous rocket aircraft, especially if they were to remain secret.

By today's aviation standards the X-1 was a very risky aircraft. Apart from the rather dubious means of escape for the pilot, it also had no backup electrical system. During one flight, test pilot Chuck Yeager found himself in a powerless X-1 due to a corroded battery just after being dropped from the carrier aircraft. He could neither ignite the engine nor open the propellant dump valves, since both required electrical power. Luckily, engineer Jack Ridley and Yeager had installed a manual system to get rid of the dangerous fluids just before that very flight, so he could still empty the tanks before landing; the X-1 had not been designed to land safely with the weight of a full propellant load.

The original X-1 aircraft, the X-1-1, made its first unpowered glide flight on 25 January 1946 over Florida's Pinecastle Army Airfield, flown by Bell Aircraft chief test pilot Jack Woolams. The first powered flight was on 9 December 1946 at Muroc using the second X-1 aircraft, with Bell test pilot Chalmers 'Slick' Goodlin ('Slick' being a flattering moniker in those days) at the controls. He also piloted the X-1-1 on its first powered flight on 11 April 1947. Two months later the Air Force, unhappy with Bell's cautious and thus slow "pushing" of the flight envelope in terms of speed and altitude, terminated the flight test contract and took over. Captain Chuck

Chuck Yeager with his X-1 [US Air Force].

Yeager, a veteran P-51 Mustang pilot of the Second World War, was selected to attempt to exceed the speed of sound in the X-1-1. After being assigned to the program, which was understood by all involved to be extremely dangerous, he was told by program head Colonel Boyd: "You know, we've got a problem. I wanted a pilot who had no dependents." Yeager responded that he was married and had a little boy, but that this would only make him more careful. This was judged sufficient explanation.

In October 1947, after several glide and powered flights, both pilot and aircraft are deemed ready to officially break the sound barrier. On the 14th, teams of technicians and engineers awaken early in order to prepare the small, bright orange X-1 for flight and install it in the bomb bay of its B-29 carrier. Then the four-engined bomber takes off and climbs to an altitude of 6 km (20,000 feet). At 10:26 a.m., the X-1-1, which Yeager has christened 'Glamorous Glennis' after his wife, is dropped at a horizontal speed of 400 km per hour (250 miles per hour). Yeager lights the four XLR11 rocket chambers one by one, rapidly climbing as he does so, and then he levels out at about 13.7 km (45,000 feet). Trailing an exhaust jet with shock diamonds (caused by shock waves in the supersonic gas flow) from the four rocket nozzles, the X-1 approaches Mach 0.85. Entering the poorly understood transonic regime, Yeager momentarily shuts down two of the four rocket chambers, holding the plane at about Mach 0.95 to carefully test the controls. As on previous flights there is buffeting and shaking due to the invisible shock waves forming on the top surface of the wings, but apart from that the plane responds well to his steering inputs. It is time. At an altitude of 12 km (40,000 feet) he levels off, reignites the third rocket chamber and watches the needle move smoothly up the Mach meter.

Suddenly the buffeting disappears and the needle jumps off the scale (which only went up to Mach 1; apparently not everyone was so confident in the X-1's supersonic capability). Yeager lets the X-1 accelerate further, and for 20 seconds flies faster than Mach 1. At supersonic speed, a strong bow shock wave forms in the air ahead of the needle-like nose, but the flow over the wings has smoothed out and he discovers that the plane behaves rather well. Not only is the X-1 able to survive surpassing the dreaded sound barrier, it is functional and controllable beyond Mach 1. Satisfied, Yeager shuts down the engine and glides back to land on the dry lake at Muroc.

The recorded peak flight speed was Mach 1.06 at an altitude of 13 km (43,000 feet), corresponding to an actual airspeed of about 1,130 km per hour (700 miles per hour). On his return to base, Yeager reported that the whole experience had been "a piece of cake". It may be that he broke the sound barrier on the previous flight when the recorded top speed was Mach 0.997, as inaccuracies in the measurements might have masked a speed slightly over Mach 1. However, no sonic boom was heard on that occasion, whereas it was on the day the sound barrier was officially broken. The loud explosion-like noise scared several people on the ground into believing that the X-1 had blown up; no one had ever heard a sonic boom before.

This first-ever officially recorded Mach 1-plus flight made Yeager a national hero and the quintessential test pilot of the new jet age. His 1985 autobiography, *Yeager*, was a multi-million-copy best seller, and he plays a prominent role in Tom Wolfe's famous book *The Right Stuff*, as well as the eponymous movie (in which he has a cameo as the old fellow near the bar in Pancho's Happy Bottom Riding Club). The introduction to the movie perfectly describes the X-1 program: "There was a demon that lived in the air. They said whoever challenged him would die. Their controls would freeze up, their planes would buffet wildly, and they would disintegrate. The demon lived at Mach 1 on the meter, 750 miles an hour, where the air could no longer move out of the way. He lived behind a barrier through which they said no man could ever pass. They called it the sound barrier. Then they built a small plane, the X-1, to try and break the sound barrier." If you desire a flavor of the rough world of the early jet and rocket plane test pilots and the first seven US astronauts, Wolfe's book and the movie are indispensable. Some of the tales may seem fictional, inserted to spice up the story, but most of it is true. Bell test pilot 'Slick' Goodlin demanding a $150,000 bonus for attempting to break the sound barrier, then being replaced by Air Force Captain Yeager willing to do the job on his government salary of just over $200 a month is true. So is the famous incident in which Yeager breaks two ribs in a riding accident, says nothing to his superiors to avoid being replaced for the historic Mach 1 flight, and then gets his close friend and X-1 engineer Captain Jack Ridley to furnish him a piece of a broom handle so that he can pull the lever to close the X-1's door using his other hand; unfortunately, the historic piece of wood has been lost to history.

Breaking the sound barrier would have been a great publicity coup for the US Air Force, which had recently gained its independence from the Army, but the flight was kept secret in the interests of national security. Then in December the trade magazine *Aviation Week* (often referred to as '*Aviation Leak*') unofficially broke the news. The Air Force did not confirm the story until March 1948, by which time Yeager and his colleagues were routinely flying the X-1 up to Mach 1.45. The National Aeronautics Association voted that its 1947 Collier Trophy be shared by the main participants in the program: Larry Bell for Bell Aircraft, Captain Yeager for piloting the flights, and John Stack of NACA for scientific contributions. They received the 37-year-old prize from President Harry S. Truman at the White House. Yeager kept the prestigious trophy in his garage and used it for storing nuts and bolts.

The original X-1-1 'Glamorous Glennis' became one of the most famous planes ever. Not only was it the first to fly faster than the speed of sound, it also attained the maximum speed of the entire X-1 program: Mach 1.45. Furthermore, it was the only X-1 to make a ground take-off (also with Yeager at the controls). On 8 August 1949, on the program's 123rd flight, Air Force Major Frank K. 'Pete' Everest Jr., flew the X-1-1 to the new altitude record of 21,916 km (71,902 feet). Like all X-1 records, it was unofficial, as according to FAI rules an aircraft must take off and land under its own power in order to be able to claim an official record (in 1961 this even prompted the Soviets to hide the fact that the world's first spacefarer, Yuri Gagarin, had landed by parachute separately from his capsule). On the next flight, on 25 August, also with Everest on board, the X-1-1 suffered a cracked canopy and the cockpit lost pressure at an altitude of approximately 21 km (65,000 feet). Fortunately Everest was wearing a pressure suit that quickly inflated to prevent his blood from boiling in the thin air, making him the first pilot to have his life saved by such a suit. The X-1-1 was retired in May 1950 after a total of 82 flights (both gliding and powered) with ten different pilots. It was given a well-earned place in the Smithsonian Air and Space Museum alongside the Wright Flyer and Lindbergh's Spirit of St. Louis, and it has recently been joined by a distant relative in the form of SpaceShipOne. Upon presenting the X-1 to the museum, Air Force Chief of Staff General Hoyt Vandenberg said that the program "marked the end of the first great period of the air age, and the beginning of the second. In a few moments the subsonic period became history and the supersonic period was born." The XLR11 engine that was used during Yeager's historic flight is on display separately at the same museum. When I first saw both the aircraft and the engine I was surprised at how crude they appear by today's standards, dramatically showing the fairly basic technology that was available to the X-1 team in tackling the challenge. The Air Force Flight Test Center Museum at Edwards Air Force Base has an X-1 replica.

Bell built three aircraft for the program: X-1-1 (serial number 46-062), X-1-2 (46-063) and X-1-3 (46-064). X-1-1 and X-1-3 were flown by the Air Force while X-1-2 was used by NACA, which had by then established a permanent presence at Edwards (initially NACA Muroc Flight Test Unit, it was renamed NACA High-Speed Flight Research Station in 1949 and then NACA High-Speed Flight Station in 1954. After the formation of NASA it became NASA Flight Research Center in 1959 and finally NASA Dryden Flight Research Center in 1976). In their original configuration, the three X-1s made a total of 157 flights between 1946 and 1951, of which 132 were under rocket power. They were flown by 18 different pilots but Yeager, with a total of 34 flights, was the most experienced X-1 pilot of the program.

The X-1-2 was essentially identical to X-1-1, and made its first powered flight on 9 December 1946 with Bell test pilot Goodlin at the controls. By October 1951 it had

NACA X-1-2 [NASA].

completed 74 gliding and powered flights, flown with nine different pilots. Then it was rebuilt as the X-1E, one of the second generation of X-1 planes.

The X-1-3 differed by having the turbopump-driven XLR11-RM-5 engine (in the XLR11-RM-3 of its predecessors high-pressure nitrogen fed the propellant into the combustion chambers). By using turbopumps, the pressures in the propellant supply lines could be kept relatively low, and metal fatigue problems diminished (concerns of which had resulted in the grounding of the X-1-2 after its 54th powered flight). The lower pressure also resulted in a considerable mass saving on the nitrogen tanks. On the other hand, the high level of complexity of the new turbopump system delayed production. When the aircraft was delivered to Muroc in April 1951 it was three years behind schedule. It gained the nickname 'Queenie' for being a Hangar Queen (an airplane that requires extraordinary preparation and maintenance time in the hanger). The X-1-3 made only one glide flight, and that was on 20 July 1951 with Bell test pilot Joe Cannon at the controls. Sadly, the aircraft was lost on 9 November whilst being de-fueled following a captive flight test mated to its B-50 carrier bomber (an improved form of the B-29). As Cannon pressurized the liquid oxygen tank a dull thud was heard, followed by a hissing sound as white vapor escaped from the X-1-3's center section. Then a violent explosion engulfed the rocket plane and its carrier aircraft in yellow flames and black smoke. Both the X-1-3 and the B-50 were totally destroyed. Cannon managed to get out of the X-1-3, but spent nearly a year in hospital recovering from severe burns on his legs, arms and body. The X-1-3 was the first (but not the last) rocket X-plane to be lost due to a violent, mysterious explosion.

Bell X-1A [US Air Force].

To follow up on the success of the original X-1 aircraft, Bell received a contract to build a second generation of X-1 aircraft with the potential to fly at speeds exceeding Mach 2. These aircraft, the X-1A to X-1E, were powered by the turbopump XLR11-RM-5 engine that was also incorporated in the X-1-3. It had the same 27,000 Newton maximum thrust of the XLR11-RM-3 and was throttled by varying the number of active combustion chambers. The X-1A resembled the X-1, but had a bubble canopy and a stretched fuselage to carry more propellant for a longer powered flight. It was delivered to Edwards on 7 January 1953. The first glide flight was made by Bell pilot Jean 'Skip' Ziegler, who went on to make five powered flights in it. Afterwards, the aircraft was handed over to the Air Force.

In parallel with the Air Force's X-1A flights, NACA initiated its own high-speed research with the Douglas D-558-2 Skyrocket (more on this later). On 20 November 1953 Scott Crossfield achieved Mach 2.005 in this aircraft, beating the Air Force to the 'magic number' of Mach 2. The Air Force promptly initiates 'Operation NACA Weep' in which a series of ever-faster flights culminate on 12 December 1953 with Yeager boosting the X-1A to a new air speed record of Mach 2.44 at an altitude of 22.8 km (74,700 feet). Moreover, Yeager achieves this speed in level flight, whereas Crossfield had required to push his Skyrocket into a shallow dive in order to surpass Mach 2. However, Yeager's elation is short lived, because soon after setting the new speed record his aircraft starts to yaw, and when he tries to compensate this causes it to suddenly pitch up violently. The aircraft enters an inverted flat spin from which Yeager is unable to recover. Bailing out is not possible at the high speed with which the aircraft is tumbling from the sky because it is not equipped with an ejection seat. Accelerations of up to 8 G throw him so violently around inside the cockpit that his helmet breaks the canopy. Only when the aircraft enters the denser atmosphere, at an altitude of 7.6 km (25,000 feet), is he able to restore control. He has literally fallen 15 km (50,000 feet). Unperturbed, Yeager glides back to Edwards and lands safely. Aerodynamicists had predicted that such 'inertia coupling' might occur when flying at high speeds but the X-1A was the first to experience it. This is a very dangerous phenomenon in which the inertia of the aircraft fuselage overpowers the stabilizing aerodynamic forces on the wings and tail. Aircraft that have low roll inertia relative to their pitch and yaw inertia are especially susceptible to it. In practice, this means that planes having stubby wings and long fuselages, and in which the mass is spread over the length of the plane rather than being concentrated near its center of gravity, will probably have problems at high speeds. With its long, relatively slender fuselage, the heavy rocket engine in the tail, and its Mach 2+ flight speeds, the X-1A matched this profile. Pilots had up to then felt that with experience and a basic flight control system, any situation in the air could be handled. But at the extreme altitudes and speeds that the new research aircraft could attain, inertia coupling would require the development of much more sophisticated flight control systems.

An attempt to surpass Yeager's record speed with the X-1A would be extremely dangerous and was never tried. However, flying the X-1A to higher altitudes was still possible. On 26 August 1954 USAF test pilot Major Arthur Murray set a new record of 27.56 km (90,440 feet). In September the aircraft was transferred to NACA High-Speed Flight Station, which returned it to Bell for the installation of an ejection seat; all of the Air Force's high-speed and high-altitude flights had been done without the pilot having a quick and secure means of escape!

Bell X-1A in NACA service [NASA].

Joe Walker gets into the X-1A [NASA].

On 20 July 1955 NACA test pilot Joseph Walker made a familiarization flight in the modified aircraft. Then, on 8 August, as he is sitting in the cockpit preparing for another drop, there is an explosion in the engine compartment of the X-1A. Flames erupt from the propellant tanks and leave a trail in the B-29's slipstream. In addition, the X-1A's landing gear has been blown down into the extended position, making it impossible to land the carrier aircraft without the X-1A touching the runway first and likely breaking apart. Walker manages to get out of the rocket plane into the relative safety of the bomb bay, grabs a portable oxygen tank to breathe, and then returns to dump the rocket plane's propellant in an effort to save both aircraft. But it is too late, and the B-29 jettisons its burning load. As the X-1A falls it suddenly pitches up and almost hits its carrier, then spirals down and smashes into the desert floor, exploding on impact. Walker and the B-29 crew return to base uninjured. The X-1A had performed a total of 29 flights (including aborts) by four pilots.

The second aircraft of the new series, the X-1B, was similar in configuration to the X-1A except for having slightly different wings (for its last three flights its wings were slightly lengthened). The Air Force used the X-1B for high-speed research from

The cockpit of the X-1B [National Museum of the US Air Force].

October 1954 to January 1955, whereupon it was turned over to NACA, whose pilots (Neil Armstrong amongst them) flew it to gather data on aerodynamic heating, a new field of study that became ever more important as aircraft speeds increased.

Aerodynamic heating occurs when the speed of the airflow approaches zero, most particularly in the strong shock waves at the leading edges of the wings and the nose of a supersonic aircraft, where much of the kinetic (movement) energy of the air is converted into heat that can transfer into the aircraft. At extreme speeds the heat can damage the structure of a plane, and even if the temperatures remain relatively low the cycles of heating and cooling that a plane goes through during each flight can still weaken its structure in the long term. Moreover, the aerodynamic heat can make life very uncomfortable for the pilot (and passengers) if no adequate cockpit or flight suit cooling system is installed. For instance, when the Concorde supersonic airliner was cruising at Mach 2.2 its nose reached 120 degrees Celsius (250 degrees Fahrenheit). When the Space Shuttle entered the atmosphere at Mach 25 on returning from orbit its nose reached a searing 1,650 degrees Celsius (3,000 degrees Fahrenheit). Special structural materials (such as the titanium alloy used on the SR-71 capable of flying at Mach 3) and thermal protection materials (like on the Space Shuttle) were required to survive the heat at extreme flight speeds.

To be able to make detailed measurements of the temperatures on different areas on the X-1B, NACA installed 300 thermocouple heat sensors over its surface. During this test campaign the aircraft was also equipped with a prototype reaction control system comprising a series of small hydrogen peroxide rocket thrusters mounted on a wingtip, the aft fuselage, and the tail to provide better control at high altitudes where there is little air for the aerodynamic control surfaces to work with. On the X-1B this system was purely experimental, as the maximum altitude was typically kept to about 18 km (60,000 feet) at which it could still rely on its standard aircraft control system; in fact, the X-1B reached its highest ever altitude of 19.8 km (65,000 feet) three years prior to the installation of the reaction control system. Subsequently a similar system was installed on the X-15, which could fly so high that it was essentially in a vacuum and unable to rely on rudders, ailerons and elevons alone. For the Mercury, Gemini and Apollo spacecraft of the 1960s, thrusters were the only means of controlling the attitude of the vehicle. The X-1B played a pioneering role in the development of such systems.

Moreover, midway through its flight test program the X-1B was equipped with an XLR11-RM-9 engine which had a novel low-tension electric spark igniter instead of the high-tension type of the earlier XLR11s. NACA flew the aircraft until January 1958, when it was decided to ground it owing to cracks in the propellant tanks. It had completed a total of 27 flights by eight Air Force and two NACA test pilots, all of which had been intended to be powered but some had ended up as glide flights due to problems with the rocket engine. In January 1959 the X-1B was given to the National Museum of the US Air Force at Wright-Patterson Air Force Base in Ohio, where it is still on display.

The X-1C was intended to test onboard weapons and munitions at high transonic and supersonic flight speeds, but while it was still under development operational jet fighters such as the F-86 Sabre and the F-100 Super Sabre were already shooting cannon and firing missiles while flying at such speeds, so the X-1C was canceled in the mockup stage.

The X-1D was to take over from the X-1B in testing aerodynamic heating. It had a slightly increased propellant capacity, a new turbopump which enabled the tanks and propellant feed lines to work at a lower pressure, and somewhat improved avionics (i.e. the onboard electrical and electronic equipment). On 24 July 1951 Bell test pilot Jean 'Skip' Ziegler made what would turn out to be the only successful flight of the X-1D. On being dropped by its B-50 carrier the aircraft made a 9 minute unpowered glide which ended with a very ungraceful landing due to the failure of the nose gear. The repaired aircraft was turned over to the Air Force, which assigned Lieutenant Colonel 'Pete' Everest as the primary pilot. On 22 August the X-1D took to the sky for its first powered flight, partly contained within the bomb bay of its B-50 carrier. But the mission had to be aborted owing to a loss of nitrogen pressure needed to feed the propellants into the turbopump of the rocket engine. Because it would be dangerous to land the B-50 with a fully loaded X-1D, Everest attempted to jettison the propellant. Unfortunately this triggered an explosion and a fire, and once again an X-1 had to be jettisoned. Luckily no one was hurt. The explosions of the X-1-3 and the X-1D were finally traced to the use of leather gaskets in the oxygen propellant supply plumbing (which had likely also caused the loss of the X-1A). The leather had been impregnated with tricresyl phosphate (TCP), which firstly becomes unstable in the presence of pure oxygen and can then explode if subjected to a mechanical shock. It was one of the hard lessons learned during the X-1 program.

After the loss of the X-1-3 and the X-1D (the crash of the X-1A would not occur until several years later) it was decided to upgrade the X-1-2 and redesignate it as the X-1E to continue the high-speed flight test campaign. It was christened 'Little Joe' in honor of its primary Air Force test pilot, Joe Walker. The most visible modifications included a protruding canopy, a rocket assisted ejection seat, and thinner wings with knife-sharp leading edges and a thickness ratio of 4% (better suited to supersonic flight). The surface of the plane was covered with hundreds of tiny sensors to register structural strain, temperatures and airflow pressures. The X-1E made its first glide flight on 15 December 1955 with Walker at the controls. He went on to make a total of 21 flights, attaining a maximum speed of Mach 2.21. NACA research pilot John McKay took Walker's place in September 1958 and made five more flights, with a maximum attained speed of Mach 2.24. It was permanently grounded in November 1958 owing to structural cracks in the fuel tank wall, and now guards the entrance of NASA Dryden Flight Research Center.

Joe Walker with the X-1E [US Air Force].

The X-1 program thus opened the door to supersonic flight, and its experimental results facilitated a new generation of military jets that could fly faster than the speed of sound. The various X-1s truly adhered to the Edwards Air Force Base motto of 'Ad Inexplorata' (Into the Unknown).

In friendly competition with the Air Force's X-1 program, the US Navy, working with NACA, initiated tests using its mixed-power Douglas D-558-2 Skyrocket. The Navy/NACA D-558 program pursued a more conservative approach to the problems of high-speed flight than did the USAF/NACA X-1. In contrast to the decision by the Air Force to go straight to supersonic rocket propelled planes, the Navy started with the transonic D-558-1 Skystreak jet-powered research aircraft. This was more in line with the careful scientific approach which NACA advocated. The D-558-1 had only just been able to surpass Mach 1 in a dive. By using rocket power in addition to a jet engine the D-558-2 was to explore the transonic and supersonic flight regimes and investigate the characteristics of swept-wings at speeds up to Mach 2. The Navy was also particularly interested in the strange phenomenon that made high-speed, swept-wing aircraft of that time pitch their nose upwards at low speeds during take-off and landing, as well as in tight turns. The original plan was to modify the fuselage of the D-558-1 to accommodate a combination of a rocket and a jet engine, but that soon proved impractical. The D-558-2 became a completely new design that had its wings swept at 35 degrees (its predecessor had straight wings) and its horizontal stabilizers at 40 degrees. The wings and the tail section would be fabricated from aluminum, but the fuselage would be primarily magnesium. For take-off, climbing and landing the Skyrocket would be powered by a Westinghouse J34-40 turbojet engine drawing its air through two side intakes on the forward fuselage and producing a thrust of 13,000 Newton. To attain high speeds, a four-chamber rocket engine with a total sea-level thrust of 27,000 Newton would be fitted. The Navy called this the LR8-RM-6 but it was basically the same XLR11 engine as used in the Bell X-1. The design called for a flush canopy similar to that of the X-1 in order to obtain a sleek fuselage, but this would have so limited the pilot's visibility that it was decided to use a normal raised cockpit with angled windows. The resulting increased profile area at the front of the aircraft had to be balanced by a slight increase in the height of the vertical stabilizer. Somewhat reminiscent of the German DFS 346 rocket aircraft, the pilot was housed inside a pressurized nose section that (as on the D-558-1) could be jettisoned in an emergency. The capsule would be decelerated by a small drag chute, and when it had achieved a suitable altitude and speed the pilot would bail out to land under his own parachute.

On 27 January 1947 the Navy issued a contract change order to formally drop the production of the planned final three D-558-1 jet aircraft and substitute instead three of the new D-558-2 Skyrockets.

The Douglas company invited its pilots to submit bids to fly the new rocket plane during the test program. However, at that time Yeager had not yet made his historic Mach 1 flight in the X-1 and trying to break the sound barrier was still seen by most test pilots as a quick and easy way to "buy the farm" (i.e. die). Rather than ignore the offer, which would have been bad for their reputations, the pilots conspired to

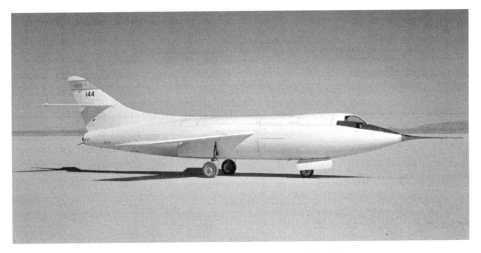

NACA 144, the second Skyrocket [NASA].

submit exceptionally high bids that would surely not be accepted by the company. However, John F. Martin was away delivering an airplane for Douglas and unaware of the plot. He submitted a reasonable bid and was promptly accepted as the Skyrocket's project pilot. On 4 February 1948 Martin took off from Muroc Army Airfield in the first aircraft (Bureau No. 37973; NACA 143) for the Skyrocket's maiden flight. At that time this aircraft employed a jet engine and was configured only to take off from the ground. It was tested in this configuration by the company until 1951 then handed over to NACA, which kept it in storage until 1954 and then modified it by removing the jet engine, installing an LR8-RM-6 rocket engine, and configuring the aircraft for air-launch from the bomb bay of a P2B (the naval version of the B-29). However, it was subsequently only used for one mission: an air-drop familiarization flight on 17 September 1956 by NACA pilot John McKay. In total NACA 143 made 123 flights, mostly in order to validate wind-tunnel predictions of the Skyrocket's performance. One interesting discovery was that the airplane actually experienced less drag above Mach 0.85 than the wind tunnels data indicated, thus highlighting the discrepancies between wind tunnel results and actual flight measurements that still existed at that time.

Skyrocket Bureau No. 37974 (NACA 144) had a much more interesting career. It also started out with a jet engine only, in which configuration NACA pilots Robert A. Champine and John H. Griffith flew it 21 times for subsonic airspeed calibrations and to investigate longitudinal and lateral stability and control. They encountered the expected pitch-up problems, which were often severe and occurred very suddenly. In 1950 Douglas replaced the turbojet with an LR8-RM-6 and modified the airframe to be carried by a P2B (B-29) bomber. The release at an altitude of about 9 km (30,000 feet) and the increased thrust compared to the turbojet enabled company pilot Bill Bridgeman to fly this aircraft up to a speed of Mach 1.88 on 7 August 1951, and on 15 August reach a maximum altitude of 24.2 km (79,494 feet) and set an unofficial world altitude record. Bridgeman flew the aircraft a total of seven times.

A Skyrocket being loaded into the bomb bay of its carrier aircraft [NASA].

NACA 144 being dropped from its carrier bomber [NASA].

During his supersonic flights he encountered a violent rolling motion due to lateral instability which was curiously weaker on his Mach 1.88 flight than on a Mach 1.85 flight that he made in June.

It was then turned over to NACA, which started its own series of research flights in September 1951 with legendary pilot Scott Crossfield. Over the next several years

Crossfield flew NACA 144 at total of 20 times, gathering data on longitudinal and lateral stability and control, aerodynamic loads and buffeting characteristics at speeds up to Mach 1.88. On 21 August 1953 Marine Lieutenant Colonel Marion Carl, flying for the Navy, set a new unofficial altitude record of 25.37 km (83,235 feet). NACA technicians then extended the rocket engine nozzle in order to prevent its exhaust gas from affecting the rudders at supersonic speeds and high altitudes (where the exhaust expands into an enormous plume). As explained later in this chapter, such additions also improve the efficiency of an engine at high altitudes; in the case of the D-558-2 increasing the thrust by 6.5% at 21 km (70,000 feet) altitude.

Meanwhile, people in the project where lobbying for the go-ahead from NACA to attempt to cross the Mach 2 boundary. They knew the Air Force was planning to try to fly faster than twice the speed of sound using the X-1A in celebration of the 50th anniversary of the first flight by the Wright brothers. The Navy and Scott Crossfield, who was a Naval officer prior to joining NACA as a civilian test pilot, were eager to claim this record. NACA preferred to focus on a steady scientific approach and leave record setting to others, but Crossfield convinced NACA director Hugh L. Dryden to consent to a Mach 2 flight attempt with the NACA 144 Skyrocket. Some years later Crossfield admitted, "It was something I wanted to do; particularly if I could needle Yeager about it."

The NACA project team knew their aircraft would need to be pushed to the very limit of its capabilities. The extra thrust from the new nozzle extension would help, but more was required. Extremely frigid liquid oxygen was put into the oxidizer tank 8 hours before the flight to cold-soak the aircraft, because this would reduce fuel and oxidizer evaporation due to the aircraft's own heat during the flight and thereby leave more propellant in the tanks for several more seconds of powered acceleration. To limit drag as much as possible they cleaned and thoroughly waxed the fuselage, even taping over every little seam in the aircraft's surface. The heavy stainless steel propellant jettison tubes were replaced with aluminum ones. In addition, these were positioned into the rocket exhaust stream so that they would burn off once the engine was ignited and were no longer required, further reducing the aircraft's weight and drag. Project engineer Herman O. Ankenbruck devised a flight plan to make the best use of the Skyrocket's thrust and altitude capabilities. It was decided that Crossfield would fly to an altitude of approximately 22 km (72,000 feet) and then push over into a slight dive to gain a little help from gravity. Despite having the flu and a head cold, Crossfield made aviation history on 20 November 1953 by becoming the first man to fly faster than twice the speed of sound; although barely: his maximum speed was Mach 2.005, or 2,078 km per hour (1,291 miles per hour). But this record stood for a mere 3 weeks, when the X-1A flew considerably faster. No attempts were made to push the D-558-2 to higher speeds; it had reached the limits of its design and there was no way that it could hope to reclaim the speed record from the X-1A.

More flights were made by NACA 144 with NACA pilots Scott Crossfield, Joe Walker and John McKay gathering data on pressure distribution, structural loads and structural heating. It flew a total of 103 missions, including the program's finale on 20 December 1956 when McKay took it up for data on dynamic stability and sound-pressure levels at transonic and supersonic speeds.

NACA 144 returning to Edwards, with an F-86 flying chase [NASA].

The third Skyrocket (Bureau No. 37975; NACA 145) could also be air-launched and was equipped with both an LR8-RM-6 rocket engine and a Westinghouse J34-40 jet engine which had its exhaust pipe exiting the belly of the plane. Taking off under its own power on 24 June 1949 this aircraft became the first Skyrocket to exceed the speed of sound, thereby proving that the design was well suited to supersonic flight; pilot Eugene F. May noted that upon passing Mach 1, "the flight got glassy smooth, placid, quite the smoothest flying I had ever known". By November 1950 NACA 145 had completed 21 flights by company pilots May and William Bridgeman, and then it was turned over to NACA. In September the following year pilots Scott Crossfield and Walter Jones began flying it to investigate the notorious pitch-up phenomenon. For this, the aircraft was flown with a variety of configurations involving extendable wing slats (long, narrow auxiliary airfoils), wing fences (long but low vertical fins that run over the wing) and leading edge chord (width) extensions. They found that whilst fences significantly aided in the recovery from sudden pitch-ups, leading edge chord extensions did not. This disproved wind tunnel tests which had indicated the contrary, and clearly demonstrated the need for full-scale tests on real aircraft. Wing slats, when in the fully open position, eliminated the pitch-up problem except in the speed range of Mach 0.8 to 0.85. The data obtained from these tests was extremely valuable when developing supersonic fighter aircraft. In June 1954 Crossfield began using NACA 145 to investigate the aircraft's transonic behavior with external stores such as bombs and drop tanks (the bombs were empty dummies, as only their shape and position were relevant). Pilots McKay and Stanley Butchart completed NACA's investigations on this, with McKay flying the last of NACA 145's 87 missions on 28 August 1956.

Together the three Skyrockets flew a total of 313 missions, both taking off from the ground on jet power as well as being air-launched from a carrier. They gathered invaluable data on the stability and control of swept-wing aircraft at transonic and supersonic speeds. The data enabled a better correlation between wind tunnel results and flights by real aircraft in the open sky, making wind tunnel tests more useful in the design of high speed aircraft. Especially benefiting from the D-558's research, as well from the X-1 program, were the so-called 'Century Series' supersonic fighters: the F-100 Super Sabre, F-101 Voodoo, F-102 Delta Dagger, F-104 Starfighter, F-105 Thunderchief and F-106 Delta Dart. The various makers of these magnificent aircraft all exploited the flight research done at Edwards, giving the US military an important edge over their Soviet counterparts.

NACA 143, the first Skyrocket, is on display at the Planes of Fame Museum in Chino, California. NACA 144, the first aircraft to fly at Mach 2, is hanging from the ceiling of the National Air and Space Museum in Washington D.C. NACA 145 can be found outside on the campus of Antelope Valley College in Lancaster, California, not far from Edwards.

In late 1944, as the design of the X-1 was getting underway, it became clear to the US Army Air Force that supersonic aircraft would greatly benefit from swept wings like those pioneered in Germany. Bell thus responded to the Air Force request for a successor to the X-1 with their Model 37D, which was essentially an X-1 that had its wings swept back at 40 degrees. However, aerodynamic and structural analyses soon demonstrated that such an upgrade of the X-1 design was not very practical, and the proposal was rejected. In September 1945, just after the Second World War ended, Bell came back with an entirely new and much bolder concept which they called the Model 52. Even although the X-1 had yet to fly, the Bell engineers told the Air Force that their new aircraft would be able to get close to Mach 3 at altitudes above 30 km (100,000 feet). The Air Force was sold on the concept and named it the XS-2 (later shortened to the X-2). This revolutionary airplane had wings that were swept back at 40 degrees (as before) but now they were mounted to the fuselage with 3 degrees of dihedral and had a 10% thickness ratio (as explained earlier, swept wings can have a greater relative thickness than a straight wing for a given critical Mach number). The wings had a bi-convex profile (a double-wedged cross section which resembled an elongated diamond) that was expected to be particularly suitable for supersonic flight as already indicated by wind tunnel experiments performed in Italy in 1940 (also the canceled British Miles M.52 would have been equipped with bi-convex wings). Like on the X-1, the horizontal tailplane was all-moveable but an innovation was that the stabilizers had the same sweep as the wings.

Where the X-1 series was to surpass the sound barrier, the X-2 was envisioned to best the 'heat barrier'. The temperatures on its exterior were expected to reach about 240 degrees Celsius (460 degrees Fahrenheit) due to severe aerodynamic heating. To survive this, the wings and tail surfaces were made using heat resistant stainless steel and the fuselage was a high strength copper-nickel alloy called K-Monel. In order to maintain a comfortable temperature in the cockpit, a cooling system weighing 225 kg (496 pounds) was installed which, under normal conditions, was sufficient to keep a room containing 300 people nice and cool.

The X-2 would be air-dropped from a B-50 bomber and land without propellant on the dry lake near Muroc, so its landing gear comprised a deployable center-line skid, a small skid under each wing, and a short nose wheel which hardly protruded beyond the fuselage. (Its peculiar attitude on the ground gave the impression that the front carriage had collapsed.) It looked very much like a manned rocket, with a

The first X-2 with its B-50 carrier, chase planes and support crew [US Air Force].

rather small cockpit capsule right at the front, housed inside a sharp pointy nose. Just as on the D-558-2 Skyrocket, in an emergency the X-2's entire pressurized nose assembly would be jettisoned and soon stabilized by a small parachute. The pilot would then have to manually open the canopy at a safe altitude and speed, and bail out. Although NACA was concerned about this system, the Air Force considered it an adequate means of escape at extreme flight speeds and altitudes and approved the design. It is another example of the more careful but slower NACA approach versus the bolder Air Force seeking faster progress in order to stay ahead in aviation (with respect to the Soviets certainly, and probably also in friendly competition with the Navy and NACA).

To propel the X-2 to Mach 3, it was equipped with an advanced Curtiss-Wright XLR25-CW-3 pump-fed dual-chamber rocket engine that ran on water-alcohol and liquid oxygen and produced a total thrust of 66,700 Newton at sea level; about two-and-a-half times that of the XLR11 used by the X-1. The upper combustion chamber could produce a maximum of 22,200 Newton and the larger, lower chamber twice that. They could be run together or separately, and each could be throttled between 50 and 100% of its full thrust level (whereas the XLR11 could only be adjusted by varying the number of chambers ignited). With full propellant tanks the X-2 weighed 11,299 kg (24,910 pounds), and its landing weight with empty tanks was 5,613 kg (12,375 pounds); both of these weights where almost twice the corresponding figures for the X-1.

The Air Force ordered two X-2 Starbuster research aircraft (airframes 46-674 and 46-675) from Bell Aircraft for the initial flight test program. NACA would initially provide advice and support, and install data-gathering instrumentation, then later use the aircraft for its own test flight campaign.

The X-2 represented a major advance in technology over the X-1. In particular, the development of the XLR25 rocket engine delayed the program by several years and many issues concerning the structure of the aircraft and its flight control system had to be overcome. The planned revolutionary fly-by-wire system where the pilot's control inputs would be interpreted by a computer and then translated into electrical signals to operate motors of the control surfaces was abandoned in 1952 because its technology was too immature. It was replaced by a conventional and much heavier hydraulic power-boosted system. This unfortunately also meant that the operation of the aircraft was completely up to the pilot, without any intervention from a computer to ensure that no maneuvers were made which would be dangerous at certain speeds and altitudes.

The Air Force purchased a Goodyear Electronic Digital Analyzer (GEDA) analog computer which NACA turned into a rudimentary X-2 flight simulator, the first ever computer simulator to be used in aviation. This machine, which could simultaneously handle the various complex interdependent mathematical equations that described the motions of the X-2, helped pilots to familiarize themselves with the aircraft and its expected handling characteristics. It also allowed detailed preparation and checking of flight plans before assignment to the real aircraft. In due course the measurements made during the actual flights helped to improve the simulator.

Consistent (although probably not intentionally) with the X-1 speed indicator only going up to Mach 1, the X-2 cockpit had a meter limited to Mach 3 and an altimeter that only went to 100,000 feet (30.5 km), even though the plane was intended to (and did indeed) fly considerably faster and higher than that! In

The second X-2 with collapsed nose gear following the program's first glide flight [US Air Force].

addition, the cockpit had a standard gyro system to indicate the plane's attitude, which the pilots found to be so inaccurate as to be unusable.

Owing to the development problems it was early 1952 before Bell concluded the captive flight tests with the X-2 remaining mated to the B-50. The first glide flight on 27 June 1952 took place at Muroc (which by then was Edwards Air Force Base) with Bell test pilot Jean 'Skip' Ziegler at the controls. The plane used on the occasion was the second X-2 (46-675) because it had been decided to leave the first aircraft at the company so that it could be equipped with an XLR25 engine as soon as one became available. Unfortunately, at the end of its first glide flight the plane was damaged by a rough landing that collapsed its nose gear. While this repair was underway, a wider central skid was installed to make landing easier. When testing resumed in October 1952, both glide flights resulted in satisfactory landings.

With the glide tests finished, the plane was returned to Bell for modifications. As the first rocket engine delivered had not yet been installed in the first (untested) X-2, it was decided to put it in the already flown one. More captive flight tests were then performed to verify the proper operation of the new propulsion system (without any ignition) at high altitude. Sadly, Ziegler, a veteran of many flights in the X-1 series, died on 12 May 1953 when this X-2 suddenly exploded during a captive flight over Lake Ontario while he was checking the aircraft's liquid oxygen system. B-50 crew member Frank Wolko also died, but the bomber managed to jettison the burning X-2 into the lake and land safely. The X-2 was never recovered and the B-50 had been damaged beyond repair. It was later found that the explosion was likely caused by the same inflammable leather gasket problem that caused the loss of the X-1-3 and X-1D, and possibly also the X-1A.

Once the remaining X-2 airframe 46-674 had been equipped with an XLR25 engine, the testing of this aircraft began with a series of glide flights. No problems were foreseen, since the glide landings with the second X-2 had been satisfactory after the wider skid was installed. The flight team was therefore surprised when 46-474's first flight ended in a very unstable landing in which the aircraft skidded sideways over the salt lakebed. After repairs, the next flight ended similarly. It appeared that the high position of the aircraft's center of gravity on the ground due to the tall landing skid booms made it wobble upon touching down. The skid's height was decreased, changing the plane's 7-degree nose-down angle to 3 degrees. This did the trick. The aircraft made perfect landings from then on. Now the X-2 was finally ready for its powered maiden flight. The first attempt took place on 25 October 1955 but because of a nitrogen leak pilot 'Pete' Everest had to complete the mission as a glide flight. The second attempt was aborted while still attached to the carrier aircraft and ended in another captive flight. On 18 November everything finally worked. As planned, only the smaller of the two thrust chamber was ignited. The maximum speed attained was Mach 0.95. However, a small fire had broken out in the tail of the aircraft. Although this did not look very severe in the post-landing inspection it nevertheless meant several months of repair. Following several more aborted attempts, Everest completed a second powered flight on 24 March 1956, this time using only the larger thrust chamber. If anything, these early flights showed the X-2 to be a complex aircraft that was difficult to fly and to maintain. Due to these problems the development and flight test program was already three years behind schedule.

When both combustion chambers were used on 25 April they enabled the X-2 to fly supersonically for the first time: it reached a speed of Mach 1.40 and a maximum altitude of 15 km (50,000 feet). Everest performed three powered flights in May that pushed the X-2's speed to Mach 2.53, making him the 'Fastest Man Alive'. Another pilot, Air Force Captain Iven C. Kincheloe, made a supersonic flight on 25 May, but a malfunction obliged him to shut the engine down early.

In a rocket, the role of the nozzle is to correctly expand the hot exhaust from the high pressure inside the combustion chamber to a considerably lower pressure but a much higher speed. For maximum efficiency (i.e. specific impulse) the expelled gas should reach the same pressure as the ambient atmosphere at the end of the nozzle. Over-expansion (in which the exhaust reaches a pressure lower than that of the air) causes a loss of thrust; as indeed does under-expansion. The higher the altitude the lower the ambient air pressure, which means that at high altitudes the exhaust can be expanded further through a longer nozzle, enabling the same engine to deliver more thrust (at the cost of the maximum thrust at lower altitudes, where the exhaust will be over-expanded). In June 1956 the X-2 received an engine nozzle extension to give it more thrust at high altitudes where there is low aerodynamic drag, thus enabling it to fly faster. Everest made a supersonic checkout of the upgraded X-2 on 12 July 1956, and on the 23rd made his final flight in the aircraft to gather data on

An X-2 igniting its engine just after being dropped by its carrier B-50. [US Air Force].

aerodynamic heating. During this mission he reached a speed of Mach 2.87 at an altitude of 21 km (68,000 feet). Kincheloe then took over as project pilot and made a series of flights in an attempt to reach the aircraft's greatest possible altitude. To achieve this, the X-2 had to make a powered ascent at an angle of 45 degrees. This was difficult to judge using the cockpit instrumentation owing to the inaccurate gyro system, so engineers simply drew a line on the windscreen with a red grease pencil: if Kincheloe kept this line parallel to the horizon while looking out to the side, he would be climbing at the required angle. After two aborted attempts he achieved the very respectable altitude of 26,750 km (87,750 feet) on 3 August 1956. On 7 September he shattered his own record by reaching a spectacular 38,466 km (126,200 feet) flying at Mach 1.7, which also marked the first time anyone had exceeded 100,000 feet altitude (corresponding to 30.5 km, but 100,000 is obviously more impressive as a 'magic number'). Since at this altitude 99.6% of the atmosphere is below the aircraft, Kincheloe was named the 'First of the Spacemen'. He later said that at the highest point, "Up sun the sky was blue-black in color and the sun appeared to be a very white spot. The sky conditions down sun, were even darker in color. This dark condition existed through the horizon where a dark gray band appeared very abruptly. This gray band lessened in intensity until eventually its appearance resembled that of a typical haze condition. Extremely clear visual observation of the ground within a 60 (degree) arc directly beneath the aircraft was noted." As expected of a military test pilot, this report was factual and devoid of any emotional response. On three occasions Kincheloe tried to go higher, but each attempt ended in an abort. His altitude record (unofficial due to the use of a carrier plane) stood until the X-15 rocket plane surpassed it in August 1960.

The X-2 was scheduled to be transferred to NACA in mid-September, which was eager to start a series of missions to investigate aerodynamic heating and study the handling characteristics of the aircraft at extreme altitudes and speeds. However, the Air Force was keen to reach Mach 3, which was the next 'magic number' in aviation,

Captain Mel Apt in the X-2. [US Air Force].

and managed to get an extension and check out another of its pilots, Captain Milburn 'Mel' G. Apt. While Apt practiced missions on the GEDA simulator, representatives from the Air Force, NACA and Bell agreed on a flight plan. It was clear the mission would involve a lot of risk, as understanding of the dynamics of a Mach 3 airplane was fairly sketchy in the 1950s. In fact, the limited aerodynamic data gathered from wind tunnel experiments for the X-2 was only valid up to Mach 2.4; what happened beyond that could at that time only be discovered by practical "cut-and-try".

On 27 September 1956 all was ready to attempt the record flight. Thanks to the grease pencil line on his cockpit window, Apt flew an almost perfect profile of speed and altitude as a predefined function of time and became the first person to fly faster than thrice the speed of sound. The maximum speed attained was an incredible Mach 3.196; equivalent to 3,369 km per hour (2,094 miles per hour). Sadly, the excitement was very short lived. As he turned back towards Edwards, Apt for some reason made too sharp a turn and lost control due to inertia coupling; the problem first suffered by Yeager in the X-1A in 1953 and which may well have been avoided if the intended fly-by-wire flight control system had been implemented in the X-2. After a series of violent combinations of roll, pitch and yaw the aircraft entered a relatively smooth subsonic inverted spin, but Apt could not get it under control. During his attempts he never unlocked the rudder, which had been manually secured prior to accelerating to supersonic speeds in order to avoid dangerous shock waves forming over the vertical stabilizer. We will never know whether unlocking the rudder would have helped to escape from the spin. Realizing that he would not be able to gain control of the plane, Apt separated his escape capsule. Unfortunately he did not manage to get out of the capsule before it slammed into the desert floor (the problem that NACA had warned of when the system was accepted by the Air Force). Ironically, the X-2, now without its cockpit, stabilized itself and continued to descend in a series of undulating glides followed by stalls, before hitting the ground and coming apart.

The most spectacular achievement of the X-2 was therefore also its last, and Apt's death cast a shadow over the program. It was a highly experimental and dangerous machine, a fact that was downplayed at the time in order to ensure continuing public support. However, the X-2 program had accomplished much of what it had set out to do: identifying the peculiarities of high-altitude flight and speeds exceeding Mach 2. Unfortunately, some of the lessons were learned the hard way. It was now clear that the safe operation of aircraft at very high speeds would require more sophisticated control systems, in particular incorporating so-called 'stability augmentation' since at Mach 3 things happen very quickly and a pilot receives little warning before inertia coupling causes loss of control. In fact, X-2 pilots found that above Mach 2.5 the safest thing to do was not to do anything at all, as any small steering correction could give rise to dangerous instabilities. One simple measure implemented during the X-2 flights was the already mentioned mechanical locking of the rudder at supersonic speeds. Everest even had a metal grab bar installed at the top of the instrument panel, on which he would place both of his hands at extreme speeds in order to force himself not to move the stick (a very difficult task for a pilot used to always being in active control of his plane). The extremely successful (and much better known) X-15 rocket plane program benefited greatly from both the good and the bad experiences of the X-2.

The dangerous nature of their X research aircraft was pretty much downplayed by both the Air Force and Bell Aircraft. The documentary movie *Flight into the Future* released by the Department of Defense in 1956 duly explained how important and challenging the research work at Edwards was, but it failed to say anything about the risks and accidents, of which there had already been many. It showed pilot Everest kissing his wife goodbye in the morning and going to work just as if he were going to spend his time at a desk. No mention was made of the considerable risks that he was undertaking on a regular basis, and that his wife was probably wondering whether he would survive to have dinner with her that evening. Many test pilots at Edwards died paving the way for the future of aviation, flying various experimental and prototype rocket planes and jet aircraft. The movie included a routine test firing of the rocket engine of the X-2 with personnel standing literally alongside the nozzle, which was a risky thing to do because rocket engine's were still not all that reliable (as an engine explosion during a ground test of the X-15 would later emphasize).

Not much of the X-2 has survived. The one that was dropped in Lake Ontario was never recovered. The one that crash-landed by itself near Edwards was salvaged, and some thought was given to reassembling the aircraft to continue the test program but this was rejected and the remains were buried (apparently nobody remembers where on the vast base). Souvenir hunters occasionally find bits and pieces at the crash site. A replica of the X-2 was constructed for the 1989 television series *Quantum Leap*, and it is currently being restored for display at the Planes of Fame Museum in Chino, California.

The X-2 also made it onto the big screen, first in 1956 in the movie *Toward the Unknown* (apparently a translation of the Latin motto of Edwards Air Force Base). It is a story about a daring test pilot trying to redeem himself after having succumbed to torture while a prisoner of war, and also win back the love of a girl. Other than using actual X-2 footage, the story has little to do with the real flight program. In 2000 the entertaining movie *Space Cowboys* featured a plane which appears to be a (computer generated) two-seat version of the X-2. In the prologue one of the pilots manages to rip a wing off the aircraft during a flight in 1958, after which both occupants (played by Clint Eastwood and Tommy Lee Jones) employ ejection seats to save themselves. So much for historical accuracy!

RUSSIAN ROULETTE

Like the Americans the British and the French, the Russians also understood that the Germans had made great advances in the development of jet and rocket technology during the Second World War. And in spite of the fascist origin of that knowledge the Soviets were not too proud to use it. At the end of the war they had captured the unfinished prototype and wind tunnel models of the German DFS 346, the advanced experimental research plane with swept wings, a pressurized cockpit, and the HWK 109-509C rocket engine. The cigar-shaped fuselage with sleekly embedded rivets was optimized for high speeds and the T-tail had all-moving horizontal stabilizers placed high on the vertical fin to prevent shock stall and disturbances by the wings. To minimize the plane's frontal cross section the pilot was prone on his stomach and viewed through a Plexiglas nose. The Germans had designed the DFS 346 to be air-launched from a bomber so that the maximum of 2 minutes at full-thrust would suffice to break the sound barrier at high altitude. The plane was to land on a retractable skid, saving considerable weight in comparison to a conventional undercarriage using wheels. For measuring the speed of the aircraft a long spike with a pitot tube projected ahead from the nose. Now standard equipment on any aircraft, this tube measured the relative air pressure, which is a function of the velocity of an aircraft through the air. Poking this pitot tube out in front of the plane ensured that its measurements were not affected by airflow disturbances closer to the fuselage. At least as important as capturing hardware was the recruitment of many of the German engineers who had developed this revolutionary plane, by offering them privileges such as additional food rations as well as the opportunity to continue their research (apparently Stalin had finally understood that positive motivation resulted in more progress than brute force when it came to developing complex technology).

The Soviets planned to use the DFS 346 in order to gain a head start in the Cold War competition for speed and altitude, and therefore converted the German Siebel Flugzeugwerke company, which during the war had been tasked with developing the DFS 346, into the OKB-2 design bureau under the direction of the German engineer Hans Rössing. Soon the factory and its staff were moved from the original location in Germany to Russia, where the team continued their work on the DFS 346. Aleksandr Bereznyak, one of the original designers of Russia's wartime BI rocket interceptor, was assigned to assist (and no doubt keep an eye on) Rössing. In order to disguise the German origin of the design, the project was renamed 'Samolyot 346' (Aircraft 346), and the Russian form of the German engine was designated ZhRD-109-510.

Wind tunnel tests showed that at high angles of attack and low speeds the angle of the leading edge of the 346's wing forced some air to flow sideways out towards the

DFS 346P.

wingtips instead of parallel to the fuselage. At the wing tips the airflow could end up flowing almost completely span-wise, sharply reducing the lift and resulting in a stall on the outer part of the wing and a loss of control of the aircraft. The solution was to add two so-called wing fences, low vertical ridges running from the leading edge to the back of the wing. This solution was later incorporated in most Soviet swept wing fighters of the 1950s and 1960s.

In 1947 the first prototype was completed. Since it had no engine installed it was designated 346P (for 'Planer', meaning glider). This version was meant to test flight stability, practice landings, and also try out the mating to and release from the carrier aircraft. It lacked a pressurized cockpit, propellant tanks and other propulsion-related equipment. In 1948 four test flights were carried out with the 346P being dropped from under the right wing of a confiscated American B-29 bomber that had suffered damage during a raid over Japan and then gone on to make an emergency landing in Soviet territory. Interestingly, during these tests the 346P was piloted by Wolfgang Ziese, who had previously been a test pilot for the Siebel company in Germany. In Russia he had prepared for the flights using a modified DFS 'Kranich' (Crane) glider that had been fitted with a prone-pilot cockpit and could be towed into the air behind a Petlyakov Pe-2 bomber.

Flying the 346 into unknown areas of aerodynamics, virtually encased in the tiny aircraft in an uncomfortable prone position and having to rely upon its complicated escape system, must have taken a lot of courage. Especially since at that time some aerodynamicists predicted that at Mach 1 an aircraft would slam into a virtual wall of air and inevitably be ripped apart by violent shock waves. The successful breaking of the sound barrier in the US by the Bell X-1 leaked by *Aviation Week* in December 1947 did tell the Soviets that faster-than-sound flight was possible, but exactly what kind of phenomena they would encounter in the 346 was still unknown; naturally, the Americans kept the X-1 flight data secret.

On one flight, Ziese forgot to check that the ailerons were in their neutral position before his aircraft was released by the B-29 carrier, so the 346P immediately flipped inverted. Only after losing almost 2,000 meters (6,600 feet) of altitude did he manage

The 346P under the wing of its B-29 carrier.

to regain control of the plane. On the whole however, the 346P drops, gliding flights and landings went very well, and it was decided to proceed with the construction of a powered prototype. This 346-1 was completed in May 1949, and had a launch weight of 3,145 kg (6,935 pounds).

On 30 September 1948 the B-29 drops Ziese in the 346-1 equipped with a dummy engine from an altitude of 9.7 km (32,000 feet). He experiences some difficulties in controlling the aircraft and is obliged to land at an excessive speed (the fact that the aircraft does not have flaps for additional lift at low speeds exacerbates the problem). After the first hard touchdown the plane bounces several meters into the air, flies a further 700 or 800 meters (2,300 or 2,600 feet) across the ground, then touches down again. At that moment the ski is pushed back into the fuselage and the plane slides along the runway on its belly prior to coming to a standstill. It is slightly damaged, and the pilot is knocked unconscious but only lightly injured when his head hits the front of the cabin (apparently his seat and safety belt system were not up to the rough landing). Investigators conclude that Ziese had not fully released the skid during his approach, probably because he was fully occupied keeping the aircraft under control.

After repairs and improvements, the plane is redesignated 346-2 and glide flight testing resumes in October 1950 with Russian pilot P. Kazmin. The plane still proves tricky to fly, and on the first flight the skid once again fails to lock when lowered for landing. However, this time the landing takes place on a snow covered field and the belly-sliding does not cause any significant damage. On its second flight the 346-2 is towed by a Tu-2 bomber to an altitude of 2 km (6,600 feet) and released for a free gliding flight. This time Kazmin lands short of the runway. The aircraft is damaged and more repairs are needed. Meanwhile Ziese has recovered from his injuries and, starting on 10 May 1951, resumes flying the engineless 346-2, and starting on 6 June also the newly constructed but still unpowered 346-3 which has thinner wings better suited to transonic flight speeds. During the 346-3 flight tests the confiscated B-29 is replaced by a Soviet copy designated the Tupolev Tu-4 (reputedly copied so literally that rivets missing from the original were omitted).

Finally Ziese and the aircraft are judged to be ready for a powered flight, and on 15 August 1951 the 346-3 is driven through the air on rocket power for the first time. For around 90 seconds Ziese is the ruler of the sky. But the flight is no treat because the plane still has a tendency to roll. And due to a malfunctioning heating regulator the temperature in the cockpit rises to 40 degrees Celsius (104 degrees Fahrenheit), all but making the pilot faint. During this mission, as well as the following flight on 2 September, only the weaker cruise chamber of the engine is used in order to hold the speed below Mach 0.9 because tests in the T-106, the Soviet's first supersonic wind tunnel have led the designers to fear that the aircraft's control surfaces will freeze up at transonic speeds. And their fears are soon proven well-founded. On 14 September Ziese is dropped for the third low-thrust flight, ignites the smaller thrust chamber and accelerates into a climb. Shortly thereafter things go wrong at an altitude of just over 12 km (39,000 feet). Ziese reports to the ground that the aircraft is not responding to his control inputs, is rolling uncontrollably and rapidly losing altitude. Evidently the rocket thrust has pushed the plane into the transonic 'no-go' zone, resulting in locked control surfaces. On falling to a lower altitude Ziese manages to regain some control and ends up in a dive from which he pulls up at about 7 km (23,000 feet). When the airplane starts to roll wildly once again, Ziese realizes that he is running out of time and altitude. The controllers on the ground tell him to bail out. For the first time he triggers the explosive bolts to separate the cockpit section from the rest of the plane. The system works perfectly. The stabilizing parachute puts the cockpit into a smooth descent, enabling him to scramble out and land safely under his own parachute. The aircraft is obviously lost, along with all the flight measurements recorded and stored inside (there was no real time telemetry link with the ground, as is standard for test flights today). Nevertheless the limited data available enables investigators to figure out what probably happened. It is concluded that when it shot up into thinner air the aircraft entered the transonic flight regime and experienced shock stall at its tailplane and wings, freezing up its controls. Once the plane started to fall it accelerated out of the transonic area and exceeded Mach 1, at which moment the shock waves at the tail moved further to the rear, releasing the elevators. And when Ziese pulled out of the dive the aircraft slowed down and again entered the transonic regime, freezing up its controls once more.

It was clear that the 346 was not well suited to transonic speeds, and the aircraft shape's aerodynamic speed limit had been achieved even without igniting the rocket engine's more powerful main combustion chamber. The 346 project was abandoned. In any case, not much valuable data was expected to be gained from further flights because by the late 1940s Soviet jet aircraft were already flying faster than Mach 1. One 'glass half full' project report stated that within the speed limits imposed by the obsolete aerodynamic design all the 346-3 hardware had functioned well, including the rocket engine, the skid landing gear, and finally the escape capsule. The German engineers involved in the 346 project were repatriated to East Germany in 1953 (this was apparently standard procedure once Russian engineers felt that they had learned everything they could from their German colleagues.)

In parallel with OKB-2 and its 346 project, OKB-256 under Pavel Vladimirovich Tsybin was working on a transonic rocket plane called the Tsybin LL (with the LL standing for 'Letayushchaya Laboratoriya', which means Flying Laboratory). Even though this aircraft was kept very simple in terms of construction and propulsion, it was meant to approach Mach 1 and if possible surpass it. After models were tested in the TsAGI wind tunnels, two prototypes were constructed. They were made almost entirely of wood, with ailerons and flaps operated by a pneumatic system powered by compressed air (the forces on the control surfaces were expected to be very high at transonic speed, and so require more than pilot muscle power to operate). The rocket engine in the tail was a straightforward solid propellant booster called the PRD-1500, and it could provide an average of 15,000 Newton for a duration of 10 seconds. The first prototype, LL-1, had conventional straight wings and an ejectable dolly take-off undercarriage similar to that of the Me 163. From mid-1947 pilots M. Ivanov, Amet-Khan Sultan, S. Anokhine and N. Rybko together completed a total of 30 flights with this prototype. After being towed by a Tu-2 bomber to an altitude of 5 to 7 km (16 to 23,000 feet) the pilot pushed it into a steep dive of 45 to 60 degrees in order to gain as much speed as possible prior to leveling off and igniting the rocket

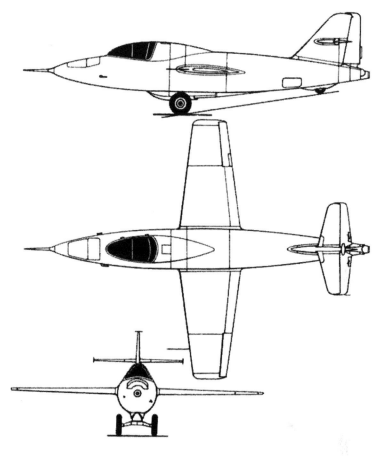

Design of the Tsybin LL-1.

motor. Then a very short, horizontal high-speed powered flight was followed by a gliding return to land on a retractable skid.

During the winter of 1947-1948 the second prototype was equipped with forward-swept metal wings, the benefits of which the Russians had learned of from German wartime research, and which had also initially been planned to be incorporated in the previously described Lavochkin 162. Water tanks were installed in the fuselage so as to be able to adjust the center of gravity of the aircraft. This was designated the LL-3. It made over 100 flights and achieved a maximum speed of 1,200 km per hour (750 miles per hour), corresponding to Mach 0.97, without any significant problems. After the LL-3 tests, the LL-1 was turned into the LL-2 by retrofitting it with swept wings, but it never flew because by then swept-winged jet fighter prototypes had already undergone extensive testing.

A more ambitious project was the Bisnovat 5 developed by aircraft manufacturer Matus Ruvimovich Bisnovat. This was intended to continue where the Samolyot 346 project had ended, providing data on transonic and low-supersonic flight speeds up

Tsybin LL-1.

to 1,200 km per hour (750 miles per hour) at an altitude of 12 km (39,000 feet), which was Mach 1.1. Bisnovat had prior experience of rocket planes because he had been responsible for the production of the Kostikov 302 prototypes by OKB-55 during the war and later had been involved in a number of missile projects. Similar to the DFS 346-based Samolyot 346, the Bisnovat 5 was an all-metal monoplane that had wings swept back at 45 degrees and augmented by fences, and a pressurized cockpit. It was also to be dropped from a carrier aircraft, in this case from under the right wing of a Petlyakov Pe-8, and then land using a simple ski undercarriage. The main ski under the fuselage was set at an angle to enable the aircraft to land with its nose slightly up to ensure sufficient low-speed lift for a soft impact. Unlike the uncomfortable prone-pilot position and complicated escape capsule of the 346, the pilot had a conventional ejection seat and sat upright, although slightly reclined in order to reduce the plane's cross section. A single Dushkin-Glushko RD-2M-3VF dual-chamber rocket engine was installed in the tail and fed nitric acid and kerosene propellants by a turbopump powered by hydrogen peroxide. This engine was similar to those of the Florov 4303, Kostikov 302P, the Polikarpov Malyutka and the MiG I-207 but the combined thrust chambers provided a maximum thrust of 16,500 instead of 15,000 Newton at sea level.

Models were tested in the TsAGI T-104 wind tunnel at up to Mach 1.45 and then one-third-scale models that were powered by small liquid propellant rocket engines were launched from carrier aircraft. After these tests had validated the aerodynamics of the new design, the first flight prototype was constructed and prepared for gliding flights. The first flight of this '5-1' aircraft on 14 July 1948 almost ended in disaster when it hit the Pe-8 carrier shortly after being released. But test pilot A.K. Pakhomov managed to keep the 5-1 under control and made an emergency landing in a rough field. This incident severely damaged the prototype but it was repaired,

Bisnovat 5-2.

and the pylon under the wing of the Pe-8 was revised to carry the Bisnovat 5 with its nose pointing slightly downward to reduce the risk of the aircraft flying up and hitting the carrier after the drop. The next glide flight showed that the aircraft had poor roll and yaw stability. This problem had not yet been resolved when the third flight was made on 5 September 1948 and caused the plane to land tilted to one side, hit the ground with a wingtip and topple over. The plane was almost broken in two and beyond repair, but Pakhomov was okay.

The '5-2' prototype was modified based on the lessons learned during the gliding tests with the 5-1. The vertical tail was swept back further aft to improve directional stability, and the simple metal wingtip bows were replaced by shock-absorbing skids better suited to dampening the impact of touchdown. The test campaign was resumed on 26 January 1949 with pilot Georgi Shiyanov taking the 5-2 on its first glide flight. Again the mission ended in a hard landing with severe damage to the aircraft, this time because the pilot had difficulty in finding the proper approach to the rather short runway and therefore came down beyond it. The 5-2 was repaired and further improvements made. The main ski, which had previously been set at an angle in the vertical direction for improved lift prior to landing, was now put horizontal and thus parallel to the fuselage to improve the pilot's view of the runway. This meant that the small ski on the tail could be removed and replaced with a ventral keel fin to further improve flight stability. No major problems occurred during the next flight but roll and yaw stability were still insufficient. This led the engineers to install downwards angled fins at the wingtips like those of the Florov 4302. The next six glide flights showed that the stability had improved, and that the plane was controllable at least up to the highest speed of Mach 0.77 that was attained in a dive.

But before the powered flight test campaign could commence, the authorities had shifted their interest to further developing supersonic jet aircraft. On 26 December 1948 test pilot I. E. Fedorov had opened the throttle on his swept-wing Lavochkin La-176 (derived from the La-168 jet fighter), pushed the plane into a shallow dive and attained Mach 1.0, marking the Soviet Union's entry into the world of supersonic flight (just over a year after Chuck Yeager made his historic flight in the X-1). Hence the authorities did not see much use for a Mach 1 rocket aircraft.

There was never a Russian equivalent to the American Douglas D-558-2 and Bell X-1 series of experimental rocket planes, and no subsequent evolution into a vehicle like the X-15. The Samolyot 346 flew until September 1951 but never managed to exceed Mach 1 and (as noted above) this project was also terminated after the loss of the 346-3 aircraft.

Nevertheless Soviets engineers continued to develop many supersonic aircraft that were as good as anything in the West, and during the Cold War proved themselves to be masters of aerodynamic theory and design. It is however clear that Soviet spies in the US aviation industry and NACA provided data that was of great assistance to the Russian designers, and at least partly made up for their lack of a supersonic research aircraft program.

7

Rocket plane spaceflight

Basic Flying Rules: "Try to stay in the middle of the air. Do not go near the edges of it. The edges of the air can be recognized by the appearance of ground, buildings, sea, trees and interstellar space. It is much more difficult to fly there."
– Anonymous

At the end of the 1950s the idea of the pure rocket fighter was dead and the role envisaged for a mixed jet/rocket interceptor already very limited. However, as far as American designers of research aircraft were concerned, the evolution of supersonic extreme-altitude rocket aircraft had barely started. After the successful X-1 and X-2 series and the D-558-2 Skyrocket, the next step was a rocket plane that could surpass all of its predecessors in terms of speed and altitude.

Whereas the early X-1 aircraft had investigated the transonic and low supersonic flight regimes, the later X-1s and the D-558-2 had explored speeds around Mach 2, and the X-2 had marginally exceeded Mach 3, the new goal was to venture into the hypersonic area of aerodynamics: Mach 5 and above. The definition of 'hypersonic' is somewhat nebulous since there is no clear and sudden change with respect to the supersonic flight regime (as occurs between subsonic, transonic and supersonic). In general, with respect to supersonic aerodynamics, what happens at hypersonic speeds is much more complex and far more difficult to model and predict. Many of the simplifications about the behavior of the atmosphere, aerodynamic heating and shock waves that can safely be used for supersonic theory are no longer valid at speeds over Mach 5. Laboratory tests for hypersonics are hampered by the fact that it is virtually impossible to generate a continuous Mach 5 + airflow in a wind tunnel. Hypersonic wind tunnels depend on extremely brief, explosive bursts of gas that only facilitate measurements on very small models during a fraction of a second. Once again, the only way to get large amounts of reliable data on this flight regime is to fly research aircraft at hypersonic speeds.

As regards altitude, with its maximum attained altitude of 38.5 km (126,000 feet) the X-2 had already reached into the upper stratosphere. But how a spaceplane or a shuttle-like vehicle would behave in a virtual vacuum, and what it would experience on returning from orbit, had yet to be investigated. Many aviation experts at the time expected the airplane to evolve into an orbital spaceplane, initially launched on top of a conventional rocket but later on capable of taking off and landing like a normal aircraft. As early rocket pioneers such as Valier had foreseen, a space plane was part of an inevitable evolution. The next step, the X-15, was therefore expected to act as a bridge between aircraft and spacecraft.

X-15, HIGHER AND FASTER

If you have any interest in aviation or spaceflight, then you'll have heard of the X-15. Entering 'X-15' into Google in early 2011 resulted in 56 million Internet hits. By comparison 'Space Shuttle' got a mere 23 million hits. The popularity of the X-15 is understandable. When this amazing machine was developed it represented the next step in rocket aircraft, flying at hypersonic speeds and reaching altitudes so high that it was considered to be outside the 'sensible' atmosphere. Eight of its twelve pilots earned the right to wear the coveted US Air Force 'astronaut wings' for achieving an altitude of 50 miles (80.5 km); strangely, such badges were not awarded to NASA pilots who flew the aircraft to such heights. Its extensive flight program took it to an altitude of 107.8 km (353,700 feet) and a speed of 7,274 km per hour (4,520 miles per hour), equivalent to Mach 6.7. The unofficial altitude record set by the X-15 on 22 August 1963 stood until the SpaceShipOne rocket plane broke it in 2004, and the unofficial aircraft speed record which it set on 3 October 1967 still stands. The X-15 was the ultimate rocket plane. It couldn't achieve orbit but in terms of altitude, flight dynamics, instrumentation, heat shields and propulsion it was very much a suborbital spaceplane.

The X-15 originated from a suggestion by Bell Aircraft's Walter Dornberger (who had been von Braun's boss in Germany during the war) in the early 1950s to develop a rocket plane to explore the realm of hypersonic flight. As with the previous rocket X-planes, the X-15 was to be carried aloft by a bomber to maximize the usefulness of the available rocket propellant for reaching high altitudes and speeds. The reason for developing the X-15 was similar to the goals that had inspired the X-1, D-558-2 and X-2: to provide experimental data on high-speed flight for improving and validating aerodynamics theories and models (which translated into sets of equations to enable the aerodynamic behavior of aircraft to be predicted). The X-15 would extend this knowledge to speeds in excess of Mach 5.

NACA (soon to become NASA), the Air Force and the Navy all had an interest in the program, but the Navy eventually opted out in order to concentrate on advanced planes for it fleet of aircraft carriers. The Air Force was in charge of the development of the X-15 and NACA was to lead the flight research campaign after the acceptance flight tests were completed. The request for proposals for the airframe was issued on 30 December 1954 based on a preliminary NACA design. Invitations for the rocket engine went out on 4 February 1955. North American, Republic, Bell, and Douglas all responded with designs that closely resembled the reference concept. In late 1955 North American won the contract to develop and build the X-15 aircraft, and shortly thereafter Reaction Motors was hired to supply the engine.

North American's design had a long, cylindrical fuselage with short stubby wings

Models of the competing designs for the X-15 arranged around the earlier Bell X-1A: clockwise North American, Republic, Bell and Douglas [US Air Force].

and a tail section that combined thin, all-moving horizontal stabilizers and the thick, all-movable wedge-shaped dorsal and ventral fins that NACA suggested would work better at hypersonic speeds than the thin fins that were commonly used on supersonic planes. The two horizontal stabilizers could be moved differentially (i.e. pointing one up and the other down) for roll control, thereby eliminating both the need for ailerons and potential shock-wave interaction problems at the wings. In unpowered flight the pilot would use a standard center-stick, but while running on rocket power he would employ a small joy-stick at the right of his seat, where his elbow would be blocked in order to prevent him from pulling the nose up as the strong acceleration forced his arm backwards.

Most of the internal volume of the aircraft was taken up by the propellant tanks and the rocket engine, prompting designers and pilots to dub it "the missile with a cockpit" and "the flying fuel tank". The retractable landing gear comprised a nose-wheel and two skids in the rear fuselage. Because the ventral fin protruded below the extended skids the pilot had to jettison part of it just before landing. The ejected fin landed by parachute to be recovered and reused. Later in the flight campaign it was discovered that the aircraft was actually more stable without the fin extension during re-entry into the atmosphere on a high-altitude mission, so from then on the ejectable part was no longer used.

The X-15 would be carried under the wing of the large and powerful B-52 jet bomber, with its dorsal fin rising through a notch in the trailing edge of the carrier

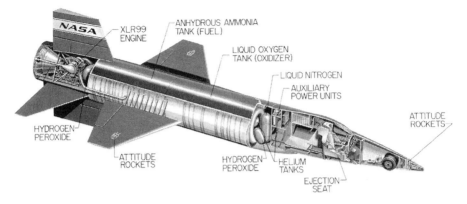

Cutaway drawing of the X-15 [NASA].

An X-15 taken up by an NB-52A bomber [NASA].

aircraft's wing and its cockpit section protruding forward from under the wing. This arrangement required the X-15 pilot to be on board his aircraft from the moment the B-52 started taxiing, but the big advantage was that in an emergency his ejection seat could be used immediately; previous rocket planes of the X series were housed in the bomb bays of their carriers and in an emergency an already seated pilot was unable to eject until after his plane had been dropped. To overcome the boil-off of the liquid oxygen supply in the X-15 during the long climb to the release altitude the B-52 was able to continuously top up the X-15's tank. After its powered flight the X-15 would glide back for a landing at Rogers Dry Lake near Edwards. The small wings were well suited for high-speed, low-drag flights, but not particularly good for gliding. To generate enough lift the aircraft had to glide rather rapidly, which meant landing at 320 km per hour (200 miles per hour). But the dry lake had so much roll-out space that the nose landing wheel did not have to be equipped for steering, simplifying the design and further lowering the X-15's weight.

The engineers developing the X-15 had to overcome many challenges. A big one was aerodynamic heating. Since shock-wave compression would heat up the air (in much the same way as air is heated in a bicycle pump) the skin of the aircraft would get very hot, not only when flying horizontally at hypersonic speeds for a prolonged time but also when re-entering the atmosphere after being boosted into the vacuum of space (albeit to a lesser extent due to the relatively brief duration of this phase of the mission). It was calculated that the temperatures of the upper fuselage would reach 238 degrees Celsius (460 degrees Fahrenheit) and the nose and the leading edges of the wings almost 700 degrees Celsius (1,300 degrees Fahrenheit). In addition to heat-resistant titanium, the project required the new high-temperature nickel alloy called Inconel X. This alloy would retain sufficient strength at such temperatures but it was a difficult material to work with (as was titanium, in fact). The wings and fuselage of the X-15 consisted of titanium frames with an Inconel X skin. The aircraft remained relatively cool by using the 'heat sink' principle: rather than applying active cooling, its structure would simply absorb the aerodynamic heat for the duration of the high-temperature phase of the flight and later radiate it away. The skin was painted black in order to maximize this heat loss. The internal structure would get fairly hot, which is why it was made of titanium rather than standard aircraft aluminum. The fact that different areas of the aircraft would have different temperatures and would therefore expand irregularly, necessitated innovative design solutions such as flexibly mounted wings which could deform span-wise and chord-wise, wing leading edges that were segmented so they could expand without buckling, and the incorporation of a variety of metals having different thermal expansion rates. The intense heating and high air pressures in hypersonic flight also meant that conventional boom-type sensors could not be used, because they would soon bend, break and melt off. An innovative 'Ball Nose' (officially a 'high-temperature flow-direction sensor') was installed instead. It was protected by a thick Inconel X skin and cooled by liquid nitrogen to prevent it from melting at high speeds. It was automatically aligned with the airflow to give the pilot data on angle of attack, sideslip (the direction of the airflow around the aircraft) and the impact pressure of the air (a measure of the velocity).

The pilot was housed in a pressurized aluminum cabin that was isolated from the aircraft's skin and insulated by heat radiation shielding and insulation blankets. He wore a full pressure suit that would instantly inflate in the event of a loss of cabin pressure or ejection from the aircraft. The cabin and the suit were both pressurized and cooled by nitrogen, an inert gas that would help to prevent fire in the cockpit but meant the pilot had to breathe from a separate oxygen supply system; when opening the visor to scratch his nose he had to be careful to hold his breath. Nevertheless, the cockpit could become rather hot and pilots usually landed drenched in sweat despite having opened the nitrogen cooling supply all the way prior to launch.

Instead of a heavy, complicated escape capsule such as on the X-2 the designers of the X-15 chose to incorporate an ejection seat. As a concept this escape system was fairly conventional but the extreme situations in which it would be required to operate were definitely not. Design specifications stated that the seat must enable a pilot to leave the aircraft whilst flying at speeds up to Mach 4, in any attitude, and at altitudes up to 37 km (120,000 feet). These were much more extreme conditions than faced by any other aircraft escape system, with the result that it became possibly the most elaborate ejection seat ever developed. When a pilot felt the urgent need to get out he drew his feet into the foot rests, his ankles striking a set of bars that activated ankle restraints and extended a set of airflow deflectors in front of his toes. He then raised the ejection handles, activating a set of thigh restraints as well as rotating elbow restraints that drew in his arms. This would protect him from the imminent onslaught of the high-speed air outside the cockpit which, depending upon the flight speed and altitude, might be several times the force of a major hurricane. At the same time the seat's oxygen supply would be activated to enable the pilot to breathe independently of the aircraft. When the ejection handles reached 15 degrees of rotation the cockpit canopy was automatically ejected and solid rocket motors boosted the seat out of the aircraft. Immediately, a pair of fins folded out and two telescopic booms extended backwards to stabilize the assembly. The seat automatically released the pilot either at an altitude of 4.6 km (15,000 feet) or 3 seconds after ejection if already below that altitude. The system jettisoned the headrest, and released the seat belt, power and oxygen lines and other restraints so that he would be free to land under his own parachute. If the automatic release system failed the pilot could release himself, and to enable him to judge his altitude his visor was kept clear of ice by a battery powered heater. The X-15 ejection seat was tested on a rocket sled track at Edwards, but neither tested or used for real in the extreme conditions for which it was designed.

Another major design issue was how the aircraft should maneuver itself when the aerodynamic controls lost their effectiveness in the near-vacuum at extreme altitudes. Attitude control required incorporating a reaction control system consisting of small thrusters that the pilot could control using a small stick placed on the left side of his console. This steering method was also in development for the Mercury spacecraft but the X-15 would be the first aircraft to depend on such reaction control (the X-1B had tested a similar system as an experiment). The assembly consisted of four 500 Newton thrusters for pitch, four 500 Newton thrusters for yaw, and four 190 Newton thrusters for roll (the roll thrusters needed less thrust because the aircraft was easier to roll than it was to pitch or yaw). Each wing had one upward and one downward pointing roll thruster near its tip (the farther a thruster was from the plane's center of gravity the more effective it was because of the cantilever torque effect), while the aircraft's nose housed the two sets of yaw thrusters and two sets of pitch thrusters. For every impulse in each direction two thrusters would fire in parallel (e.g. for pitch there was one pair to push the nose up and one pair to push it down), with the system continuing to function if one thruster of each set malfunctioned. The thrusters ran on the gas (super-heated steam and oxygen) provided by the decomposition of hydrogen peroxide.

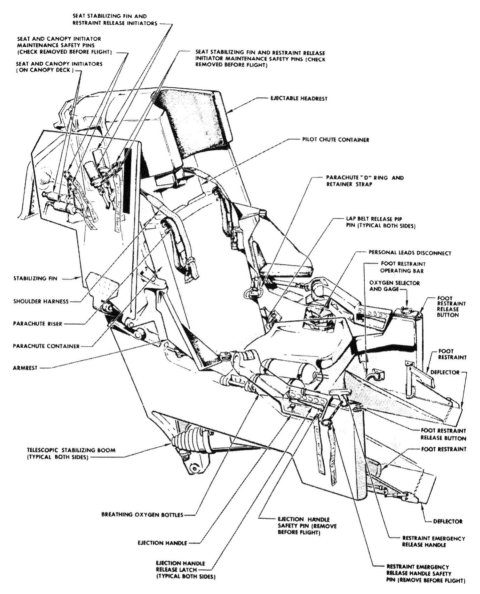

SEAT STABILIZING FIN AND
RESTRAINT RELEASE INITIATORS

SEAT AND CANOPY INITIATOR
MAINTENANCE SAFETY PINS
(CHECK REMOVED BEFORE FLIGHT)

SEAT AND CANOPY INITIATORS
(ON CANOPY DECK)

SEAT STABILIZING FIN AND RESTRAINT RELEASE
INITIATOR MAINTENANCE SAFETY PINS (CHECK
REMOVED BEFORE FLIGHT)

EJECTABLE HEADREST

PILOT CHUTE CONTAINER

PARACHUTE "D" RING AND
RETAINER STRAP

LAP BELT RELEASE PIP
PIN (TYPICAL BOTH SIDES)

PERSONAL LEADS DISCONNECT

FOOT RESTRAINT
OPERATING BAR

OXYGEN SELECTOR
AND GAGE

FOOT
RESTRAINT
RELEASE
BUTTON

STABILIZING FIN

SHOULDER HARNESS

PARACHUTE RISER

PARACHUTE CONTAINER

ARMREST

FOOT
RESTRAINT

DEFLECTOR

FOOT RESTRAINT
RELEASE BUTTON

FOOT RESTRAINT

TELESCOPIC STABILIZING BOOM
(TYPICAL BOTH SIDES)

BREATHING OXYGEN BOTTLES

EJECTION HANDLE

EJECTION HANDLE
RELEASE LATCH
(TYPICAL BOTH SIDES)

EJECTION HANDLE
SAFETY PIN (REMOVE
BEFORE FLIGHT)

DEFLECTOR

RESTRAINT EMERGENCY
RELEASE HANDLE

RESTRAINT EMERGENCY
RELEASE HANDLE SAFETY
PIN (REMOVE BEFORE FLIGHT)

Diagram of the X-15 ejection seat [North American Aviation].

Reaction control in a space-like environment is very different from maneuvering an aircraft using normal aerodynamic control surfaces. If you make a turn in an airplane in the atmosphere, all you have to do to return to straight flight is to push the controls back to neutral. It is just like on a boat. This is called 'static stability' because the airflow around the aircraft ensures that it automatically assumes a stable attitude when the pilot lets go of the stick and foot pedals. However, in space there is no such thing. In a (near) vacuum, if you use a bit of rocket thrust in the plane's nose to push it to the left, the aircraft will not stop turning after you cease thrusting. The thrust has accelerated the nose and thus given the aircraft a leftward rate of rotation that will remain constant if nothing interferes with it. To point the nose in a certain direction you have to start it rotating in that direction and then, when the moment is right, fire the thrusters on the other side of the nose to cancel the rate of turn. Thus if you ignite the thrusters on the right side of the nose for 2 seconds to start a turn, you will then require to fire the thrusters on the left side for the same duration (presuming that they have the same thrust) to stop the rotation. Of course, while the thrusters are firing either to start or stop a rotation the rate will be either increasing or decreasing, making pointing the aircraft in a certain direction using reaction control thrusters a very delicate and difficult task. The dynamics are completely different from a normal aircraft with air flowing over its wings and tail, and they do not come naturally even to an experienced pilot.

The first two X-15s to be delivered had conventional hydraulically actuated flight controls, aided by a simple 3-axis stability augmentation system that would weakly counteract any unintended motions. But the X-15-3, which was specifically meant to fly at extreme altitudes, had 'fly-by-wire' adaptive flight control. This would monitor the pilot's movements of the stick and rudder pedal and adjust them prior to passing the actions to the aerodynamic control surfaces and the reaction control system, thereby making the plane handle in a similar manner in all flight regimes. At higher flight speeds it reduced the sensitivity of the controls and seamlessly integrated the reaction control thrusters with the aerodynamic controls: the lower the ambient air density the more the thrusters would be called upon. It was believed that at extreme altitudes the X-15 would not be controllable without this adaptive control system, until pilot Pete Knight experienced a total electrical failure in X-15-3 during a high-altitude mission and still managed to land safely.

Powering the X-15 would be a Reaction Motors XLR99 rocket engine, generating an awesome 227,000 Newton of thrust (the equivalent of half a million horsepower) at sea level, and 262,000 Newton in a near-vacuum. As the weight of a fully fueled X-15 was 15,400 kg (34,000 pounds) this meant the engine's thrust was about twice the plane's weight at the moment it was dropped from its B-52. The X-15 could thus fly straight up and still accelerate. When it ran out of propellant the aircraft's weight was a mere 6,600 kg (14,600 pounds), which meant that just before shutting down its engine it had a tremendous thrust to weight ratio of 4 at high altitudes; accelerating at 4 G! At that time the mighty XLR99 was the most powerful, most complex yet safest man-rated rocket engine in the world. It could be throttled from 50 to 100% of thrust, shut down and restarted in flight. The restart capability was useful if the engine failed to ignite upon the aircraft's release from the carrier aircraft but other than during the early demonstrations intentional stops and restarts were deemed unnecessary and too risky.

The XLR99 used ammonia and liquid oxygen as propellants, and the turbopumps were driven by hot steam produced by the decomposition of hydrogen peroxide using a silver catalyst bed. A kind of spark plug ignited propellant in a small combustion chamber, which then acted as a blow torch for an instant start of the

Reaction Motors XLR99 rocket engine being installed in the engine test stand [NASA].

rocket engine itself. Ammonia is toxic and expensive but gives better performance than the alcohol used in for instance the XLR11, whilst not burning as hot as for instance kerosene. It proved a good compromise combining a high performance with a relatively simple and therefore reliable engine cooling system. The standard X-15 carried sufficient propellant to run the XLR99 at full power for 85 seconds but the modified X-15A-2 with two external drop tanks could fly at maximum thrust for just over 150 seconds. The XLR99 was a big unit weighing 415 kg (915 pounds) and required an overhaul after every accumulated hour of operation, so a standard X-15 could make about 40 missions before the engine needed to be replaced. For electrical and hydraulic power the aircraft relied on a pair of redundant auxiliary power units driven by steam from decomposed hydrogen peroxide, just like the engine turbopumps.

Because of delays in the development of the XLR99, early X-15 flights used two XLR11 engines (running on ethyl alcohol and liquid oxygen) similar to that which had powered the X-1, and they provided a combined thrust of only 71,000 Newton.

X-15-1 was rolled out from North American Aviation's plant outside Los Angeles on 15 October 1958 applauded by some 700 spectators, amongst them Vice President Nixon. Here was (part of) America's answer to the Soviet Sputnik satellite, which had beaten the US into orbit a year earlier and caused them to suddenly realize that they were in a technological race for space supremacy with the USSR. The aircraft's first glide flight was made on 8 June 1959 piloted by Scott Crossfield, who had left NACA to become chief test pilot at North American. His job was to demonstrate the rocket plane's airworthiness at speeds up to Mach 3, which needed to be verified

The dual XLR11 engine setup used for the early X-15 flights [NASA].

before the aircraft could be handed over to the government. As an aeronautical engineer as well as a test pilot, Crossfield had also played a major role in the design and development of the aircraft.

The first free flight of the X-15-1 came close to disaster shortly prior to landing. Crossfield pulled the nose up to slow his descent, then found he had to push the stick

Scott Crossfield gets ready for the X-15's first captive-carry flight during which it was not released from its carrier plane [NASA].

forward again because the nose had come up too far. This was the start of a severe divergent pitching oscillation. The more he tried to correct the motion the worse it got. Only his superb piloting skills enabled him to smack the X-15 onto the desert airstrip at the bottom of a cycle without damage to the plane or injury to himself. Afterwards it was found that the aircraft's pitch controls had been set at too sensitive a level, resulting in 'pilot-induced oscillations', a situation in which inputs from the pilot tend to overcorrect and cause a pendulum motion of increasing magnitude.

Crossfield was also at the controls of the second aircraft, the X-15-2, when it made the program's first powered flight on 17 September. Once clear of the B-52 carrier he ignited first one XLR11 and then, when satisfied, added the second engine. The flight plan had called for a 'safe' maximum speed of Mach 2 but even with the air brakes fully extended he could not keep the X-15 from creeping up to Mach 2.1. He ended this promising, brief powered phase of the flight with a lazy barrel role for the benefit of the two jet planes that were flying 'chase'. Later missions soon had the X-15 going much faster, especially when flights with the XLR99 engine were started in November 1960.

North American built three X-15 aircraft, with the second being specifically set up for high-speed missions and the third for high-altitude missions. The X-15 program made a total of 199 powered flights over a period of nearly 10 years (a planned 200th flight in November 1968 was canceled due to technical problems and bad weather). Thirteen flights were to altitudes exceeding 50 miles (80 km), earning eight pilots the right to wear US Air Force 'astronaut wings', but only two of these qualified as true space flights by the rules of the International Aeronautical Federation and they were by Air Force pilot Joe Walker of the X-1 series and Skyrocket fame. The FAI defines spaceflight as occurring above 100 km (62 miles) altitude; Walker flew up to 105.9 km (347,000 feet) on 19 July 1963 and to 107.8 km (354,000 feet) on 22 August. At such altitudes 99.9% of the atmosphere was below the aircraft (the decision on where the atmosphere ends and space begins is pretty arbitrary because there is still atomic oxygen even above the altitude at which the Space Shuttle orbits). In terms of speed records, the X-15 enabled Air Force pilot Robert White to become the first person to fly at Mach 4, 5 and 6. It had taken 44 years for aircraft to reach Mach 1 but White increased his maximum achieved velocity from Mach 4 to Mach 6 in the span of only 8 months.

Generally an X-15 mission would fall into one of two categories: high-speed or high-altitude, but for most phases of the flight they adopted similar procedures. The pilot followed a strict pre-determined flight plan defining exactly what combinations of thrust, speed, altitude and heading were required as functions of time. This depended on the type of vehicle tests or experimental measurements to be made and therefore was different for each flight, but a typical X-15 mission would proceed as follows:

After mating of the X-15 to the B-52, propellant loading and pre-flight checks, the ground crew disconnects the servicing carts and the big bomber with its heavy load taxies for several miles along the dry lake bed at Edwards to the start of the runway. The X-15 pilot is already fully enclosed in the cockpit of his rocket plane, doing his own checks in preparation for the mission. Initiating the take-off run, accompanying ground support vehicles are soon left behind as the B-52 rapidly accelerates. On a hot day in the Mojave Desert, when the air density is relatively low, more than 3.7 km (12,000 feet) of runway is required to get the combination into the air. Slowly the jet bomber climbs to an altitude of about 14 km (45,000 feet), where it continues to fly at around 800 km per hour (500 miles per hour). The underside of the X-15 builds up a coating of frost at the location of the liquid oxygen tank due to the intense cold of that propellant. The cruise to the release point takes up to an hour, during which the cryogenic liquid warms up and evaporates; if it were not constantly replenished from a tank on the B-52 the X-15 would boil off 80% of its oxygen. The sleek rocket plane is carried a suitable distance so that it can be launched straight into the direction of its intended landing site (Rogers Dry Lake), eliminating the need for the rocket plane to make any turns during powered flight. Mission planners have made sure that there are enough dry lake beds along the X-15's flight route for emergency landings in the event that the rocket engine does not ignite or extinguishes in flight. Twelve minutes prior to launch the X-15 pilot starts the auxiliary power units, which then produce an exhaust trail behind the aircraft. He also checks all onboard systems, tries the flight controls, tests the reaction control system, sets all the switch positions, activates the main propulsion system, and powers up the data recorders and cameras. The X-15 is accompanied by several jet fighter planes during the various phases of the mission to help and advise its pilot, who is unable to see any part of his own aircraft through the tiny windows (although otherwise the view was not too bad because his helmet was very close to the windows). Several ground stations along the flight path are ready to relay radar measurements of the rocket plane's location, speed and direction of flight, as well as telemetry data from the X-15, to the ground control center, which is in turn in contact with the pilot. This enables the control center to help the pilot to verify that the instrumentation is giving accurate information, and also help him to keep up with the often complex flight plan and offer advice if something goes awry.

Release from the B-52 is sudden, since the X-15 is aerodynamically a very poor glider at low speeds and, moreover, very heavy with a full propellant load. It drops like a streamlined brick, falling clear of its carrier in seconds. The rocket engine must be ignited promptly; if it does not start after two attempts then the pilot has barely enough time to dump the propellant and prepare for an emergency landing. But when the mighty XLR99 ignites, the X-15 rapidly accelerates and leaves the B-52 and the chase planes for that part of the mission far behind. Climbing at nearly 1,200 meters (4,000 feet) per second at an angle of 42 degrees it shoots up into the thin atmosphere at full power.

As the X-15 lightens due to its voracious propellant consumption, the acceleration gradually increases from 2 G to 4 G. This subjects the pilot to a peculiar sensation in which, although he is holding a steady pitch and climb attitude, he feels he is pulling G in a sharp pitch maneuver that is increasing the climb angle and even rotating the aircraft over onto its back, as if looping. The instruments in the cockpit tell him it is an illusion but even an experienced test pilot like Robert White once could not help himself from momentarily pushing the nose down to check whether the horizon was still in the right place; because of this little maneuver he actually failed to reach his planned maximum altitude on that flight.

The relatively long high-G acceleration was pretty uncomfortable; Milt Thompson once said that the X-15 was the only airplane he ever flew where he was glad when the engine quit. A G-suit integrated into his flight suit would help a pilot to cope with the acceleration by inflating bladders to press on his abdomen and legs and prevent a black-out from blood draining away from the brain into the lower parts of the body.

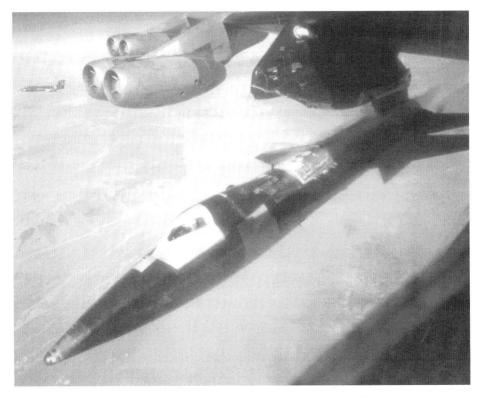

An X-15 is dropped from its carrier plane [NASA].

If the flight is a high-speed mission, the pilot levels off at an altitude below 30 km (100,000 feet) so that the X-15 can employ its standard aerodynamic controls and fly as a conventional airplane. The remaining propellant is used to accelerate to the top speed required, with the pilot varying the thrust and flight angle to control speed and altitude. But if a high altitude is the objective the X-15 continues to climb until the engine exhausts its propellant supply, some 85 seconds after launch and at an altitude of about 50 km (160,000 feet). The aircraft then continues to climb unpowered for up to 2 minutes until gravity reduces the vertical speed to zero, at which point the plane has achieved its maximum altitude of up to 108 km (354,000 feet). With an optimum flight profile the X-15 was capable of flying even higher, but the re-entry would have been too fast and too steep for the structural limits.

During the unpowered ascent the pilot is in a zero-drag, zero-thrust 'free fall' (in effect falling upwards). Although essentially weightless, he remains strapped firmly into his seat. The view is spectacular, as described by Robert White: "My flights to 217,000 feet and 314,750 feet were very dramatic in revealing the Earth's curvature. At my highest altitude I could turn my head through a 180 degree arc and wow! The Earth is really round. At my peak altitude I was roughly over the Arizona-California border in the area of Las Vegas, and this was how I described it: looking to my left I felt I could spit into the Gulf of California. Looking to my right I felt I could toss a dime into San Francisco Bay." But an X-15 pilot had little time for sightseeing, not so much because the flight was brief but because keeping the plane under control required the utmost attention.

After reaching the top of its ballistic arc the aircraft falls back to Earth in zero-gravity conditions until the deceleration by the increasing aerodynamic drag of the atmosphere becomes noticeable at lower altitudes. During the time the air density is too rarefied for the aerodynamic controls to work, the pilot orients the X-15 using the reaction control system. Initial penetration of the atmosphere has to be done holding the plane's nose high up, presenting the broad underside to the air to create a strong shock wave that slows the vehicle down and deflects the resulting heat away from its skin (the Space Shuttle Orbiter would adopt a similar re-entry attitude when coming back from space). This requires very precise steering, as too high an angle of attack will put the plane into a flat spin that is extremely difficult to escape from, whilst too low an angle will plunge the X-15 into the denser atmosphere too fast and result in pressures and temperatures that will destroy it. The weightless ballistic part of the flight lasts at most 5 minutes. Together with its extreme altitude this makes the X-15 very similar to a spaceplane, albeit one that cannot achieve orbit.

As the thickening air slows the falling X-15, the pilot experiences a maximum of 5 G of deceleration for about 15 seconds. Electronic stability augmentation helps to keep the aircraft in a proper attitude during re-entry, preventing inertia coupling such as killed Mel Apt in the X-2. The aerodynamic control surfaces are banging against their stops and sending loud noises reverberating through the empty propellant tanks. Once the speed stabilizes, the pilot pulls out into level flight and initiates a shallow supersonic gliding descent to the landing site. He adjusts the glide path by extending or retracting the air brakes: the further these are deployed, the greater is the drag, the lower is the speed, the lower is the lift, and thus the steeper is the rate of descent.

The round trip of up to 640 km (400 miles) has brought the X-15 back to where it started. At 11 km (35,000 feet) altitude the pilot guides the aircraft into an approach pattern for a landing on Rogers Dry Lake, banking to visually check the landing site. He dumps any remaining propellants to make sure he is not too heavy for landing, jettisons the ventral rudder if it is present (as otherwise it would dig into the ground), lowers the landing flaps and undercarriage and closes the air brakes to avoid running out of necessary flight speed so near the ground. Lowering the aircraft gingerly at a sink rate of about 0.6 meters per second (2 feet per second) he touches down with a forward speed of 320 km per hour (200 miles per hour). When pilot Joe Walker was asked whether he thought it would be possible to land the X-15 very accurately while coming out of a very steep gliding approach he responded, "There's no question of where you're going to land, it's how hard." Generally X-15 pilots managed to touch down gently and very close to the intended landing spot.

Because the main skids are located far back on the fuselage, once they touch the ground the rest of the aircraft slams down fairly hard onto the single nose wheel. The X-15 then skids on the dry lake bed surface for about 2 km (1 mile) before stopping, with the high friction of the skids eliminating the need for active braking. While the

X-15 just before touchdown [NASA].

jettisoned ventral rudder (which landed under a small parachute) is retrieved, ground support personnel drive up to the X-15, assist the pilot in getting out of the cockpit, and prepare the aircraft for transport back to the hangar. The pilot, now relaxed after the tensions of the flight, enters the transportation van to have his flight suit removed and post-flight physiological checks. The B-52 roars overhead at low level and then makes a 180 degree turn while climbing (a so-called chandelle maneuver) in order to celebrate 'mission accomplished'. So ends another X-15 mission that has added more data points to the collection of aerodynamic data on flight at hypersonic speeds and extreme altitudes.

Of course not all missions went according to plan. On the fourth powered test flight of the program, Scott Crossfield had to make an emergency landing in the second X-15 due to a small fire in the engine compartment. Because he did not have enough time to dump all of the propellant he had to land with a much higher angle of attack and hence a nose-high attitude to generate sufficient lift. Once the skids hit the ground, the nose wheel smacked down hard, since it was impossible to keep the nose up with the skids being all the way at the rear of the plane (normal aircraft have their main wheels under the wings, near the center of gravity in order that the nose can be lowered slowly after main gear touchdown). Because of this, as well as the weight of the propellant, the airframe buckled just aft of the cockpit. Crossfield was unharmed but the plane needed extensive repairs. He also survived a fuel tank explosion during a test of the third X-15's XRL99 in 1960. He was sitting in the cockpit wearing his normal clothes when suddenly he was blasted forward. The aircraft was engulfed by a fire but because of the hermetically sealed cockpit Crossfield survived unscathed. The remainder of the test team had been safe inside a control bunker, so nobody was harmed.

Many missions failed in less dramatic ways, often due to the XLR99 not igniting or quitting early and other malfunctions of onboard equipment. Even on what were deemed successful flights not everything always worked perfectly. On several high-speed flights hypersonic air penetrated the X-15 via small gaps between access doors and panels, burning tubes and wires and allowing smoke into the cockpit. Typical of the less safety-conscious manner in which experimental programs were run in those days, quick fixes were implemented with minimal disruption to the schedule.

The second X-15 was badly damaged in a crash landing by pilot John McKay on 9 November 1962 due to failing wing flaps and the weight of unjettisoned propellant braking a landing skid on touchdown. McKay suffered several cracked vertebrae and the aircraft was virtually destroyed. However, both lived to fly another day. McKay's injuries healed and he returned to flight status. The plane was rebuilt and modified by North American for flying even faster than previously. A big improvement was the installation of attachment points for two large drop tanks (one for liquid oxygen and the other for ammonia). The propellants in these tanks were to be used for the initial phase of a high-speed mission, then the empty tanks would be discarded to lower the aircraft's aerodynamic drag. The X-15-2's fuselage was also slightly lengthened to accommodate an additional liquid hydrogen tank intended to power a small prototype ramjet engine that was to be placed on the ventral fin. A dummy engine was carried to determine how it affected the aerodynamics, but the X-15 program finished before a real ramjet could be installed.

For protection from the extreme temperatures of the high-speed flights a special ablative heat shield material was applied to the upgraded X-15's surface. This would slowly burn off, removing heat so that it would not reach the aircraft's structure.

X-15-2 after its crash in 1962 [NASA].

Launch of the X-15A-2 with its white painted thermal protection and dummy ramjet [NASA].

One issue was that the melted material formed an opaque coating on the windows. The simple solution was to cover one window with protective doors during the powered phase of the flight. It would be uncovered for landing, so that the pilot at least had one clean window to look through. A smaller issue was that the ablative material was pink. No self-respecting test pilot was willing to fly in a pink aircraft, but luckily a protective white coating was also necessary to protect the ablative material from liquid oxygen.

The improved aircraft was renamed X-15A-2 and first flew with the ablative coating on 21 August 1967, when it achieved a speed of 5,419 km per hour (3,368 miles per hour). A new layer of coating was then applied in preparation for the next, much faster flight. On 3 October of that same year Pete Knight flew the aircraft to a maximum speed of 7,274 km per hour (4,520 miles per hour), Mach 6.72. It was the highest speed of the X-15 program and still represents the highest speed achieved by any aircraft except the Space Shuttle. However, after landing, the plane was found to be in a sorry state. Some of the skin of the ventral fin was burned and excessive heat had also damaged the nose and the leading edges of the wings and equipment inside the ventral fin, particularly the dummy ramjet. In fact, Knight didn't need to eject the ramjet prior to landing, it fell off by itself due to the heavy damage to the pylon onto which it was mounted (it was later discovered that the ramjet created a shock wave that impinged on the pylon, locally causing extremely high temperatures). Clearly the limit of the thermal protection system, and as such the aircraft's speed limit had been reached. The X-15A-2 was repaired but never flew again.

A total of a dozen test pilots flew the X-15, including Neil Armstrong, who would become the first man to walk on the Moon, and Joe Engle, who would command a Space Shuttle mission. One pilot, USAF test pilot Major Michael J. Adams, lost his life flying this challenging machine on 15 November 1967. He flew the X-15-3, the one specifically built for high-altitude missions and the aircraft in which seven pilots had already earned their 'astronaut wings'. During the climb aiming for an altitude of 81 km (266,000 feet), an electrical disturbance from an onboard experiment caused the reaction control system to function only intermittently. The glitch also caused the inertial system and boost-guidance computers to display incorrect data on the cockpit instruments. As the X-15 began to deviate from its proper direction of flight, Adams, possibly disoriented and confused by the false instrument data, made control inputs which actually increased the heading error. Soon the aircraft was flying sideways, a situation that was not serious while in near-vacuum but would spell disaster once the plane fell back into the atmosphere. Adams reported to the ground control team that the aircraft seemed "squirrely", then said "I'm in a spin." It sent a chill up the spine of the control engineers. Since no pilot had ever experienced a hypersonic spin, there was little advice they could offer.

Adams managed to recover from the spin, but then found himself in an inverted (upside-down) dive. But this attitude was stable and there was sufficient altitude for

Pilot Neil Armstrong and X-15-1 [NASA].

Adams to regain control of the aircraft. Next the fly-by-wire control system began to try and correct the erroneous attitude, resulting in a violent out-of-control oscillation. With the flip of a single switch Adams could have shut off this runaway system, but no one thought to suggest it as the plane rapidly plummeted into the ever denser air. At an altitude of 19 km (62,000 feet) and falling at almost 6,400 km per hour (4,000 miles per hour), the X-15-3 was ripped apart by the rapidly increasing aerodynamic pressures and forces which exceeded 8 G. Adams did not eject, probably because he lost consciousness or was otherwise incapacitated, and was killed when the aircraft's forward section struck the desert floor near Johannesburg, California. Wreckage was found scattered over an area of 130 square kilometers (50 square miles). Adams was posthumously awarded 'astronaut wings' for this flight.

There were several concepts for an even more advanced version of the X-15 with delta wings, uprated engines, increased propellant volumes, and structures that could withstand higher temperatures. A plan to launch such an X-15A-3 from the top of a Mach 3 high altitude XB-70 Valkyrie bomber in order to achieve even higher speeds and altitudes came to nothing, primarily due to a lack of funding. In any case, by then the priority was switching to achieving orbital flight by launching capsules or winged vehicles on top of expendable ballistic missiles.

The many accomplishments of the X-15 program include the first application of hypersonic aerodynamics theory and wind tunnel data to an actual flight vehicle, the first use of a reaction control system in space, the first application of a reusable high-

A delta-winged X-15 launched from an XB-70 Valkyrie [North American Aviation].

temperature alloy structure, and the development of the first practical full pressure suit for flying in space (the direct ancestor of the suit the Mercury astronauts would wear). The X-15 pilots showed that it was possible to safely land an unpowered plane that had a very poor lift to drag ratio, time after time, and this greatly influenced the Space Shuttle Orbiter concept. Many technologies developed for the X-15 were later incorporated into airplanes, missiles, and spacecraft. Experience gathered during the development of the reusable XLR99, for instance, was extremely useful developing the Space Shuttle Main Engine. The Shuttle also incorporated some key parts made of Inconel X, the 'super' alloy that formed the skin of the X-15. The idea of a ground control center actively assisting a pilot during his flight was picked up by the orbital space program, laying the foundation for the famous NASA Mission Control Center that played such a vital role in the Mercury, Gemini, Apollo and Shuttle projects.

The X-15 was also the first aircraft to make extensive use of a 'man-in-the-loop' simulator, the so-called 'Iron Bird' that allowed pilots and flight planning engineers to test and evaluate flight procedures and explore the aircraft's behavior whilst safely on ground. The simulator was initially set up with calculated, theoretical figures for the X-15's flight characteristics but once real flights began it was constantly updated with actual measurements. Nowadays such simulators are used in any new aircraft project and enable designers and pilots to 'fly' it long before any hardware leaves the factory. An X-15 pilot typically trained for weeks in the simulator prior to his flight. This was necessary because each mission had its own unique set of requirements (in terms of speed, altitude, attitude, durations of different flight phases, etc) to ensure that between them the missions covered the full flight envelope that the program was meant to explore. Once every movement that was planned for the nominal flight had become second nature, the simulator would run a pilot through strings of unexpected emergencies. A pilot would typically fly 200 simulated missions before taking to the air. Mike Adams actually got so bored with his training sessions that he started to fly them upside down! As the free-flight time was only 8 to 10 minutes, the preparations lasted much longer than the actual mission (as is typical for any type of crewed space flight). Apart from rehearsing in the simulator, new pilots would also make several unpowered X-15 flights to familiarize themselves with the demanding procedures for making a landing.

Arguably, no other aircraft in aviation history has expanded our knowledge about high-speed flight as much as the X-15. During its 199 powered flights it accumulated a total flight time of 30 hours and 14 minutes, of which 9 hours were spent flying faster than Mach 3 (powered as well as gliding) and 82 minutes at speeds over Mach 5. Data was gathered by an array of sensors, telemetered to the ground during flight and recorded for detailed analyses. In addition, the pilots were closely monitored by various sensors, providing the US with the first biomedical data on the effects of weightlessness on the human body. Whether the X-15 was an aircraft able to reach space or a spacecraft with wings remains a matter of opinion, but arguably it was the world's first reusable spacecraft.

The X-15 program proved key elements of hypersonic theory as it was understood at the time, but also showed several inconsistencies. This led to improved theories for the prediction of lift, drag, stability, control, and temperatures that were fundamental to developing the Space Shuttle. The data that the three X-15 aircraft gathered is still being used today in the development of new spaceplanes and hypersonic missiles. In fact, because the program provided such a wealth of information, and aerodynamics change little between Mach 6 and orbital velocities, there has been no need for a new X-plane capable of flying faster than the X-15 in the atmosphere. The X-15 data goes so far beyond what is required for the development of a normal aircraft that the data such a successor could yield has not been thought worth the cost up until today.

In addition, the experience gained in the development and flying of the X-15 was of tremendous value for the fledgling US manned space program; in the words of NACA scientist and X-15 advocate John Becker, the project led to "the acquisition of new manned aerospace flight 'know how' by many teams in government and industry. They had to learn to work together, face up to unprecedented problems, develop solutions, and make this first manned aerospace project work. These teams were an important national asset in the ensuing space programs." The experience that North American acquired in developing and building the X-15 helped it to win the contract for the role of prime contractor for the Space Shuttle two decades later.

Because it could fly so incredibly high and fast, the X-15 was also a very useful platform for carrying research experiments not specifically related to aerodynamics. These could be mounted in the cockpit, in a wing-tip pod, in a tail-cone box, or in a special skylight compartment with protective doors just behind the cockpit (giving a free view into space at high altitudes). Many types of experiments were flown, such as micrometeoroid collection pods, astronomical instruments, radiation detectors, star tracker sensors and ablative heat shield samples for the Apollo program, an electric side-stick controller, a landing computer, and high-temperature windows. Especially in the last six years of its operation the X-15 was more used as a platform to support a variety of technology programs than it was for the aerodynamic research for which it was conceived.

The two X-15s that survived the flight program can both be seen in museums: the X-15-1 (56-6670) is in the National Air and Space Museum in central Washington, D.C., hanging from the atrium ceiling close to the X-1. The X-15A-2 (56-6671) is in the Air Force Museum in Dayton, Ohio. There are also mockups at the Dryden Flight Research Center at Edwards, at the Pima Air Museum in Tucson, Arizona, and at the Evergreen Aviation Museum in McMinnville, Oregon.

THE NF-104A AEROSPACE TRAINER

With the advent of manned spaceflight in the early 1960s, the US Air Force foresaw an important role for its pilots in space, not only as part of the X-15 program but also as astronauts on Gemini and Apollo missions, manning an orbital USAF outpost, and flying military shuttle-like vehicles starting with the X-20 Dyna-Soar (which will be described later). Thus in 1962 the USAF Experimental Flight Test Pilot School at Edwards Air Force Base became the Aerospace Research Pilots School (ARPS) and its training for armed forces test pilots was expanded beyond the traditional aviation curriculum to include an 8-month aerospace course involving spacecraft operation. In line with the school's change in scope, it soon required a high-performance but low-cost training aircraft that would be able to fly far up into the stratosphere. With such a plane the flight profiles of the X-15 and X-20 could be rehearsed in order to enable test pilot students to familiarize themselves with flight in very rarefied air, where the standard aerodynamic control surfaces become ineffective and control thrusters are needed. Re-entry is a particularly dangerous phase that leaves little margin for pilot error. For the X-15 and indeed any spaceplane, the correct orientation for re-entry into the atmosphere is of paramount importance because only the properly shielded part of the aircraft is able to protect it from the extreme temperatures that it will experience. In the early 1960s there were no computers that could take care of such delicate maneuvering and were also small enough to be accommodated inside an aircraft. The X-15 was flown manually all the way, and it would be the same for the X-20 once this had been inserted in orbit by its launch rocket. The ARPS training planes would also familiarize students with zero-gravity, since during the unpowered ascent and descent in a near-vacuum they would effectively be weightless.

North American proposed a modified version of their X-15 design, adding another cockpit for an instructor and a conventional undercarriage with wheels instead of the usual skids to enable the trainer to take off under its own power. But the X-15 was a complex, expensive machine. A more cost-effective solution was found in the shape of an F-104 Starfighter jet modified for mixed propulsion. An additional benefit of a modified F-104 was that it would enable students to rehearse reigniting a jet engine after a high-altitude parabolic flight during which it would either be deliberately shut down or left to flameout by being starved of oxygen.

The ARPS had already been using standard production Starfighters to simulate the very steep, low-lift and high-drag glide approaches of the X-15 and the planned X-20 Dyna-Soar. This involved climbing to an altitude of 3.7 km (12,000 feet), throttling the jet engine back to 80%, 'dirtying up' by extending the flaps, speed brakes and undercarriage to maximize the aerodynamic drag, and then establishing a 30 degree dive. This resulted in a very steep, low lift-over-drag descent similar to that of a spaceplane. (The lift-over-drag, or L/D, ratio is a measure of a plane's aerodynamic efficiency: a 'dirtied-up' Starfighter had a ratio of 2.2, meaning the lift force its wings provided was only 2.2 times the amount of drag the aircraft caused. In contrast, a 'clean' Starfighter had an L/D of 9.2. The L/D of a typical airliner is 17, and that of the X-15 was about 4 during its glide phase.) The pilot would pull the nose up for the landing flare a mere 460 meters (1,500 feet) above the ground, which left very little room for error. It was a risky profile but it did prepare pilots for landing future spaceplanes.

The school's new F-104 space trainers would need to be equipped with a reaction control system similar to that tested by the X-1B and used operationally by the X-15, to control the aircraft at high altitudes and allow students to rehearse attitude control for spacecraft. NASA already had experience in this, having modified an F-104A in 1959 to use a hydrogen peroxide reaction control system. Following zoom climbs to altitudes up to 25 km (83,000 feet) this gave the NASA Starfighter controllability in the rarefied upper atmosphere. To achieve higher altitudes the ARPS trainer would have an auxiliary rocket engine in the tail.

In 1962 the ARPS awarded Lockheed (the Starfighter manufacturer) a contract to modify three F-104A single-seat fighters for the dedicated role of aerospace trainer. Three existing aircraft were subsequently taken out of long-term storage, relieved of all unnecessary equipment (such as the cannon) to reduce their weight, then equipped with improved instrumentation for high-altitude flight. The reaction control system was based on that of the X-15 and consisted of four 500 Newton thrusters for pitch, four 500 Newton thrusters for yaw and four 190 Newton thrusters for roll. As on the X-15, two thrusters would fire in parallel for every impulse in each direction. The thrusters ran on hydrogen peroxide from a dedicated tank and the pilot controlled them using a handle mounted on the instrument panel. The wings were extended at their tips both to make room for the roll thrusters and to increase lift in order to compensate for the modified aircraft's greater weight. The standard vertical fin and rudder were substituted by the larger versions of the two-seat F-104 to increase their effectiveness in the thin air of the high stratosphere, and the fiberglass nose radome and its radar were replaced by an aluminum cone that housed the pitch and yaw thrusters. A battery was added to run the onboard systems when the jet engine cut out. A long nose probe was installed to measure the angle of attack and sideslip of the plane with respect to the airflow without disturbance from the flow around the body.

A normal jet engine only operates with subsonic air flowing into it. On the basic Starfighter, shock cones were installed in front of the air intakes to guarantee that at supersonic speeds a shock wave formed. This slowed the air passing through it into the intakes to subsonic speed. The inlet shock cones on a standard Starfighter created a proper shock shape up to Mach 2 but extensions were fitted to these cones because the new trainer would fly faster than that. Because the jet engine would not operate in very thin air and thus could not provide 'bleed' air to pressurize the cockpit at high altitude, an additional pressurization system was needed. The pilot would be wearing a pressure suit that would inflate in an emergency but an inflated suit would seriously hamper the precise control required from the pilot during normal operation, so it was decided to fully pressurize the entire cockpit using nitrogen from an added gas tank (oxygen would have offered the benefit that the pilot could breathe it, but would have been a serious fire hazard).

The F-104As were equipped with a standard J79 jet engine that gave the plane a normal maximum thrust of 43,000 Newton at sea level, which could be increased to 67,000 Newton on afterburner. For the required extra boost a compact Rocketdyne AR2-3 (LR121-NA-1) rocket engine was installed at the base of the vertical tail, just above the jet's exhaust. It was canted slightly so that its thrust was aimed through the plane's center of gravity and thus would not continuously push the nose down. This engine ran on a mixture of standard JP-4 kerosene jet fuel drawn from the aircraft's standard fuel tank and 90% concentrated hydrogen peroxide oxidizer. It provided a thrust of 27,000 Newton and could be restarted and throttled in the 50% to 100% range using a specific throttle lever on the left side of the cockpit. The aircraft carried sufficient oxidizer for about 90 seconds of full-thrust rocket operation. With the need

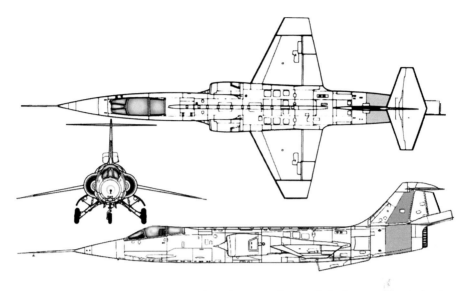

An NF-104A. Note the long nose probe, wing extensions and the large rocket engine in the tail [William Zuk].

to replace some of the AR2-3's parts after an hour of operation some 40 flights could be made before a minor overhaul was required (a major overhaul was required after 2 hours and the total life of an engine was 4 hours). The heavily modified Starfighter design was designated the NF-104A (with the 'N' standing for 'Nonstandard') AeroSpace Trainer (AST).

Every aircraft has a so-called service ceiling, which is the maximum altitude at which it can operate under normal conditions and whilst in steady, horizontal flight. To reach altitudes that are well beyond their service ceiling for brief periods of time jet fighters use so-called zoom climbs in which the plane accelerates horizontally to great speed and then pitches up into a steep climb. Zoom climbs enable aircraft to exploit the good performance of their jet engine at relatively low altitudes to build up speed which can then be traded for height on the way up (the thrust rapidly diminishes due to a lack of air as the altitude increases). The NF-104A pilots used a similar approach but they had the benefit that their additional rocket engine would not lose thrust in the thin air of the stratosphere. An experienced NF-104A test pilot would typically use his jet engine with afterburner to accelerate to Mach 1.9 at an altitude of 11 km (35,000 feet), then ignite the rocket engine to full thrust. Shortly afterwards he would reach Mach 2.2, whereupon he would pitch the aircraft sharply up into a 70 degree climb at 3.5 G. As the amount of cooling air flowing through the engine dropped with the decreasing atmospheric density the jet engine's temperature would gradually increase. To prevent this from reaching levels that would impair the engine's structure the pilot would start to throttle down the afterburner at an altitude of about 21 km (70,000 feet) and completely shut down the engine at around 26 km (85,000 feet). After the rocket engine had depleted its hydrogen peroxide the aircraft would continue to climb ballistically farther into the stratosphere until its vertical

A publicity photo of an NF-104A with its rocket engine ignited at low altitude [US Air Force].

rate reached zero. Then it would fall back into the denser air, where the jet engine would be restarted.

During the ballistic climb and descent the aircraft and pilot would effectively be weightless. The pilot remained strapped in but would feel an absence of body weight against his seat, and loose objects would fly through the cabin. However, he had little time to enjoy this experience or the view because during this phase he would need to use the reaction control system to nose the aircraft over through 140 degrees in order to cancel the climb attitude and push the nose 70 degrees down for re-entry into the atmosphere whilst maintaining zero roll and yaw. This ensured a stable position in the thicker atmosphere and also that sufficient air would flow into the jet's intakes to windmill its compressor and thereby enable it to be restarted. A poor entry attitude would result in a pilot soon finding himself in an out-of-control aircraft and unable to restart the engine (as occurred to Chuck Yeager during a particularly hairy NF-104A flight, as we will see later).

The Bell X-2 had reached altitudes very similar to those planned for the NF-104A but unlike the long, thin Starfighter with its tiny wings the X-2 was an excellent and stable glider designed to land without engine power. An additional complication that did not affect the all-rocket X-2 was the gyroscopic effect of the still rapidly turning turbine and compressor of the extinguished jet engine: this effect resisted the reaction control thrusters, and required careful compensation by the pilot whilst executing the pitch over maneuver.

The F-104 itself was, even by today's standards, an impressive high-performance aircraft and with the additional rocket engine it truly became a near-spaceplane. But it was definitely something for experienced pilots (as all ARPS students were). The normal Starfighter was so difficult to handle that it had gained the nickname 'Widow Maker'. The West German Air Force bought 917 of them in the early 1960s and by 1976 had lost 178 in accidents. The added rocket engine and the need to operate the reaction control system in the high stratosphere made it even more dangerous to fly. During the steep climb the plane's attitude and the pilot's rigid helmet meant that the horizon could not be seen, so all critical stability control and maneuvering had to be performed on cockpit instrumentation only. Furthermore the NF-104A was a single-seat aircraft so there was no instructor in the back ready to take over if things went wrong. To ready a student for this challenging aircraft he would first make a flight in a standard F-104 with his pressure suit fully inflated to familiarize himself with the movement restrictions this would impose in an accidental cockpit depressurization. Next he would make a zoom flight in a conventional Starfighter trainer supervised by an instructor in the rear seat. Following 4 hours of rehearsal in a flight simulator he would execute three solo zoom flights in a standard F-104 while being coached by an instructor in the back of an accompanying two-seat trainer. During these flights he would gradually build up the climb angle: 30 degrees on the first flight, 40 degrees on the second, and finally 45 degrees. Only then would he be deemed ready to zoom the rocket-equipped NF-104A, and even then the maximum allowed climb angle was 50 degrees (the optimum for reaching extreme altitudes was 70 degrees).

The first NF-104A (56-0756) was tested by Lockheed's test pilot Jack Woodman and Major Robert W. Smith of the Test Division of the Air Force Flight Test Center at Lockheed's factory near Palmdale Airport, prior to its formal handover to the Air Force. During this phase Smith broke the standing altitude record by zooming up to an astonishing altitude of 36,230 meters (118,860 feet) on 22 October 1963. During this flight he managed to keep the aircraft under control even though all three axes of the reaction control system had accidentally been wired incorrectly! After acceptance by the Air Force, and during the next phase of testing at Edwards, Smith surpassed his record by achieving an altitude of 36,800 meters (120,800 feet) on 6 December, and the unpowered parabolic arc provided no less than 73 seconds of weightlessness. Nearly half a century later, Smith's achievement still stands as the highest altitude ever achieved by a US aircraft taking off from a runway. Although the plane left the ground under its own power (unlike the X-planes) and the altitudes achieved were accurately recorded by ground stations equipped with radar and powerful telescopes, both records remained unofficial because the Air Force had not requested the flights to be monitored by the International Aeronautical Federation.

The second NF-104A (56-0760) was delivered 25 days after the first, and the third and final aircraft (56-0762) on 1 November 1963. Unfortunately the third plane was lost barely a month later, on 10 December, when it crashed during a flight piloted by Chuck Yeager, who was Commander of the Aerospace Research Pilots School at the time. According to him, he was unable to push the nose back down once he reached the zenith of the zoom climb, possibly due to a malfunction of the reaction control system, causing the Starfighter to go into a disastrous flat spin at an altitude of 33 km (109,000 feet). During a normal descent the aircraft would be orientated (using the reaction control system) so that air would flow into the jet engine's intakes and make its compressor windmill, thus providing hydraulic pressure for activating the control surfaces and enabling the engine to be restarted. However, in the flat spin no air was flowing into the engine, so there was no power to control the ailerons, elevators and rudder and no means to restart the engine. As the aircraft plummeted from the sky, Yeager had no option but to abandon the aircraft. He ejected just 2.6 km (8,500 feet) short of hitting the ground, was struck by his own discarded ejection seat on the way down and was badly burned by its glowing solid rocket motor, but managed to land by parachute. This is depicted in the movie *The Right Stuff*, although it shows him flying a standard F-104G without a rocket engine.

The ensuing Air Force investigation cleared Yeager of responsibility for the crash, blaming the accident on an aircraft malfunction. However, the NF-104's primary Air Force test pilot Major Robert Smith, who had trained Yeager to fly the profile, insists that Yeager simply did not perform a proper zoom climb, pulling up to the full 70 degree climb angle too slow and too late. This meant he ran out of speed before reaching the intended maximum altitude, was too late trying to nose the aircraft down, and hence began to fall with the nose still 70 degrees up. Yeager may have believed he was still climbing because the rocket was still operating, but by then the thrust of the rocket alone was insufficient to prevent the plane from falling back essentially tail first. Still in relatively dense air, no amount of reaction control thrust could have maneuvered him out of the attitude in which it was impossible to restart the jet engine. According to Smith, Yeager's fame and influence meant that the investigation ruled in his favor and unjustly labeled the NF-104A a dangerous aircraft. The differing accounts of this incident, Yeager in his famous autobiography and Smith on his highly informative NF-104A website, don't even agree on the purpose of the disastrous flight: according to Yeager it was part of his investigation of a known pitch-up problem of the aircraft (caused by the T-tail being masked by the wings at high angles of attack, preventing the airflow from reaching the horizontal stabilizers; the same problem that troubled the British SR.53) while Smith insists Yeager was merely trying to break the altitude record and had been assigned by the Air Force to fly the NF-104A solely for this purpose.

After the accident a restricted flight regime was enforced to ensure safe flights for the ARPS students using the two remaining NF-104As, part of which was imposing a limit of 50 degrees on the climb angle (again, unjustly according to Smith, which in his opinion left the students with little opportunity to experience the peculiarities of reaction control at really extreme altitudes).

The remaining aircraft were used to train students, but not very often and only for zooming flights to relatively low altitudes. The dangerous hydrogen peroxide caused some trouble, as was to be expected from experiences with earlier hydrogen peroxide powered rocket aircraft. Once a tail tank ruptured on the ground and another time a small explosion occurred in a wing while in flight, both caused by hydrogen peroxide reacting with metal aircraft parts. Modifications where made to prevent recurrences but the 56-0756 suffered an inflight rocket motor explosion in June 1971 owing to a hydrogen peroxide leak. The rocket engine and most of the rudder where blown off but the student was able to land safely. The seriously damaged aircraft was scrapped.

By then, however, the Air Force's human spaceflight ambitions had withered: the X-15 program was over, the X-20 had long since been canceled, and the task of the planned Manned Orbiting Laboratory (MOL), namely military reconnaissance, could be done better and at far lower cost by unmanned spy satellites; MOL was canceled in 1969. The Space Shuttle would not fly for another decade and it would be run by NASA. The remaining NF-104A was therefore retired, mounted on a pole and placed outside the Air Force Test Pilot School where it can still be seen today.

Various parts of this aircraft, including the extended wing tips and the metal nose cone, were loaned to Daryl Greenamyer for his civilian aviation record attempts with a highly modified Starfighter that was based on equipment from various F-104s. In 1977, after a practice zoom flight working up to his altitude record attempt, one main wheel of the plane's undercarriage did not completely deploy and Greenamyer had to eject. The NF-104A parts were lost along with the rest of his aircraft.

Several test pilots who flew the NF-104A before the planes were handed over to the ARPS experienced severe difficulties in controlling the aircraft. To achieve high altitudes and to allow sufficient time in near-vacuum for the very large (140 degree) change in pitch angle required the pilot to fly a very precise zoom maneuver. The initial speed, the fast pull-up maneuver to the 70 degree climb angle, and maintaining this angle, were all extremely important. Near the peak of its parabolic trajectory the NF-104A moved from the aerodynamic control region into the space control region and back in less than a minute, giving the pilot little time to transition from the well-known aerodynamic controls to the less familiar reaction controls and back again.

Nevertheless, according to the Air Force's primary test pilot, Robert Smith, it was not a particularly dangerous aircraft to fly as long as the pilot flew the proper zoom trajectory, had sufficient understanding of the peculiarities of reaction control at such altitudes, and did not attempt to push the aircraft beyond its established boundaries. Indeed, X-15 test pilot Bob Rushworth flew the NF-104A to the impressive altitude of 34 km (112,000 feet) without trouble on his first and only flight in it. In total some 50 pilots flew the NF-104A during 302 flights and accumulated a total of 8.6 hours of rocket engine operation.

ROCKET PROPELLED LIFTING BODIES

In preparation for the development of a reusable space glider or space shuttle, in the early 1960s NASA started to investigate so-called 'lifting body' aircraft. As the name implies, such planes have fuselages shaped to provide all or most of the necessary lift, dispensing with the need for large wings. For a glider or spaceplane returning from orbit such a configuration was thought to be ideal: wings need to be relatively thin, making it hard for them to handle the brutal aerodynamic forces of hypersonic flight and the extreme heat of re-entry, whereas a lifting body can be very robust and have a large volume relative to the surface that is heated up. Keeping a lifting body cool and in one piece is therefore theoretically simpler than preventing protruding wings from melting or being ripped off at hypersonic speeds. A lifting body basically combines the robustness and structural simplicity of a ballistic capsule with the maneuverability, flight range and landing accuracy of an aircraft. The questions to be answered were whether a lifting body was sufficiently controllable at all speeds and could be landed safely.

In 1962 the NASA Flight Research Center at Edwards (now called Dryden Flight Research Center) approved a program to build a very simple, unpowered lifting body prototype for low-speed flight tests on a shoe-string budget of $30,000; equivalent to about $220,000 in 2011. It had to be so lightweight that it could be towed into the air behind a car for the initial take-off and low-speed flights. In this mode it would also be used to train pilots, before progressing to towing to higher altitudes with Dryden's C-47 transport aircraft and release for gliding trials. The resulting contraption, which looked like a horizontal cone cut in half, was designated the M2-F1 with the 'M' referring to 'Manned' and 'F' indicating it concerned a 'Flight' version. However, its unusual shape quickly earned it the nickname 'Flying Bathtub'.

The M2-F1's structure consisted of a tubular steel frame made by Dryden, which was then covered over with a plywood shell by the Briegleg Glider Company, a local glider manufacturer. The fixed undercarriage was taken from a Cessna sports plane. The little aircraft had a maximum take-off weight of about 570 kg (1,250 pounds).

For towing the M2-F1 over the hard, flat surface of Rogers Dry Lake a powerful and fast, but not very expensive car was required. Members of the flight test team bought a Pontiac Catalina convertible with the largest engine available, which was then fitted with a special gearbox and racing slicks by a renowned hot-rod shop near Long Beach. With these modifications the Pontiac could tow the M2-F1 into the air

The M2-F1 in tow behind a C-47 [NASA].

in 30 seconds at a speed of 180 km per hour (110 miles per hour). The first car-tow test run in April 1963 did not go all that well because the M2-F1 started to bounce uncontrollably on its two main wheels the moment that NASA research (and X-15) pilot Milt Thompson raised the nose off the ground. The problem was quickly found to be in the rudder control and was fixed. About 400 successful car-tow tests were made, all with Thompson at the controls. They produced enough flight data about the aircraft to proceed with flights behind the C-47 starting in August of that same year. For this the M2-F1 was equipped with a simple ejection seat as well as a small solid propellant rocket motor in the rear base. If required, this "instant L/D rocket" could provide a thrust of 300 Newton for 10 seconds; just enough to keep the plane in the air for a little bit longer if the pilot were to find himself descending too rapidly just prior to touchdown. The tow plane typically released the M2-F1 at an altitude of 3.7 km (12,000 feet) for a 2 minute glide down to Rogers Dry Lake at a speed of 180 to 190 km per hour (110 to 120 miles per hour). Apart from Thompson, who flew most of the glide tests, several other pilots took the controls of this strange aircraft, among them the famous test pilot Chuck Yeager.

A total of 77 aircraft-tow flights were performed, the success of which convinced NASA and the Air Force of the feasibility of the lifting body concept for horizontal landings of atmospheric entry vehicles. The solid rocket motor was only used once, on the last flight when USAF pilot Captain Jerauld Gentry accidentally rolled the relatively unstable M2-F1 onto its back just after take-off. Flying inverted behind the C-47 a mere 100 meters (300 feet) above the lakebed he released the tow line, finished the barrel roll into level flight, fired the rocket and made a perfect landing. The roll instability was caused by a lack of wingspan.

The single solid propellant rocket motor of the M2-F1 can be seen in the rear base of the plane. [NASA].

The success of the M2-F1 encouraged NASA to put some real money into lifting body research and gave rise to both the M2-F2 by NASA Ames Research Center and the HL-10 by NASA Langley Research Center (with 'HL' for 'Horizontal Landing' and '10' referring to the design number). Unlike the lightweight M2-F1 glider these new aircraft were all-metal and were fitted with an XLR11 rocket engine (previously used in the X-1 series, the D-558-2 and early X-15 missions) for testing their lifting body shapes at high speeds and high altitudes. As with the rocket propelled X-planes the XLR11 was used for a short powered flight phase, after which the plane would glide to a landing. In emergencies the engine could be reignited just before landing, eliminating the need for the solid propellant rocket of the M2-F1.

To be able to attain high speeds and long flight times the rocket propelled lifting bodies were carried to about 14 km (45,000 feet) under the wing of a B-52 and then released at a speed of about 720 km per hour (450 miles per hour). In fact they used the same carrier aircraft as the X-15, whose flight program was running concurrently. Special adapters were designed to enable the lifting bodies to use the wing pylon that was developed to carry the X-15. Both the M2-F2 and the HL-10 were equipped with pressurized cockpits and ejection seats.

The HL-10 and M2-F2 [NASA].

The M2-F2 was built by Northrop and was similar in shape to the M2-F1. It made its first unpowered free flight on 12 July 1966 with Milt Thompson in the cockpit. Another 14 successful glide flights followed and revealed that, just like the M2-F1, the aircraft had a stability problem which often caused it to violently oscillate in roll during the ascent. The last planned unpowered flight in May 1967 ended in disaster when, just prior to landing, Bruce Peterson suffered from a pilot-induced oscillation problem. The situation was similar to that experienced by Scott Crossfield during his first X-15 flight, except that rather than a pitch oscillation the M2-F2's problem was roll control. With the craft rolling from side to side and also having to avoid a rescue helicopter that was in his way, Peterson fired the XLR11 rocket engine to prolong the landing approach but nevertheless smashed onto the lake bed before the landing gear was fully down and locked. The aircraft skidded

The M2-F3 with test pilot John Manke [NASA].

across the ground in a cloud of dust, rolled over six times and came to rest upside down. Peterson was severely injured; he later recovered but had permanently lost the vision in his right eye. Some of the film footage of the crash was used by the 1970s' TV series *The Six Million Dollar Man*. It took three years to rebuild and improve the M2-F2, which was redesignated the M2-F3. The most important modification was the addition of a third, central vertical fin to improve low-speed roll control. NASA pilot Bill Dana made the M2-F3's first glide flight on 2 June 1970 and found it to possess much better lateral stability and control characteristics than the M2-F2. After only three glide flights the aircraft was taken on its first powered flight on 25

November. The M2-F3 subsequently made a total of 24 flights under rocket thrust, during which it attained a top speed of Mach 1.6 and, on its last flight in December 1972, a maximum altitude of 21.8 km (71,500 feet). This unique vehicle is now hanging in the Smithsonian Air and Space Museum in central Washington D.C.

Northrop also built the HL-10, which made its first glide flight in December 1966 with Bruce Peterson at the controls. Peterson found the plane to be very unstable and only managed to maintain control by keeping the speed up: once again actual flight testing had proven its worth as proof-of-concept of aerodynamics and control theory. The HL-10 was grounded while NASA engineers studied data recorded during the flight, as well as from additional wind tunnel tests. It was found that so-called flow separation at the outboard fins was the culprit: because the air was not flowing over them properly they were ineffective in providing lateral stability. The leading edges of the outboard fins were modified and the ensuing glide tests established that this fix worked. After 11 unpowered flights the first powered flight was made on 23 October 1968 by Jerauld Gentry, the pilot who made the last (and unusually exciting) flight of the M2-F1. During this first powered flight the XLR11 engine malfunctioned shortly after launch, forcing him to jettison the remaining propellant and make an emergency landing on a conveniently located dry lake bed. On 13 November NASA pilot John Manke made the first successful powered flight. The HL-10 was flown a total of 37 times, logging (on different flights) the highest speed and altitude in the entire lifting body program: a top speed of Mach 1.86 and a maximum altitude of 27,440 meters (90,030 feet). Although it had its share of teething problems, pilots eventually found the HL-10 to be more stable and easier to fly than the M2-F3 and in this respect also better than the later X-24A (see below). On its two last flights the XLR11 engine was replaced with three small hydrogen-peroxide engines that provided a continuous low thrust of 4,000 Newton in total during the final approach for landing, reducing the glide angle from 18 down to 6 degrees. However, it was found that this provided few benefits over a completely unpowered glide landing and led to the conclusion that adding low-thrust landing engines to aircraft with relatively low lift over drag ratios didn't make much sense. It added complexity and weight, offered little assistance to the pilot, and the weak thrust did not enable him to abort a landing and fly around for another try. This HL-10 result later helped engineers to decide not to put any landing engines on the Space Shuttle Orbiter.

In his book *Wingless Flight: The Lifting Body Story*, HL-10 engineer Dale Reed describes a plan that he proposed for sending this vehicle into orbit. It would require to be fitted with reaction control thrusters for attitude control in space and an ablative heat shield for re-entry. It would be launched unmanned on a Saturn V moonrocket along with a manned Apollo capsule; the HL-10 would basically take the place of the lunar module in the adapter of the upper stage. Once in orbit, one of the astronauts would make a spacewalk from the Apollo capsule and enter the cockpit of the lifting body. On the first of two such missions the pilot would make in-orbit checks of the vehicle and return to his spacecraft, whereupon the HL-10 would return to Earth automatically. On the second mission an astronaut would actually pilot the vehicle all the way back to Edwards. This very adventurous plan was never

The Northrop HL-10 with flyby of the B-52 carrier plane [NASA].

implemented, although Wernher von Braun was apparently enthusiastic about using his Saturn V for these missions.

The HL-10 can be found guarding the entrance of NASA Dryden Flight Research Center, mounted on a pedestal as if coming in for a landing. *The Six Million Dollar Man* also used footage of the HL-10.

The M2-F2 and HL-10 were followed by the joint USAF/NASA X-24A. It was built by the Martin Aircraft Company and looked somewhat like a fat version of the M2-F3 with a curved back, three vertical fins and an XLR11 engine. Jerauld Gentry piloted this "potato with three fins" on its first unpowered flight on 17 April 1969, as well as on its first powered flight on 19 March the next year (it thus flew before the wrecked M2-F2 reappeared as the M2-F3). During its flight test phase this aircraft was flown 28 times, achieving a top speed of Mach 1.6 and a maximum altitude of 21.8 km (71,400 feet).

The Martin Marietta Corporation (which the Martin Aircraft Company became) went on to strip the X-24A down and rebuild it as the X-24B, which looked rather different. Whereas the X-24A was round and fat the X-24B had a triangular 'double delta' shape with a flat bottom and pointed nose, and was affectionately called the

The Martin Aircraft X-24A [NASA].

'Flat Iron'. The double-delta planform meant it had delta wings which incorporated a bend (as on the Space Shuttle). This shape, derived from a study by the Air Force Flight Dynamics Laboratory of possible future re-entry vehicles, resulted in a more stable aircraft with a much better lift over drag ratio owing to its greater area of lift-generating surface. The reuse of much of the X-24A's equipment and airframe saved the Air Force a lot of money in comparison to what it would have cost to build a new vehicle from scratch.

The first to take the X-24B up (or rather, down) for a glide flight was NASA pilot John Manke on 1 August 1973, and he was also at the controls on the first powered flight on 15 November. During its total of 36 flights the X-24B managed to reach a speed of Mach 1.76 and a maximum altitude of 22.6 km (74,100 feet). The X-24B also made two landings on the main concrete runway of Edwards to demonstrate that accurate runway landings were possible for a lifting body glider with a low lift over drag ratio (it had nose-wheel steering, unlike the X-15 and the other lifting bodies).

In 1975 Bill Dana made the final flight of the X-24B, drawing to a close not only the lifting body test program but also rocket aircraft flying at Edwards in general. For the occasion the team prepared a sign depicting the X-1 and the X-24B and the text "End of an era; Sept. 23, 1975; Last rocket flight." It truly was the end of one of the most interesting periods in aviation history, with rocket planes pushing technology, flight speeds and altitudes to levels that had been mere dreams when the X-1 broke the speed of sound almost three decades earlier.

There were various proposals for an X-24C, including one by the Lockheed

The Martin Aircraft X-24B with test pilot Tom McMurtry [NASA].

Skunk Works for an aircraft that would use scramjets (able to function at higher speeds than a ramjet) to reach Mach 8, but in the post-Vietnam era the military had little money to spare and NASA was developing the Space Shuttle. The lifting body program thus ended with the last flight of the X-24B, after which the aircraft found a home in the National Museum of the US Air Force at Wright-Patterson Air Force Base. Although the original X-24A no longer exists, a very similar vehicle was put next to the X-24B in the museum to represent it. It is not a replica, but actually a conversion of a never-flown, jet-powered version of the X-24A lifting body called the SV-5J.

The lifting body program taught NASA invaluable lessons for the Space Shuttle, which it started to develop in the early 1970s. Although it was decided that the Space Shuttle Orbiter would have a relatively conventional fuselage with wings, rather than a lifting body shape, its low lift to drag ratio would produce a similarly steep gliding descent for landing. The lifting body flights showed that accurate and safe landings could be made with such a vehicle, without the need for a means of propulsion. The earlier planned jet engines for the landing were discarded, simplifying the design and lowering the vehicle's weight. The Orbiter went on to routinely land on the runway at Kennedy Space Center, and its pilots never found themselves wishing they had the jet engines available.

In the 1990s the shape of the X-24A returned in the form of the X-38 technology demonstrator that was expected to lead to a Crew Return Vehicle to enable astronauts aboard the International Space Station to return to Earth in an emergency. The X-38 made several unmanned glide test flights after release from a B-52 but the program was canceled in 2002. Currently the SpaceDev company is developing the somewhat similar Dream Chaser mini-shuttle partly funded by NASA under the Commercial Orbital Transportation Services program intended to encourage private companies to develop space transportation vehicles for servicing the International Space Station. If introduced, the Dream Chaser lifting-body vehicle will be launched atop an Atlas V rocket to take crew and cargo into low orbit.

SHUTTLES

Even before the X-15 took to the air the Air Force, NACA and North American were making plans for an orbital version. This X-15B would be launched using a multi-stage rocket derived from the launch booster of the SM-64 Navaho missile; a project that had just been canceled and had left North American with a warehouse of rocket boosters. But the X-15B was canceled when Project Mercury was approved, a much simpler capsule concept that promised early results in the developing space race with the Soviets. Then the national goal of being first to land a man on the Moon gave rise to the Gemini and Apollo capsule-style spacecraft.

However, the Air Force, regarding a capsule as merely a step on the way to more routine access to space, saw the benefit of a reusable shuttle-type vehicle for manned missions. During the 1950s Werner von Braun had proposed a reusable canard space glider to be launched atop a rocket using two expendable stages. This was explained in *Collier's* magazine in 1952, in one of a series of articles on future spaceflight that von Braun wrote with Willy Ley between 1952 and 1954. *Collier's* printed 4 million copies per issue, so these articles did much to spur enthusiasm for spaceflight in the US. The articles were enlivened by beautifully detailed illustrations by leading space artists, including Chesley Bonestell, that effectively dramatized von Braun's manned spaceflight development blueprint for the general public. Further publicity came soon afterwards with von Braun presenting a hugely popular three-part Disney television show on the future of space travel, which also featured his designs. At one point von Braun presents his winged spacecraft design: "Now here is my design for a four-stage orbital rocketship. First we would design and build the fourth stage and then tow it into the air to test it as glider. This is the section that must ultimately return the men to the Earth safely." His 'Ferry Rocket' is basically an upper rocket stage with multiple engines and long, slightly swept-back wings, each with an elongated vertical stabilizer mid-way. Large horizontal stabilizers were fitted to the nose, resulting in a canard design. The launcher's first stage had huge fins as well, not for flying but for counteracting the imbalance caused by the rocket glider's wings on top of the rocket, as otherwise a slight wind during take-off or buffeting while ascending through the atmosphere would easily blow the whole assembly off course.

By 1959 the Air Force was promoting a new program as the means of performing military manned space missions. Like von Braun's concept, this 'Dyna-Soar' would not be a real rocket plane but a reusable space glider (its name was a contraction of Dynamic Soaring) bolted to the nose of a conventional launch rocket. It would only use its wings during the unpowered descent back through the atmosphere and make a controlled landing near its launch site. There it would be quickly readied for its next launch on a new expendable rocket. In its operational form Dyna-Soar would be able to perform all kinds of missions and even put things into orbit because it would have a cargo bay; it was basically to be a small, early version of the Space Shuttle.

In contrast, Mercury and the later Gemini and Apollo spacecraft were completely single-use; very little of what was launched would come back, and with their ablative heat shields that burned away during re-entry the capsules were not reusable. They also had very little means of maneuvering once they started their ballistic fall back to Earth and at best could be expected to land within a radius of several kilometers of a specific point. Because of this uncertainty in the landing spot, as well as to ensure a soft landing, they had to come down in the ocean, which meant a fleet of search and recovery ships was required. The X-20 pilot would fly his craft back to its airbase, making the return much more economical. Of course a controllable, winged machine was also much more appealing to the Air Force than a 'spam-in-a-can' capsule using parachutes. It facilitated a dignified landing on a runway, rather than an inglorious splash into the ocean.

The Air Force forecast numerous versions and missions for the X-20, involving payloads for gathering aerodynamic flight data, satellite inspection, electronic and photographic intelligence, and even dropping nuclear bombs with greater precision than was possible using a ballistic missile! Dyna-Soar was to be a research aircraft, spy plane, orbital bomber and transportation shuttle all in one.

In June 1959 Boeing was awarded the development contract. Boeing's design was a 5,200 kg (11,400 pounds) delta-winged vehicle with large vertical winglets instead of a more conventional tail for lateral (yaw) control. Most of the internal structure as well as the upper surface was to be made from René 41, a 'super alloy' that was able to withstand extreme temperatures. However this material would not suffice for the areas that would see the highest temperatures during re-entry, when slamming into the atmosphere at some 28,000 km per hour (17,000 miles per hour), or Mach 28. Capsule spacecraft solved this issue by using expendable ablative heat shields, but as the X-20 was to be a completely reusable space glider its underside would require to be made by placing molybdenum sheets over insulated René 41, while its nose-cone was to be made from pure graphite incorporating zirconia rods.

The X-20 would be controlled by a single pilot. Behind him was an equipment bay that could contain either data-collection equipment, reconnaissance equipment, weapons, or seats for up to another four astronauts. The X-20 would be connected to a small rocket stage to enable the craft to shoot itself away from the Titan III booster during launch in case of an abort or, once in space, to change its orbit. At the end of a mission this 'transition stage' would fire its main rocket engines against the velocity vector so that the X-20 would fall back to Earth. It would then be jettisoned and the aircraft would descend through the atmosphere and use aerodynamic drag to further slow down ('aerobraking'). During the initial, high-temperature re-entry, the pilot's windows would be protected by an opaque heat shield that could be jettisoned once the aerothermodynamic onslaught was over. Because rubber tires would burn during re-entry, the X-20 would land using wire-brush skids made of René 41. The Titan III

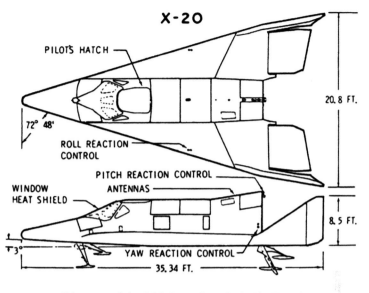

Diagram of the X20 Dyna-Soar [US Air Force].

launcher would be based on the Titan intercontinental ballistic missile and be fitted with exceptionally large stabilizing fins to compensate for the disturbances caused by the X-20's wings on top of the assembly during ascent through the atmosphere.

In April 1960 seven astronauts were secretly chosen to fly the X-20, among them (future) X-15 pilots Neil Armstrong, Bill Dana, Pete Knight and Milt Thompson.

But the X-20 Dyna-Soar never made it out of the factory, let alone into orbit. In the early 1960s it was already becoming evident that many of its foreseen missions could be performed sooner and more cheaply by unmanned spacecraft or manned capsules. The high cost and questionable military utility led to the cancellation of the program in December 1963. This made sense at the time in terms of immediate technological and military priorities, but in hindsight it is definitely a pity. Although the X-20 was a vertically-launched reusable glider with some orbital maneuvering capability rather than a real rocket plane, it was a logical next step in the evolution of rocket aircraft. The X-15 was a true rocket plane, but suborbital. It could reach orbital altitudes but lacked the propellant to accelerate to orbital speed. With the X-20 the Air Force was developing the technology needed to fly a plane from orbit back to Earth, with all the aerodynamic and thermodynamic complications, and some new materials capable of withstanding extreme temperatures. Had the X-20 continued, it would have delivered valuable experience that could have made the later Space Shuttle a more economical launch vehicle and perhaps opened the way to fully reusable spaceplanes.

Conceptually, a follow-on to the X-20 could have been to replace the expendable launching rocket with a reusable rocket powered carrier aircraft to create a two-stage rocket plane (similar to the X-15B/Valkyrie combination but with the carrier aircraft able to fly higher and faster than the Valkyrie, and with the secondary vehicle able to

The X-20 Dyna-Soar launched on a Titan III booster [US Air Force].

boost itself into orbit). That would have been considerably simpler than developing a single-stage rocket spaceplane, whilst retaining the benefits of an all-reusable system with (hopefully) aircraft-like operations. Prior to the cancellation of the X-20 there were many outline designs for orbital rocket plane and shuttle-type vehicles, most of them involving multiple stages. There were all kinds of combinations: orbital gliders with or without integrated rocket stages; with expendable or reusable, winged or non-winged stages using either liquid or solid propellants; and with horizontal or vertical take-offs.

Martin offered the very ambitious 'Astroplane', a horizontal-take-off, horizontal-landing, single stage spaceplane which, powered by "nuclear magnetohydrodynamic engines", sounds like something straight out of *Star Trek*. This intriguing propulsion system would extract nitrogen from the atmosphere, rapidly cool it, then accelerate the resulting liquid using powerful electromagnetic forces generated by an onboard nuclear reactor. The vehicle had a long, slim shape that would have been perfect for low drag at high speed but unsuitable for horizontal runway landings at reasonable speeds. The designers therefore envisaged deployable wings that would be extended at low speeds to provide additional lift. The bat-like airfoils would be composed of rigid ribs with a flexible membrane stretched between them. What would happen if a wing failed to deploy properly and this flying radiation hazard fell out of the sky was obviously something that would have to be worked on a bit.

In general, however, it was understood that without resorting to exotic, far-fetched propulsion systems, a reusable launcher would have to be a multi-stage vehicle. For conventional multi-stage expendable launchers the useful payload that can be placed into orbit typically represents only 3% of the total weight of the rocket that leaves the pad, and around 18% in terms of hardware (i.e. without propellant). A single-stage vehicle cannot shed the weight of empty tanks and no-longer-needed rocket engines on the way up, so its payload is necessarily considerably less than for a multi-stage launcher. The wings, heat shields and control systems that are required to enable a single-stage vehicle to return to Earth can all too easily reduce its payload to zero, or even to negative values implying the vehicle will not even be able to reach orbit. To compensate, Single Stage To Orbit (SSTO) vehicles must carry more propellant per unit of hardware, which leads to larger vehicles. All this makes the development of a single-stage spaceplane with a reasonable payload extremely challenging, expensive and risky, and certainly beyond the technology of the 1960s and 1970s (and indeed beyond today's technical capabilities; more on this later in this book).

The Martin 'Astrorocket' concept studied in the early 1960s therefore consisted of two vertically launched, winged rocket stages mated belly to belly. The task of the larger vehicle was to get its smaller sibling to high altitude and high speed, release it and then glide back to Earth while the second stage continued into orbit. The orbital vehicle would later make a gliding return (very much like the Space Shuttle) and be reunited with the first stage and prepared for another mission. This VTHL (Vertical Take-off and Horizontal Landing) concept would be proposed many times in various forms and by several different companies.

Douglas offered the 'Astro' design. This was based on the same idea but with the orbital vehicle mounted on the nose of the carrier booster rather than attached to its belly, and with both vehicles having lifting body configurations.

Lockheed's 'Reusable Orbital Carrier', which was studied at about the same time, also involved two winged rockets but this combination would take off horizontally using an accelerated sled. The orbital vehicle of this HTHL (Horizontal Take-off and Horizontal Landing) launcher would ride on the back of its large carrier stage, which would use jet engines to return to its launch base. Interestingly NASA had stated in the specifications for this carrier stage that it "should offer a potential commercial application in the late 1970s, such as operating the vehicle over global distances for surface-to-surface transport of cargo and personnel". In other words, it must be able to be converted into a hypersonic, rocket propelled airliner; a kind of super-Concorde similar to that proposed by Max Valier in the 1930s.

Lockheed and North American both had similar concepts for three instead of two winged stages, which meant more complicated operations but less stringent vehicle mass-minimization requirements. A proposed further development of Lockheed's

'System III' would involve a giant first stage booster aircraft with ramjet propulsion and was called 'System IV' (System I was the Apollo Saturn IB combination as later used on the Apollo 7 mission, while System II would consist of a Saturn IB rocket with a reusable 10-man spaceplane that would also form the basis for the System III and System IV orbital stages).

In the mid-1960s NASA, along with the Department of Defense, adopted a similar phased approach involving a 'Class III' vehicle based on the Lockheed's System IV design but with a hypersonic booster aircraft powered by combined turbofan-ramjet-scramjet engines. Such engines would run as regular turbojets at subsonic and low supersonic speeds, and as ramjets at supersonic velocities where the airstream could still be slowed down to subsonic speeds; at hypersonic flying velocities the airflow through the engine would be supersonic and so the engines would go into scramjet mode. A scramjet (meaning 'supersonic combustion ramjet') is a ramjet in which the combustion takes place in a supersonic airflow so that the air coming into the engine does not need to be slowed down to subsonic speeds (as is required for ramjets). The big benefit is that the incoming air does not lose useful energy due to deceleration, energy that would otherwise be converted into heat that would require the engine to be actively cooled. The combined engine system NASA envisaged represented an enormous technological challenge: at the time the only existing hybrid jet engine was the J58 of the SR-71 Blackbird, whose turbojet and ramjet modes enabled the aircraft to reach about Mach 3.2. The hypersonic stage of the Class III vehicle would require to fly much faster and its engine would be considerably more complicated. The joint Aeronautics and Astronautics Coordinating Board Subpanel on Reusable Launch Vehicle Technology, formed by NASA and the Department of Defense, concluded that the required technology could not be expected to be operational before 1982. (In fact, it still does not exist today.) Preliminary calculations for the Class III carrier vehicle indicated it would weigh 306,000 kg (675,000 pounds) at liftoff, if required to carry a 132,000 kg (290,000 pound) orbiter plane with a 16,000 kg (35,000 pound) low-Earth-orbit payload. This meant the total liftoff weight would have been about a quarter of that of the Space Shuttle for a payload only about one-third less. Of course the Space Shuttle actually flew whereas this Class III vehicle remained a paper study so we must be wary of such comparisons. Nevertheless, the numbers do indicate the potential mass benefits to be gained by using airbreathing engines.

A concept called 'Mustard' (Multi-Unit Space Transport And Recovery Device) was studied by the British Aircraft Corporation (BAC) in the mid-1960s. It involved three similar winged vehicles that would be launched vertically as a single stack. A number of configurations were examined: a belly-to-belly-to-back sandwich, triangular belly-to-belly, two belly-to-belly and the third inline on top. Two of the vehicles would serve as boosters to put the third one into orbit, with the boosters pumping their remaining propellant into the orbital vehicle to top up its tanks prior to separation. In this way all of the engines could fire simultaneously for the first phase of the flight, yet the orbital shuttle would still initiate its solo flight with a full propellant load. The concept would potentially have enabled the orbiter to reach the Moon, which is a unique capability for a winged reusable design. All three stages would have flown back to base after their mission, landing horizontally to be readied for the next launch. Around this same time the Warton Division of BAC also studied several launch vehicle concepts in which a very large hypersonic aircraft carried on its back a small shuttle mounted on top of an expendable rocket. The 'European Space Transporter', which originated as the Nord Aviation Mistral and was studied by French and West German aerospace companies, was rather similar; it looked like a giant airbreathing fighter with enormous ramjet intakes and a rocket propelled lifting-body shuttle strapped onto its belly. Dassault's 'Aerospace Transporter' was also based on essentially the same idea, involving an orbital 'space taxi' shuttle carried under a turboramjet Mach 4 aircraft that resembled the Concorde. Dropping a rocket vehicle rather than (as envisaged for the X-15A-3) launching it from the back of a carrier aircraft reduces the risk of the shuttle striking its motherplane, but the downside is a longer and more complex undercarriage on the carrier in order to give the shuttle enough clearance above the ground during take-off and aborted landings of the aircraft combination.

In 1967-1968, the US Air Force awarded several study contracts for an 'Integral Launch and Re-entry Vehicle' (ILRV) which would consist of a reusable single-stage VTHL that would jettison relatively inexpensive propellant tanks. This represented a compromise between a fully reusable single-stage launch vehicle (which would be large, heavy and extremely challenging and expensive to create) and a fully reusable two-stage vehicle (that would require the development and operation of two separate vehicles). The McDonnell-Douglas ILRV design featured a VTHL lifting body that resembled a steam iron, mated to four jettisonable tanks: two large ones for liquid hydrogen and two smaller ones for liquid oxygen. It would fold out small wings to increase its lift over drag ratio in the final descent and landing. Lockheed's proposed 'Starclipper' was a VTHL lifting body design in which a single integrated A-shaped propellant tank would 'wrap around' the reusable lifting body. General Dynamics proposed a 'Triamese' solution with three reusable rocket stages which had fold-out wings and jet engines to land using a standard airport runway. The concept did partly satisfy the ILRV concept of a single vehicle design because the three winged stages would have been virtually identical.

The primary mission of all these reusable spaceplane concepts dreamed up in the 1960s was to transport crews to a large Earth orbiting space station. However, other missions such as the launching, inspection, repair and retrieval of satellites were also envisaged. The reusability was expected to translate into very low operating costs: for instance Lockheed claimed that their Starclipper would have a turn-around time (the time necessary to prepare the vehicle for its next mission) of only 24 hours and would reduce the costs of launching cargo into space to less then $100 per kg in 2011 terms. In fact, launching satellites into low orbit currently still costs $10,000 per kg for expendable launchers (and double this for the Space Shuttle when it was retired that year).

By mid-1969 NASA was planning an extremely ambitious and financially rather unrealistic, manned space exploration program to follow up on Apollo. The wish list included space stations, interplanetary 'space tugs' and eventually bases on the Moon and Mars. The first task was to build a 12-person space station in orbit by

1975 and to expand this into a 50-person 'space base' by 1980. Smaller 'way-stations' would then be deployed in geostationary Earth orbit and around the Moon and Mars. Naturally all this infrastructure would require an efficient, dependable and inexpensive means of transportation in the form of a space shuttle.

In a meeting in April 1969 Maxime Faget, the renowned designer of the Mercury capsule and by then director of engineering and development at NASA's Manned Spacecraft Center, presented a balsa-framed, paper-skinned airplane model that had short, straight wings and a shark-style nose. It looked remarkably similar to Sänger's Silbervogel concept of the Second World War. When introducing the design, Faget explained, "We're going to build America's next spacecraft. It's going to launch like a spacecraft; it's going to land like a plane."

Back in January NASA had awarded four Phase-A study contracts to McDonnell-Douglas, North American Rockwell, Lockheed and General Dynamics to begin the development of the shuttle launch system. Martin Marietta, whose bid had been rejected, decided to participate using its own funds in order not to be left out of what promised to be a very lucrative project. The NASA requirements called for a vehicle able to launch 12 people as well as some 11,300 kg (25,000 pounds) of supplies to a space station in low orbit. Soon these became even more demanding and complex when it was decided that the vehicle must also be capable of launching satellites and interplanetary probes. Then the Air Force demanded that the stated payload capacity be doubled.

NASA had by then decided that the shuttle was going to be a fully reusable VTHL Two Stage To Orbit (TSTO) vehicle, since that was expected to produce the lowest cost per flight and highest payload capability in combination with good operability and mission flexibility. This meant the next generation launch vehicle was not going to be a real single-stage spaceplane but rather a combination of vertically launched winged rocket stages, each capable of gliding back to Earth and making a horizontal, unpowered landing. In an article optimistically titled 'The Spaceplane That Can Put YOU in Orbit' in the July 1970 issue of *Popular Science*, Wernher von Braun (at the time NASA Deputy Associate Administrator) commented on the decision: "It would be ideal, of course, if we could build a single-stage-to-orbit shuttle, which, without shedding any boost rockets or tanks, could fly directly up to orbit and return in one piece to the take-off site for another flight. Although we may well know how to build such a vehicle some day, most studies unfortunately show that with the present state of propulsion and structural technology this objective would be just a shade too ambitious."

Out of the 120 different concepts studied, five emerged as the most promising. 'Concept A' of North American as well as 'Concept C' of Lockheed involved a huge winged, vertically launched booster stage carrying a smaller orbital shuttle stage on its back. 'Concept B' of McDonnell-Douglas and 'Concept D' of General Dynamics both had a winged booster stage and a shuttle fitted belly-to-belly. Martin Marietta's 'Concept E' involved a shuttle vehicle mated to a twin-body winged booster stage. For the orbital elements, North American, General Dynamics and Martin Marietta favored conventional, fuselage-with-wings shapes, whereas McDonnell-Douglas and Lockheed proposed lifting bodies. However, McDonnell-Douglas soon changed their shuttle design to a more conventional shape which would re-enter with folded wings. General Dynamics' shuttle was based on a similar idea but with retractable switch-blade wings. Not having the wings extended during re-entry would save on thermal protection weight, but of course the extension system would be complicated, heavy, and involve a fair amount of risk (for example if they did not open).

For the ensuing Phase-B step of the project Lockheed teamed up with Boeing to further develop their concept, but did not win a contract. North American Rockwell and General Dynamics did, and joined forces to work on the concept of a North American shuttle stage to be launched on the back of a large winged booster designed by General Dynamics. McDonnell-Douglas and Martin Marietta, the other winning team, jointly studied a concept that was very similar but employed re-entry thermal protection based on metal alloys rather than the silica tiles favored by North American Rockwell and General Dynamics. Both teams considered two basic orbiter designs (NASA having not expressed a preference either way): one involved Maxime Faget's design with straight wings and lower re-entry heating but a relatively small cargo capability and a limited cross-range, and the other a delta-winged design with a greater cross-range and a larger cargo bay but higher heat loads. All Phase-B concepts were for a fully reusable launch system.

In addition to the primary Phase-B study contracts, in June 1970 NASA also let three contracts to Grumman/Boeing, Chrysler, and Lockheed to study an 'Alternate Space Shuttle Concept' (ASSC). This was intended as a backup plan in case the fully reusable concepts turned out to be too expensive to develop (as would soon prove to be so). The Grumman/Boeing team received the most important contract (which was later upgraded into a full Phase-B contract) for studies of a shuttle with expendable propellant tanks and of a reusable orbiter boosted by existing liquid propellant rocket engines and expendable solid rocket motors. For its ASSC contract, Lockheed further refined its Starclipper concept, which already envisaged drop tanks. The ASSC study results showed that the use of expendable tanks greatly reduced the size and weight of the orbital vehicle as well as the booster vehicle, whilst also significantly lowering the total system's development costs.

While Phase-B was underway, NASA's original grand plan involving large space stations, lunar bases and manned missions to Mars evaporated. This left the Space Shuttle "a project searching for a mission", as critics in the US Congress derided it. NASA and the Air Force then began to focus on the shuttle as a stand-alone project, as a low-cost 'space truck' for launching, repairing and retrieving satellites, and for flying all kinds of onboard experiments. Based on what we now know to have been wildly optimistic assumptions, NASA's plan was for a total of 445 flights during the 10 year period 1979 through 1988 (the Space Shuttle as we know it actually flew only 135 missions during its 30 year lifetime).

In May 1971 it also became clear that the government was not going to allocate sufficient funding to enable NASA to develop a fully reusable Space Shuttle. NASA and its contactors then spent the next 6 months frantically revising their concepts by incorporating expendable propellant tanks, solid propellant rocket boosters, and even modified forms of the first stage of the Saturn V moonrocket. At the end of 1971 the manned, winged flyback booster options were discarded on the basis of the expected development cost and system complexity (both for development as well as in-service operations). By mid-1972 NASA had finally decided on the general concept for the Space Shuttle as a reusable, winged orbiter with a single expendable propellant tank and two huge solid propellant rocket boosters which would be retrieved by parachute and refurbished for reuse. Based on the expected sizes of future military spy satellites and the orbits they would require, the Orbiter had to have a huge payload bay as well as the ability to put large satellites into polar orbit from Vandenberg Air Force Base in California (from which no Shuttle ever launched, even although a nearly complete launch facility was created there). The Orbiter also required a fairly large cross-range during its descent to Earth, since its landing site in California would be rotating away from under it when returning from a reconnaissance mission involving flying a single polar orbit. Consequently the Orbiter had to be equipped with large delta-wings, and Faget's straight-wing concept was discarded. In mid-1972 the Phase-C/D contract for the design and construction of the Orbiter was awarded to North American Rockwell. Its Rocketdyne subsidiary would supply the main engines that would burn hydrogen and oxygen, Martin Marietta would supply the external propellant tank and Thiokol would supply the solid rocket boosters. The concept illustrations of the time show a vehicle very similar to the Space Shuttle we are familiar with. Early in 1974 NASA also decided that the Orbiter would not require jet engines, because the X-24 and other lifting bodies had proven that 'dead-stick' glide landings would be sufficiently safe. Discarding the (deployable) jet engines further simplified the design and resulted in considerable weight and volume savings. It did mean that Shuttle pilots had only one chance to land, but the decision was correct because not a single Orbiter ever crashed during landing.

The selected configuration had a major impact on the test flight philosophy. Up to then all crewed launch systems had been unmanned on their maiden flights. For the early Shuttle concept, when it involved a reusable winged booster stage, von Braun said in his aforementioned article that the first stage and the orbiter would each be tested separately. Although the system as a whole would be launched vertically when operational, both winged stages would initially make a series of horizontal take-offs and subsonic flights using their own jet engines. In the next step, each vehicle would make individual vertical take-offs and supersonic flights using their rocket engines. Only after these test flights had been satisfactorily concluded would the orbiter be strapped to the booster and the combination launched into space. This logic could not be applied to the Space Shuttle. The Orbiter was able to make unpowered, subsonic glide tests from the back of a Boeing 747 but the first powered flight simply had to be an all-out launch with two pilots on board because it was not designed to fly in an unmanned role. The Orbiter was not able to take off horizontally since it lacked jet engines. There was no way to flight test the large External Tank and the two Solid Rocket Boosters separately because the Orbiter's computers provided the commands. When astronauts John Young and Bob Crippen first flew the Space Shuttle in April 1981 NASA took an enormous risk with a new system that was so radically different from the previous manned capsules (for their later Buran shuttle the Soviets actually flew their orbiter with jet engines installed for flight tests, launched the large Energia carrier rocket separately to verify its performance, and only then launched and landed the entire Buran system unmanned).

The chosen Space Shuttle design was an awkward compromise between mission requirements, technology, costs, schedule, the need for the project to provide jobs all across the US, and the political wish for something ambitious to reassert the nation's technological leadership. Nevertheless, NASA still believed its partial reusability and great payload capabilities would revolutionize spaceflight. Access to space would be relatively cheap, opening Earth orbit and beyond for all manner of exciting scientific, industrial and even commercial activities. In a speech to the National Space Club on 17 February 1972 NASA Deputy Administrator George Low said that now the great challenge was to develop a "productive Space Shuttle, one that performs as required, can be developed at a reasonable cost, and is economical to operate. If we meet the first two of these objectives, but not the third, we will have developed a white elephant."

During its 30 year life the Space Shuttle performed an amazing array of missions but it never fulfilled its promise of regular, safe, economic flights into space. Instead of weekly launches NASA managed to fly an average of fewer than five missions per annum. Two of the Shuttle flights have ended in disaster; an accident rate of 1 in 68, which is rather dismal compared to the fewer than two crashes per million flights for commercial airliners in the US. Instead of costing around $40 million per flight as predicted in 1975 (equivalent to about $160 million in 2011), actual costs were closer to $500 million per launch. When also taking into account the development costs and the production costs for the five Orbiters, the actual average cost per flight was well over $1 billion!

During the 1970s NASA claimed the Orbiter's spacious cargo bay would relax the constraints on the size and weight for its payloads, resulting in lower costs. However in reality the stringent safety constraints imposed on spacecraft and equipment meant to be transported on crewed space vehicles generally resulted in higher development and production costs than when launching on an unmanned expendable rocket.

There are several reasons why the Space Shuttle cost far more to operate and took much longer to 'turn around' than expected. An important reason is the maintenance of the Orbiter. For a start it had 35,000 brittle thermal protection tiles, and these had to be individually inspected after every flight and often replaced. Structural elements, instrumentation, and electrical wiring all had to be thoroughly inspected after every flight; a task that is only done periodically for normal aircraft. The Orbiter required a team of some 90 people working more than 1,030 hours in total on maintenance and refurbishment after each mission, costing about $8 million per flight in 2011 prices. Not included in this was the heavy maintenance work on the three very complicated Space Shuttle Main Engines that was done separately from the vehicle. Even though the engines only operated for 8 minutes per flight they were not nearly as reliable as airliner jet engines; they suffered more rapid wear and required a lot of checking and maintenance after each mission. In contrast, jet engines run for hundreds of hours before preventive maintenance is needed. In addition the two Solid Rocket Boosters, which landed in the ocean by parachute and were retrieved by specialized ships, had to be completely disassembled, cleaned and repaired before they could be filled with solid propellant and reassembled for reuse. Most of the things that on a conventional rocket are thrown away, with the Shuttle had to be retrieved, refurbished and than put together again. Only the large External Tank was discarded once the Shuttle had almost reached orbit and left to burn up upon falling back into the atmosphere. But that meant a new one was needed for each flight.

The idea that the Space Shuttle would launch any and all US payloads, whether from NASA, the military or commercial users, never materialized. It was simply too expensive, and it was deemed too risky to have the nation's access to space depend on a single launch system. To stimulate the diversity of launch systems following the loss of Challenger, President Reagan even decided that commercial satellites would no longer be launched using the Shuttle. This meant a dramatic decline of the number of Shuttle launches relative to the rate advertised in the 1970s (although on the other hand that rate was unattainable). Another important reason for the reduced number of launches required was the increased lifetime of satellites, yielding lower replacement rates than were assumed during the development of the Shuttle. Another mission that was claimed for the Shuttle, retrieving satellites, was undermined by the fact that the Shuttle could not fly high enough to retrieve geostationary communications satellites (potentially the biggest market); it could only reach satellites in low orbit. In any case the retrieval of a malfunctioning satellite was usually more expensive than building a replacement. The relatively low flight rate in itself also increased the cost per flight, because the so-called fixed costs for facilities and personnel had to be shared across fewer missions. This was foreseen by several people even before the Shuttle started flying, as by, for instance, Gregg Easterbrook in the very insightful article 'Beam Me Out Of This Death Trap, Scotty' published by the *Washington Monthly* in 1979.

Roger Launius, former Chief Historian for NASA and now Senior Curator at the Smithsonian Air and Space Museum, has perfectly summarized the Shuttle program: "In 135 missions, with two catastrophic failures, the US Space Shuttle proved itself a vehicle filled with contradictions and inconsistencies. It demonstrated on many occasions remarkable capabilities, but always the cost and complexity of flying the world's first reusable space transportation system ensured controversy and difference of opinion."

The Soviets also investigated the development of reusable spaceplanes in the early 1960s. In 1962 the Mikoyan Design Bureau presented its '50-50' concept, so-named since it would have comprised a reusable, air-breathing, hypersonic aircraft carrying a two-stage expendable rocket with a small orbital space glider. A first version of the carrier aircraft, using kerosene as fuel, would have accelerated to Mach 4 prior to releasing the rocket/spaceplane combination at an altitude of about 23 km (14 miles). A variant for the longer term would have reached Mach 6 at 30 km (19 miles) using hydrogen fuel. The orbital spaceplane was to be a lifting body with extendable wings for use during the subsonic part of the return flight. It was officially named the MiG

The 50-50 spaceplane system [NASA].

MiG 105-11 prototype.

105 'Spiral' but unofficially was the 'Lapot' (meaning 'flat shoe') for its somewhat awkward appearance. The MiG 105 would only have been able to accommodate a single cosmonaut-pilot, housed in an emergency escape capsule that could return to Earth independently even after being ejected in orbit.

The first flight of the complete 50-50 system was planned for 1977, and a group of cosmonaut-pilots started training in 1965 to fly test models and eventually the real spaceplane. But it proved difficult to maintain financial and political support because it was expensive, the technology was difficult to master and the lengthy development did not promise the quick results the Kremlin preferred. In 1973 the

The Buran shuttle landing after its only mission in space.

cosmonaut team was disbanded and in 1976 the seriously underfunded project was finally halted. The large 50-50 carrier aircraft was never built, but a turbojet-powered prototype of the MiG 105 for subsonic tests flew eight times between 1976 and 1979, both taking off under its own power and being dropped from a Tupolev Tu-95 bomber. Today this MiG 105-11 test vehicle languishes in a muddy field at the Monino Soviet Air Force Museum outside Moscow.

Despite the 50-50 project's demise the flight test program of the MiG 105 was continued to collect data for the new but less ambitious 'Buran' project. The Buran (Russian for 'Snowstorm') was the Soviet equivalent of the US Space Shuttle. It was developed purely to keep up with the US in technology and capability, and the Buran orbiter ended up looking remarkably similar to its American competitor. But unlike the Space Shuttle it did not use its own rocket engines for launch, it was mounted on the side of a giant Energia rocket. The Buran made only one, completely automatic, orbital flight without crew in 1988 and then the financial problems of the crumbling Soviet Union killed both the Buran and Energia projects.

The only Buran vehicle flown in space was destroyed in 2002 when the roof of its hangar collapsed due to poor maintenance, but a decrepit ground-test prototype is an attraction in a Moscow park. The atmospheric flight and landing test prototype with jet engines for a runway take-off found a more dignified home in the Technical Museum Speyer in Germany.

Being similar to the Space Shuttle, it is very unlikely that the Buran could have lowered launch costs relative to those of reliable expendable Russian launchers like Soyuz and Proton. In fact, the Russians regarded the optimistic Space Shuttle flight cost figures that NASA published during the 1970s as misinformation to disguise the Shuttle's real purpose as an orbital bomber and "space pirate ship" for destroying or abducting Soviet satellites.

In the 1979 James Bond movie *Moonraker*, highjackers steal a Space Shuttle Orbiter during a transfer flight on top of its Boeing 747 carrier. They ignite its main engines and fly off, with the rocket exhaust blowing up the 747 (in reality the Orbiter was of course only transported with empty tanks for its maneuvering thrusters, and could not carry propellant for its main engines). From 1988 until 1991 the Russians briefly worked on a shuttle that would do something similar, without the destruction of its carrier. A concept called MAKS (Russian abbreviation for 'Multipurpose Aerospace System') involved launching a fairly small shuttle mated to a large expendable propellant tank from atop a giant, subsonic Antonov An-225 airplane (which had already been built for transporting the Buran orbiter). The standard version of the MAKS shuttle would have had a crew of two and a payload of 6,600

Artistic impression of the MAKS system and mission.

Full-scale mockup of the MAKS shuttle [NPO Molniya].

kg (14,500 pound) to low orbit. It was supposed to reduce the cost of transporting materials into orbit by a factor of ten. But the collapse of the Soviet Union and the ensuing poor Russian economy killed this project as well. In June 2010 Russia nevertheless announced that it was considering reviving the MAKS program.

In the early 1980s the US Air Force studied a very similar concept. It was initially called the 'Space Sortie Vehicle', then the 'Air Launched Sortie Vehicle', and finally the 'Air Force Sortie Space System' (AFSSS). The carrier was to be a souped-up version of the modified Boeing 747 used to transport Space Shuttle Orbiters. Several companies came up with designs based on the initial specification which involved a lifting-body mini-shuttle about 15 meters (50 feet) in length, either unmanned or with a single pilot, and powered by RL-10 rocket engines whose liquid oxygen and liquid hydrogen propellants would be in large external tanks that would be discarded when empty. Some consideration was given to wrapping these expendable tanks around the shuttle in order to create an aerodynamic assembly that would be capable of a lifting ascent, meaning generating lift to help it attain altitude after release from its carrier (in which case it would have been a real rocket plane). Rather than having its shuttle take off gently from a horizontal flying carrier (as was envisaged for MAKS) the Air Force Rocket Propulsion Laboratory had something more spectacular and effective in mind: the 747, normally not an aerobatic aircraft, would have the thrust of its jet engines augmented by the installation of afterburners: liquid hydrogen from tanks on the Boeing (also used for topping up the propellant tanks of the mini-shuttle shortly prior to launch) would be injected into the hot exhaust of the four otherwise standard turbojet engines to produce a massive increase in power of up to 400%. This would enable the giant airliner to zoom up at a 60 degree angle to reach a launch altitude of 15 to 17 km (50,000 to 55,000 feet). An alternative idea was to install a single Space Shuttle Main Engine or a cluster of RL-10 rocket engines in the tail of the 747. When standing on 'alert status', this rapid response system was to facilitate a flyover of any point on Earth within 75 minutes of the moment the 747 started to taxi. The shuttle was also to deliver small payloads into orbit, rendezvous with satellites or space stations, and fly "low altitude penetration of target area" missions to drop bombs before reigniting the engines to regain altitude and return to its base. As the cryogenic propellants would boil off whilst standing on alert, constant replenishment from tanks on the ground would have been required to keep the system ready for take off at a moment's notice. This project never progressed beyond the preliminary design stage, in part because it was judged too expensive but primarily because there was no urgent need for it.

The French space agency (CNES) and the European Space Agency worked on a small, manned space shuttle named Hermes. The concept was similar to that of the X-20 Dyna-Soar, with a small space glider launched on top of a European Ariane 5 rocket. The project was approved in November 1987 but canceled in 1992 when it became apparent that neither the cost nor performance goals could be achieved. The Ariane 5 development continued as a conventional, expendable launch vehicle and is now very successful in the geostationary satellite market.

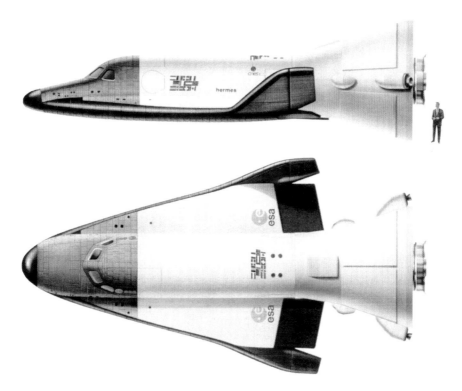

The canceled Hermes shuttle with its expendable Resource Module attached [ESA].

X-34 on the tarmac [NASA].

In the US, work on a successor for the Space Shuttle has not been very successful. During the 1990s NASA and prime contractor Orbital Sciences worked on the X-34, an unpiloted, experimental rocket plane powered by an inexpensive, non-reusable 'Fastrac' engine running on RP-1 kerosene and liquid oxygen. The X-34 was to test reusable launch vehicle technology. Like the X-15 it would have been dropped from a carrier airplane, but would have been able to reach Mach 8. But NASA canceled it in 2001 after the company refused to incorporate significant design changes without additional funding.

The most ambitious reusable launcher test project, NASA's X-33 developed by Lockheed Martin, involved a suborbital, single-stage, unpiloted VTHL vehicle with a wedge-shaped lifting-body. It was to lift off vertically without making use of the lift generated by its shape, fly at Mach 15 and then land horizontally like an airplane. It was to be powered by a 'linear aerospike' engine that consisted of a series of small rocket motors along the outside edge of a wedge-shaped protrusion. The aerospike is essentially an inside-out bell-shaped rocket nozzle in which the 'unwrapped' bell (or ramp) serves as the inner wall of a virtual nozzle along which the expanding hot gas flow produces thrust. The other side of the nozzle is effectively being formed by the outside air. The advantage is that the expansion of the rocket exhaust automatically adjusts itself to the ambient pressure of the atmosphere, preventing thrust losses due to underexpansion or overexpansion. While a conventional rocket nozzle can only be optimized for a single altitude, and hence only one point in a rocket's trajectory, an aerospike engine runs efficiently at all altitudes as well as in the vacuum of space.

Normal Bell-Nozzle
Rocket Engine

Linear Aerospike
Rocket Engine

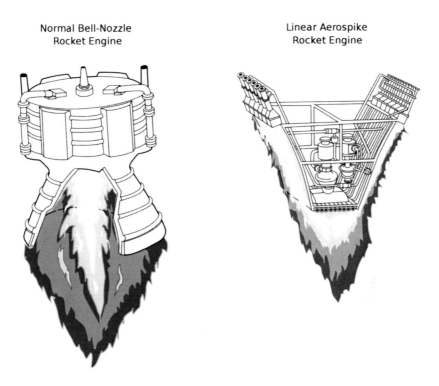

Comparison of a conventional rocket nozzle and a linear aerospike [NASA].

Sadly the X-33 project was scrubbed in 2001 owing to major problems with the development of this aerospike engine and with the lightweight composite-material hydrogen tanks (which required complex shapes to fit the lifting-body curves). The cost of the project exceeded the $1.2 billion budget limit even before any test flights could be made. The X-33 was to have led to the development by Lockheed Martin of the operational single-stage reusable launch vehicle named 'Venture Star'. However, it would probably have been very difficult to scale up the X-33 test vehicle without putting on too much weight, which (as explained earlier) is a frequent problem in the design of reusable single-stage launchers.

NASA's Orbital Space Plane program to develop a crew transportation vehicle for the International Space Station was initiated in 2002. Four competing concepts from different industries emerged, three of which involved mini-shuttle designs and one a capsule spacecraft, all of which were to be launched on top of an expendable rocket. But by 2004 they had been superseded by the Orion Multi-Purpose Crew Vehicle, an Apollo-style capsule that is still under development.

In April 2010 the Air Force finally launched a small robotic demonstrator shuttle on top of an Atlas V rocket. The payload and operations of this X-37 'Orbital Test Vehicle' were classified, but after 7 months in low orbit it made a gliding return. A second X-37 was launched in March 2011. The X-37 was intended to be carried into orbit inside the Space Shuttle cargo bay, but once it was realized that a Shuttle flight

Artist impression of the X-33 [NASA].

would be uneconomic and moreover that the Shuttle would be retired in 2011, it was redesigned for an expendable launcher and in order to prevent its wings steering the rocket off course the mini-shuttle was covered by an aerodynamic shroud. The X-37, which was developed by Boeing, would appear to be a test vehicle for an operational military, reusable, robotic spacecraft rather than a sub-scale precursor for a crewed shuttle system.

The Space Shuttle taught us that to really lower the price of transporting people and cargo into space, we require a fully reusable launcher that is easy to maintain. Ideally, it would be a single integrated vehicle without expendable tanks or boosters that require retrieval and refurbishment. It should return to its launch site to preclude complicated and expensive transportation; if the Orbiter landed anywhere other than at Kennedy Space Center it had to be mated with a specially equipped Boeing 747, flown to Florida, then de-mated from its carrier. Instead of thousands of fragile heat resistant tiles, a limited number of readily replaceable metallic shingles ought to be employed. The propellants should be non-toxic, safe and relatively easy to handle to avoid complicated tanking and propulsion system maintenance procedures (although the various rocket planes have yielded a lot of experience in handling dangerous liquids such as hydrogen peroxide). Future reusable rocket planes may carry computers and sensors that constantly check the health of all subsystems and components in flight, both to warn the crew/flight operators of any problems during a mission and to make it easier for the turn-around team to determine when and what kind of maintenance is required. The reusable rocket engines should last longer than the Space Shuttle Main Engines, require less maintenance, and be easier to repair. The RS-25 engines of the Space Shuttle can only be operated for about 10 minutes before major maintenance, which involves time-consuming (and hence expensive) tasks and the replacement of a lot of equipment. In contrast, jet engines as used in modern airliners can operate for months of accumulated flight time with only very limited checking and maintenance. Ideally, a future space plane would have some kind of combined rocket/jet engine that can use oxygen from the atmosphere while flying at relatively low altitudes. This would mean less onboard propellant, smaller tanks, and therefore a smaller, lighter vehicle. However, such engines tend to be complex, which potentially makes them hard to maintain.

8

Future spaceplanes

"Flying can never be a success until it ceases to be an adventure." – Sir Alan Cobham, aviation pioneer

It does not require a leap of imagination to see that aircraft are a more economic and practical means of transportation than the ballistic rocket launchers that we are currently using. In the 1960s, orbital rocket planes with airliner-like characteristics were therefore generally expected soon to render expendable rockets obsolete, and also to lead to cost reductions of several orders of magnitude: where the flight of an expendable launcher has the cost of building a new vehicle as its price starting point, a reusable spaceplane ideally would only incur costs for propellant, maintenance and flight operations. But the availability of existing, relatively mature ballistic missile technology, the limited market for launching satellites (currently about 70 flights per annum worldwide), and the major investments and risks associated with spaceplane development have resulted in the world's launch vehicles remaining fully expendable (with the notable exception of the partly reusable but operationally very complicated Space Shuttle, which has now been retired). As a result, for the last few decades spaceplanes have always been rather futuristic concepts, doubtless sure to supersede expendable rockets but not just yet. Launching things into space has thus remained extremely expensive and only affordable for government space agencies and several commercial applications, primarily telecommunications. As we shall see though, this is not for a lack spaceplane concepts and projects.

SILBERVOGEL

The basic, modern rocket spaceplane concept was born in the 1930s. It involved a hypersonic rocket aircraft launched by a rail with the assistance of a rocket sled, and capable of flying almost around the globe by using a suborbital trajectory in which it repeatedly bounced off the upper layers of the atmosphere, like a flat stone skipping over the surface of a pond. This 'rocket glider' idea was way ahead of its time, and in fact is still ahead of our time. The 'Silbervogel' (German for 'Silverbird') was the brainchild of Austrian-German aerospace engineer Eugen Sänger. He first published the idea in his 1933 book *Raketenflugtechnik* (*Rocketflight Technology*). In fact, this work was an elaborated version of his original plan for a doctoral dissertation, but his university had advised him to find a more classical topic and he subsequently got his degree for a study of aircraft wings. Rocket pioneer Hermann Oberth had shown that a rocket plane could only hope to achieve a long range by starting with a very steep ascent, leveling off at high altitude in order to use the remaining propellant to attain its maximum speed in the thin air, and then glide back down for the remainder of its flight. Sänger's calculations led him to the same conclusions. His initial rocket plane design involved an aircraft shaped like an elongated bullet with large but thin double-wedged wings (like those of the X-2 subsequently built in the US), a small single tail fin and two small horizontal stabilizers. It bore little resemblance to the blunt-shaped biplanes of that time.

Together with Austrian-German mathematician and physicist Irene Bredt, who he would marry in 1951, Sänger further iterated the design during the 1930s and early 1940s. Initially they intended their plane to transport passengers at hypersonic speeds all over the globe but he also tried to pitch the idea as an intercontinental bomber to the Austrian military in 1933. Not surprisingly, in an age where transonic jet airliners were still science fiction the proposal was rejected, primarily because of the risk of the rocket engine exploding.

The German Luftwaffe, however, did see value in a 'RaketenBomber' (Rocket Bomber), as well as the basic rocket engine development and testing Sänger was already conducting at the University of Vienna. Not wanting to be outdone by the Wehrmacht (Army) and its rocket development by von Braun's team, the Luftwaffe in 1936 invited Sänger to establish a secret rocket propulsion research institute in Trauen, Germany. It was named "aircraft test site Trauen" to hide its true nature, and Sänger was even given a fake job at the University of Braunschweig so that people wouldn't wonder where he was working. Like Wernher von Braun, Sänger funded his work via the military as this was the only source available in Germany during the war. Under his leadership, the Trauen site quickly became an impressive laboratory with a wind tunnel, a rail launch track test rig, and a large rocket engine test stand. Sophisticated experimental rocket engines that ran on fuel oil and liquid oxygen were developed and tested, including components for a 1,000,000 Newton engine intended to propel the Silbervogel. An innovative feature of Sänger's engine designs was the cooling of the combustion chamber by circulating propellant around it. This cooling strategy, sometimes called the 'Sänger-Bredt design', was actually implemented in early rocket engines such as that of the A4/V2, and in most liquid propellant rocket engines since then (including, as we have seen, those used to propel aircraft). For the engine of the Silbervogel, Sänger proposed to use a water loop to cool the massive nozzle and combustion chamber because the propellants he intended to use (fuel oil and liquid oxygen) were poor coolants. The water would be turned into superheated steam by the high temperatures in the engine and then be fed into a turbine to run the propellant pumps and the water coolant pump. After the turbine, the water would be condensed and pumped back into the cooling loop. But having three types of liquids (fuel, oxidizer and coolant) and all the associated pipes, valves and tanks made it a relatively complicated engine.

Although the Nazi government was interested in the Silbervogel as a possible 'Amerika Bomber', for which a number of more traditional aircraft designs were also developed, it was clear that the concept was much too advanced for the technologies available in the 1940s and that the bomber would never be ready in time to affect the war. Even Sänger himself predicted that it would take at least 20 years to make his machine operationally useful. The institute at Trauen was closed down in 1942 when the military gave priority to the V1 pulsejet bomb and A4/V2 rocket developments at Peenemünde, because these would be able to be introduced sooner and hence make a greater contribution to the war effort.

Sänger and Bredt transferred to the DFS institute in order to continue their studies and experiments on ramjet technology. But the Silbervogel never attracted sufficient support and funding, and they ended up spending most of the war developing basic rocket engine technology and a design for a ramjet fighter that never got to fly either. They nevertheless always kept thinking about their favorite concept in private and in 1944 summarized all their design and experiment work in a secret report titled 'Über einen Raketenantrieb für Fernbomber', subsequently translated and published in the US as 'A Rocket Drive For Long Range Bombers'. The concept they finally arrived at was a 100,000 kg (220,000 pound) rocket plane that would be 28 meters long with a wingspan of 15 meters. It featured a pressurized cockpit for a single pilot near the nose, a central bomb bay, and the main rocket engine as well as two smaller auxiliary engines in the tail. The rest of the fuselage would mostly consist of two parallel rows of tanks containing a total of 90,000 kg (200,000 pounds) of liquid oxygen and fuel oil. Two short wedge-shaped wings with swept-back leading edges and a horizontal tailplane that had a small fin on each tip give the Silbervogel a conventional-looking rocket plane shape. However, the design incorporated the innovative idea of having the fuselage itself generate lift through its 'flat iron' shape, hence acting as a partial lifting body.

A powerful rocket sled would be used to help the Silbervogel take off from a long rail track. The sled's rocket engines would not need to be very efficient, since they would only operate for a brief time and would not leave the track with the plane. For the sled, trading efficiency and thus propellant weight for more brute power in order to get the Silbervogel going would involve little penalty. Sänger proposed to use von Braun's A4/V2 rocket engines. The starting speed given by the innovative rail launch system resulted in a smaller, lighter spaceplane requiring a less powerful engine and smaller wings in comparison to a similar vehicle that had to take off on its own (at higher speeds the same amount of lift can be provided with smaller wings). The rail system also meant that the Silbervogel would only need a light undercarriage for its glide landing with empty tanks, thereby further reducing its weight. Furthermore, the rail sled would make detailed knowledge of transonic flight behavior unnecessary, as it would push the plane through this mysterious region of aerodynamics whilst it was still firmly connected to the track (as mentioned above, transonic aerodynamics were not properly understood until well after the war). The rail launch system would limit the possible launch azimuths, but Sänger suggested that it could be used with either end as the starting point. As Germany's enemies were primarily to the west (the US) and the east (the USSR), a single launch track

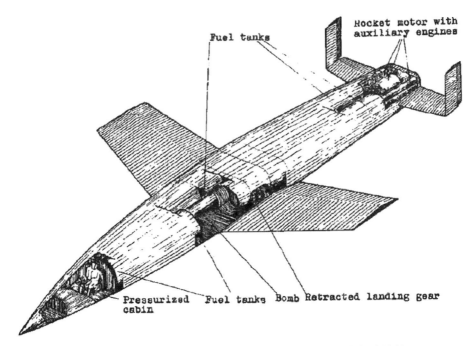

Fuel tanks

Rocket motor with
auxiliary engines

Pressurized
cabin

Fuel tanks

Bomb

Retracted landing gear

Illustration of the Silbervogel in a translated version of the original 1944 report.

aligned east-west would probably have been adequate. The spaceplane would be able to alter its flight direction after leaving the track, but the larger the maneuver the more costly it would be in terms of energy and thus range. The aircraft, the rocket sled, the track and the other infrastructure on the ground would all be reusable.

The 1944 Silbervogel design was to use a 6,000,000 Newton thrust rocket sled to accelerate to a speed of 1,800 km per hour (1,100 miles per hour) on a rail track 3 km (2 mile) long. As this would occur in only 11 seconds it would expose the plane and pilot to a substantial but certainly manageable acceleration of 4.6 G. Leaving the first stage sled behind (which would have brakes to rapidly decelerate to avoid flying off the track) the Silbervogel would climb unpowered to an altitude of 1.7 km (5,600 feet) while turning into the intended direction of flight, then continue gliding up at an angle of 30 degrees. At an altitude of 3.7 km (12,000 feet), 25 seconds after take-off, the onboard 1,000,000 Newton rocket engine would fire and in about 8 minutes push the aircraft up to an altitude of 150 km (500,000 feet). Leveling off, the Silbervogel would then continue to accelerate to a speed of 22,000 km per hour (14,000 miles per hour). At the end of its powered run the aircraft would have consumed almost all of its propellant (90% of its take-off weight) and under the constant thrust of its rocket engine reached its peak acceleration of 10 G, which is about the tolerance limit for a trained pilot. In the climbing phase of the flight the G would build up gradually, as gravity would partially counteract the rocket's thrust, but in level flight the increase from 7 G to 10 G would occur in a mere 20 seconds. Nevertheless, the pilot would sit upright with his back to the rocket engine so that his heart would not have to pump

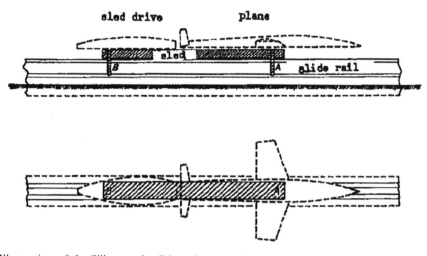

Illustration of the Silbervogel rail launch system in a translated version of the original 1944 report.

blood into his brain against the direction of the acceleration. Knowledge gained from Luftwaffe acceleration experiments on human volunteers and drugged primates led Sänger to believe that such acceleration would not cause a well-trained pilot seated in this position to black out.

The ultimately attained speed would be some 6,000 km per hour (3,800 miles per hour) short of that required to enter orbit, so after reaching its highest altitude the Silbervogel would gradually descend into the stratosphere. However, at an altitude of 40 km (130,000 feet) the increasing air density would generate enough lift to cause the plane to "bounce" back up to about 125 km (410,000 feet) altitude. Part airplane, part satellite, the Silbervogel would repeat this profile a number of times, with each successive bounce getting shallower and covering less distance horizontally owing to the continuous loss of speed resulting from the aerodynamic drag. The aircraft would heat up each time it hit the atmosphere but have time to cool down (by radiating the heat away) during each hop back into space.

The pilot would not have much window space to admire the view because to keep the external shape smooth, and probably also the difficulty of producing windows for extreme temperatures, Sänger foresaw the pilot being provided with side view slits and "optical aids" (presumably periscopes). However, for landing, he noted, "a kind of detachable windshield can be used, since then the pressurization of the cabin and maintenance of the bullet-shape are unimportant". Exactly what this would have involved is not entirely clear, but it seems to suggest the pilot was to open the cockpit and stick his head out when approaching the airfield; a rather awkward and archaic way to land a hypersonic spaceplane.

Sänger's baseline hypersonic bomber, equipped with a rocket engine capable of a specific impulse of 300 seconds, would have dropped a rather small bomb of 300 kg (650 pounds) on New York, then continued its unpowered hop-fly over the continent to land in Japanese-held territory in the Pacific. The Silverbird could only lay a small explosive egg, but the psychological and hence political impact of a German bomb of any size dropped on US soil would have been substantial (as was the first US raid on Japan). With a more efficient engine and/or closer targets, the bomb load could have been increased to several metric tons. After landing as a glider (like the Space Shuttle Orbiter) the aircraft would be re-launched back to Germany, dropping another bomb on the way. From point to point, the range would be a spectacular 19,000 to 24,000 km (12,000 to 15,000 miles). Sänger proposed that it would be possible to stretch this distance even further and enable a landing back at the launch site if the efficiency of the rocket engine could be increased. One possibility to achieve this would be to add metal particles to the fuel in order to increase the exhaust temperature, improving the specific impulse to 400 seconds and so either producing higher thrust with the same propellant consumption or running the engines at the same thrust for longer without increasing the tank volume. The basic idea was sound, because metal particles were later added to solid propellants to increase the performance of boosters such as those used by the Space Shuttle. However, for liquid propellant engines it turned out to be impractical since particles can only be kept well mixed with liquid propellants if they are turned into a gel, and doing that leads to all manner of complications.

The feasibility of bouncing off the atmosphere was inadvertently demonstrated in 1962 by Neil Armstrong when flying the X-15. He came down from a high altitude with an overly steep angle of attack (i.e. with the aircraft's nose too far up), skipped back up and as a consequence severely overshot his planned landing site at Edwards. In fact, he flew all the way to Los Angeles. However, he managed to make a gliding turn with a huge radius and thus fly back to base. Although he bounced only once, he proved the extended range of a Silbervogel-type trajectory. Crewed space capsules returning to Earth can use similar 'skip re-entry' trajectories, typically bouncing only once. The Apollo capsules were capable of flying such a 'skipping' trajectory, but it was never an operational necessity. However, the Soviet unmanned Zond prototype lunar mission capsules did demonstrate it successfully during several flights.

A postwar analysis of the aerothermodynamics of the Silbervogel discovered that Sänger and Bredt had made an error in the heat load calculations: during the first re-entry into the atmosphere the structure of the spaceplane would have become much hotter than they predicted and consequently would have needed additional protection in the form of heat shield material which would have made the vehicle much heavier than the original design, leaving little room in terms of weight for people or bombs.

After the war Sänger and Bredt were recruited by the French government to work on missiles and ramjets. In 1949 Sänger founded the Fédération Astronautique, and in 1951 he was made the first president of the International Astronautical Federation that hosts the annual International Astronautical Congress. Meanwhile in the Soviet Union, Stalin had become fascinated by the Silbervogel concept, which had come to his attention through three captured copies of Sänger and Bredt's highly secret 1944 report. Stalin saw it as a possible means of attacking the United States because the advent of nuclear weapons meant that the puny bomb load of Sänger's design could now yield a very powerful punch. In 1947 Stalin sent his son Vasily together with aviation experts Serov and Tokayev to either win over or kidnap Sänger and Bredt and bring them to Russia to further develop their concept. However, Vasily overly enjoyed his playboy life in Paris and Tokayev defected to the British and blew the operation, so nothing came of it. By then, however, the design bureau NII-1 NKAP, established by Mstislav Keldysh in 1946 specifically to work on a Soviet version of the Silbervogel, had further elaborated on the idea presented in the captured report. By 1947 they had concluded that with the rocket engines and structures which either already existed or were likely to become available in the near future, some 95% of the initial weight of the spaceplane would have to be propellant; Sänger and Bredt had calculated 'only' 90%. This would leave very little weight for the structure, tanks, engines, and other essential equipment. The poor ratio between take-off weight and structural weight is a well known issue for any spaceplane concept, even today, and particularly so for a pure rocket propelled vehicle.

The Russians decided that ramjets using atmospheric oxygen instead of internal oxidizer would make the design more feasible: the weight limitation of the vehicle without propellant (i.e. the dry weight limit) would increase from 5 to 22%, while the lower maximum speed in comparison to the Silbervogel would still give the plane an intercontinental range of 12,000 km (7,500 miles). This so-called 'Keldysh Bomber' would have a take-off weight of about 100,000 kg (220,000 pounds). It would start with an 11 second run along a track that was 3 km (2 mile) long, pushed by a rocket sled powered by five or six liquid oxygen/kerosene RKDS-100 rocket engines with a combined thrust of 6,000,000 Newton. The bomber would climb to an altitude of 20 km (66,000 feet) and accelerate to Mach 3 using a single RKDS-100 and a pair of wingtip-mounted kerosene ramjets. The ramjets would then flame out due to a lack of oxygen and possibly be jettisoned to shed their weight. Next, the rocket engine would push the vehicle to a speed of 18,000 km per hour (11,000 miles per hour) and up into space, after which there would be a series of atmospheric skips just like the Silbervogel. It was estimated that developments in rocket and aircraft technologies would make it possible to start serious development of the design in the mid-1950s. But by then the concept of the intercontinental ballistic missile, which was easier and cheaper to develop and did not require a far-away landing site, had made the Keldysh Bomber obsolete.

Walter Dornberger, the former head of A4/V2 development in Germany, tried to interest the US military in the Silbervogel concept when he joined Bell Aircraft in 1950, carefully calling it the 'Antipodal Bomber' rather than its wartime moniker of the 'Amerika Bomber' ('antipodal' meaning two sites exactly opposite each other on the Earth's surface; more or less the Silbervogel's maximum flight distance). But the sheer technical complexity of advanced materials, hypersonic aerothermodynamics and guidance accuracy, together with the doubtful military value of such an aircraft, meant that a real project never materialized. Nevertheless Dornberger's lobbying did eventually lead to the X-15 rocket plane, the X-20 Dyna-Soar concept and ultimately the Space Shuttle. (The Shuttle is not a true rocket plane and it doesn't use a skip re-entry trajectory but it does resemble Sänger's design in that it has a flat base, straight fuselage sides, a rounded top, and a series of massively powerful rocket engines in the back.) The X-15 and the Shuttle incorporated several features that Sänger had not foreseen but whose inclusion would have increased both the weight and complexity of the Silbervogel. One example is a reaction control system to control the spaceplane outside the atmosphere and to properly orientate it for re-entry. Another is the need for precise guidance control sensors and electronics. Although one X-15 pilot managed to fly a manual re-entry when his guidance control system failed, for a safe re-entry both the X-15 and the Shuttle required guidance computers that were not available in the early 1950s. Such assistance would certainly have been required for the complicated bouncing trajectory of the Silbervogel.

As a passenger transporter, the original role Sänger intended for his rocket plane, the Silbervogel would have been much too complex and expensive because of its huge propellant consumption, its elaborate take-off and landing infrastructure, and its payload mass and volume allowing only a few passengers. These considerations still plague recent concepts for hypersonic airliners such as the US National Aerospace Plane (NASP) of the late 1980s. The fact that even the less complex and more economical Concorde never reached its financial break-even point makes it difficult to see how a hypersonic passenger transport vehicle could ever be a financial success.

Sänger returned to Germany in 1954 and three years later became director of the newly-created Institute for the Physics of Jet Propulsion in Stuttgart. However, when in 1961 the German government found out that Sänger was secretly making trips to Egypt, presumably to assist a group of rogue German engineers to develop missiles that could be used to attack Israel, they forced him to resign. The loss of his position, and scant prospect that his Silbervogel would be built during his lifetime, sent him into a state of depression and his health rapidly deteriorated. Sänger was unofficially rehabilitated when it was realized he had only acted as a consultant for an Egyptian meteorological sounding rocket project and had lectured at the university of Cairo. He then continued his work by assisting with a new spaceplane project at the revived Junkers company. Based on decades of research, he proposed a 200,000 kg (450,000 pound) reusable, two-stage vehicle consisting of a delta-wing carrier rocket plane and a smaller delta-wing orbital spaceplane. A steam-rocket sled was to provide a starting velocity of about 900 km per hour (560 miles per hour) at the end of a track 3 km (2 mile) long. The steam rocket would be relatively simple, using only heated water, but not very efficient in terms of performance. However, as with the original Silbervogel's sled, that would not be particularly important for the overall spaceplane system's performance. Its simplicity would save development effort and cost, whilst making it reusable would be relatively straightforward due to the easy-to-handle and cheap propellant (water) and the absence of extremely hot exhaust gases to eat away at the nozzle.

The orbital vehicle, called 'HORUS' (Hypersonic ORbital Upper Stage), would separate from the carrier at an altitude of 60 km (200,000 feet), then proceed into a 300 km (980,000 feet) orbit while the carrier plane returned to Earth. Having fulfilled its mission, HORUS would de-orbit and glide back to Earth very much like the Space Shuttle Orbiter. Both planes would be powered by similar rocket engines, three on the carrier and one on the orbital plane, each engine delivering 2,000,000

Newton of thrust at sea level and burning the powerful combination of liquid oxygen and liquid hydrogen. This is another similarity with the later Space Shuttle, whose three main engines used those propellants and each had a sea-level thrust of 1,800,000 Newton. In fact, the entire concept has similarities to some of the Space Shuttle designs of the early 1970s. Designated the Junkers RT-8, this reusable launcher was to be able to place a 3,000 kg (7,000 pound) payload into low orbit. An improved version would later use a single-stage spaceplane with an integrated rocket-ramjet engine.

In October 1963 Sänger also accepted a professorship at the Technical University of Berlin where in February 1964, aged 58, he died while giving a lecture. Shortly before his death he wrote that he believed the US and the USSR would direct their full technological capacity towards an 'Aerospace Transporter' similar to the RT-8 as soon as the 'Moon Race' ended and that "there is therefore at the moment a unique but only a short-lived opportunity for Europe, with its great intellectual and material resources, to become active in a sector of spaceflight in which the major space powers have not yet achieved an insuperable lead". His last spaceplane design, which after his death evolved into the 'Sänger-I' RT-8-02, would take off vertically without the need for a launch sled and rail, but it never left the drawing board. Junkers was unable to secure enough support from the government since the ambitious, expensive and economically risky project was deemed to be a step too far.

Like Max Valier, but unlike most other rocket and spaceflight enthusiasts of the time, Sänger was a real rocket plane pioneer: he fully intended his rocket engines to propel rocket aircraft rather than ballistic rockets such as the A4/V2. Sänger was certainly a genius but, like Wernher von Braun, his willingness to work for the Nazi military in order to move his dream machines closer to reality, even if it meant they would be used to bomb innocent civilians, has made him a rather controversial figure (his 1944 report contains a map of New York displaying the expected efficiency of a bombing campaign by a fleet of Silbervogel planes). Irene Bredt worked for the institute in Stuttgart until 1962 and survived her husband by 19 years. In 1970 she was honored with the Hermann Oberth Gold Medal for her scientific contributions.

Sänger and Bredt's ideas have certainly influenced rocket plane design since the end of the war: the idea for the Space Shuttle can be traced back (via Dyna-Soar and the Bell 'Antipodal Bomber') to their revolutionary design. In his 1963 book, which was published in 1965 in English as *Space Flight; Countdown for the Future*, Sänger cites all of the reasons why, in his opinion, "aerospace planes" hold the future of spaceflight instead of expendable missiles: lower costs per flight, the ability to flight-test each type of production vehicle, the possible use of airfields in densely populated countries, easy self-transportation of the vehicle from the production plant to launch site, and the ability to use them for both orbital missions and long-distance flights. These are still the principal reasons for pursuing spaceplanes although nowadays the need for (and the advantages of) such vehicles seem less obvious. Certainly nothing resembling the Silbervogel has yet been built. In the same book he writes, "military aerospace planes will be used as reconnaissance planes, fighter planes against Earth satellites and extraterrestrial space stations, satellite inspections planes etc" and also "we can possibly count space fighter planes among the most important of all space weapons". At that time work on the X-20 Dyna-Soar was still ongoing, but after its cancellation work on such military orbital vehicles all but ceased (with the exception of the Space Shuttle, which was partially based on military requirements). However, considering the current renewed interest by the US Air Force in hypersonic vehicles like the X-51 and the Falcon Hypersonic Technology Vehicle, as well as automated orbital shuttles like the X-37, Sänger may well have been right on this after all.

FICTION

While on the subject of future spaceplanes let us have a look at how these were (and still are) depicted in fiction. The existence of spaceplanes as an efficient means of transportation between Earth and space is generally taken for granted in science fiction, much like airliners are used in the real world. Apart from ideal aircraft-like operations, one thing that most of the spaceplanes in science fiction books, movies and television series appear to have in common is an amazingly efficient propulsion system because most of the volume is available for passengers and cargo rather than being taken up by bulky propellant tanks.

Take for instance the Orion III 'Pan Am Space Clipper' of the famous movie *2001, A Space Odyssey* of 1968. We are shown a spaceplane that operates much like an airliner, even with stewardesses. No take-off is shown in the movie but it appears that the Orion is a single stage vehicle with double-delta wings (similar to those of the Space Shuttle, which had not yet been designed when the movie was produced). Windows cover only a short stretch of the middle of the fuselage, which makes sense for a spaceplane requiring large volumes of propellant. Still, as a relatively small single-stage-to-orbit spaceplane (judging from the size of the windows) *2001*'s Orion must have fantastically efficient engines; jet engines, if the series of holes that can be seen on the leading edges of the plane's wings and in front of the engine module are air intakes. However, having intakes on the wing leading edges would be rather poor engineering: tapping off the air before it has a chance to flow over the wings would seriously reduce lift. Nevertheless, Orion is probably the most believable spaceplane ever depicted in any block-buster movie.

The most famous space fighters of the movies, the *Star Wars*' X-wing and *Battle Galactica*'s Viper, have wings and very apparent air intakes but spend most of their time in the vacuum of space. Even so, they seem to require only very small amounts of propellant as there are no large tanks to be seen. This does not seem to make sense: if you can fly and maneuver in space for hours without any external airflow, why bother with air intakes and wings for the brief periods in an oxygen-rich atmosphere? Moreover, these vehicles seem to maneuver in space as if they were flying through air, banking into nice round turns, flying in loops etc. In space there are of course no aerodynamic lift and drag forces, making such maneuvers only possible with the help of a serious set of reaction control thrusters which aren't apparent in these fictional designs. Moreover, such airplane-like actions are rather useless in space (but not in a movie, as it certainly looks more exciting than real orbital mechanics).

Fireball XL5, a spaceplane shown in the British 1962-1963 television series of the same name also lacks apparent propellant tank volume, but according to the series it is powered by a "nutomic reactor" that enables it to have a range of many lightyears. Although the Fireball is a fictional vehicle without any regard for the limitations of real spaceflight, it is interesting to note that it takes off using a rocket powered sled and a mile-long launch rail that culminates in a 40 degree "sky ramp", whereupon the spacecraft uses its own propulsion to ascend into space; very similar to the take-off mode of Sänger's Silverbird.

In reality, even although serious work on orbital spaceplanes has been underway since before the Second World War, they still only fly in the realms of fiction. Like most of the hardware depicted in *2001, A Space Odyssey*, nothing like an Orion III was available in the real year 2001; nor in 2011 for that matter. Only the videophone used the movie is actually better in today's reality. In general, science fiction often overlooks the complexity of spaceplane transportation, and especially the limitations of propulsion technology. Of course, with engines working on fantasy and built out of unobtainium, anything is possible. What the spaceplanes of the movies do show us, however, is the ultimate goal of the real-world launch business: airline-like flights into space.

SPACEPLANE REVIVAL

By the mid-1980s it had become clear that a spaceplane based on rocket propulsion only was very difficult to achieve, and that introducing airbreathing engines for at least part of the flight would be necessary. Whilst a single-stage orbital rocket plane must consist of at least 90% propellant, for an efficient airbreathing spaceplane this drops to less than 70%. In other words, for the same payload and propellant weight the structure weight of an airbreathing spaceplane can be more than tripled to over 30% instead of 10% of the overall weight. The dramatic relaxation of this constraint just might make a single-stage spaceplane possible, although it would require a propulsion system significantly more complex than a pure rocket design.

In Germany in 1985, Messerschmitt-Bölkow-Blohm (MBB, with which Junkers had been merged in the meantime) revived the idea for a two-stage, horizontal take-off and landing (HTHL) spaceplane. They named the new concept the 'Sänger-II', and its goal was to lower the launch price of satellites and other cargo by a factor of three to ten. As with the Sänger-I it would involve a large, delta-winged hypersonic carrier aircraft and a smaller orbital vehicle, but the carrier would take off from a regular runway and instead of rocket engines it would use airbreathing turboramjets with liquid hydrogen as fuel. Since no oxidizer would be carried, this would result in great mass and volume savings. However it required five very complicated engines that would work as turbojets at speeds up to Mach 3.5 and then as ramjets at higher velocities (the Lockheed SR-71 Blackbird had turboramjet engines but the Sänger-II carrier would be required to fly much faster than the SR-71's top speed of Mach 3.2). It would be 84 meters (275 feet) long and have a wingspan of 41 meters (135 feet). After taking off from a runway in Europe the carrier aircraft would fly at Mach 4 to

Model of the Sänger-II concept [MBB].

the appropriate latitude for its intended orbit (typically south of Europe, closer to the equator) at an altitude of 25 km (82,000 feet) in order not to pollute the critical ozone layer of the atmosphere. Next it would accelerate to Mach 7 and fly up to an altitude of 31 km (100,000 feet) to release its upper stage, which would use its own rocket engine (burning liquid oxygen and liquid hydrogen) to continue into orbit while the carrier glided back to base. In a modified form the first stage would also be able to function as a hypersonic airliner capable of flying 230 passengers at a top speed of Mach 4.4 for a distance of 11,000 km (6,500 miles); e.g. Frankfurt to Cape Town in under 3 hours. The upper stage would either be a non-reusable, unmanned 'Cargus' with up to 15,000 kg (33,000 pounds) of payload, or a HORUS reusable shuttle with a two-person crew and either 36 passengers or 3,000 kg (7,000 pounds) of cargo. The maximum lift-off weight of the Sänger-II would be 350,000 kg (770,000 pounds).

The German government funded a concept study in order to refine the design, as well as a technology development program that led to ground runs of Europe's first turboramjet engine at MBB in 1991. However, it was concluded in 1994 that full development would be much too costly and the operational launch cost savings in comparison to those of expendable launchers too uncertain, so the entire project was canceled. All that seems to be left are a large scale model of the spaceplane and a laboratory-sized ramjet demonstrator, both of which are on display in the German Technik Museum Speyer (close to the jet-engined Buran shuttle).

The Sänger-II project was part of a kind of spaceplane revival that began in the mid-1980s and ended abruptly in the mid-1990s. This renewed interest was prompted by the rationale that something new was required in order to cut the costs of access to space in comparison with uneconomical expendable rockets and the Space Shuttle, and also that the technology necessary for this was now within reach. In parallel with Germany's Sänger-II the US devoted a lot of effort to the aforementioned NASP, the British worked on their HOTOL spaceplane, and France, Japan and Russia were all independently working on reusable aircraft-like launch vehicles. It appeared that the days of the expendable launch vehicle were finally numbered.

Star-Rakers loading cargo at an airport [Rockwell International].

Even as the Space Shuttle was being developed, in the US in the 1970s there were many studies of a possible successor in the form of a single-stage-to-orbit reusable launch vehicle. The intrinsic weight issues led to the conclusion that complicated tri-propellant rocket engines would very likely have to be created for rocket propelled spaceplanes. These would initially use low-energy but high-density kerosene as fuel to generate high thrust for take-off with a limited tank volume, and then low-density but high-energy liquid hydrogen to efficiently accelerate up to orbital velocity. Sled-launch systems such as Sänger envisioned were also seen as a potential solution; for instance the Reusable Aerodynamic Space Vehicle single-stage spaceplane proposed by Boeing, which envisaged using two Space Shuttle Main Engines for propulsion. Rockwell International offered an alternative concept called the 'Star-Raker', a delta-winged HTHL SSTO with ten (!) "supersonic-turbofan/air-turbo-exchanger/ramjet" engines, three large rocket motors and an undercarriage that would be jettisoned and recovered by parachute. The company issued colorful illustrations showing several Star-Rakers at a commercial airport with their hinged noses open to load cargo, thus emphasizing airline-like operations.

The various studies by NASA, the US Air Force and contracted industries in the 1970s led to a classified military program called Copper Canyon (as with most secret military programs the meaningless name was intended to mask what it was about) that ran between 1982 and 1985, and out of which the US National Aerospace Plane (NASP) emerged as the less classified follow-on announced by President Reagan in his State of the Union of 1986. NASP was to lead to an air-breathing scramjet HTHL spaceplane prototype designated the X-30, operational derivatives of which would be able to function as a single-stage-to-orbit launch vehicle, a hypersonic airliner called the Orient Express (somewhat similar to the dual use the Germans had in mind for the first stage of their Sänger-II), or a military Mach 12 reconnaissance plane and/or strategic bomber offering the response speed of a ballistic missile and the flexibility, accuracy and 'recallability' of a bomber. (Recallability means that you can change your mind about blowing some place to bits when the bomber is already on its way, something impossible with a ballistic missile.) Early artistic impressions showed an elegant Concorde-like design which looked as sleek and fast as it was intended to become.

The program was jointly run by NASA and the Department of Defense. In 1990 Rockwell International became the prime contractor for its development. By then the spaceplane had grown considerably in weight and size in comparison to the original design of 1984. Having lost its resemblance to the Concorde, it now had a wedge-shaped aerodynamic configuration called a 'waverider' (essentially a hypersonic surfboard) with most of the lift being generated by a shock wave compressing the air below the plane. This shock wave, created by the forward fuselage, would also compress the air before it entered the engines, effectively

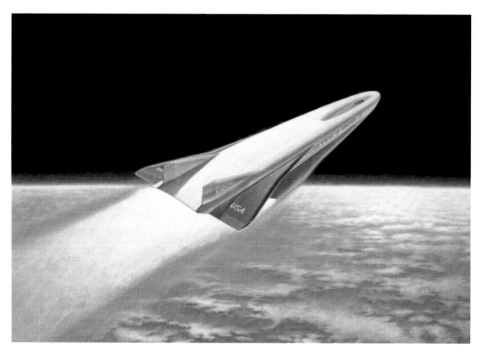

NASP as NASA imagined it in 1990 [NASA].

supplying the scramjets with more oxygen. The aft fuselage formed a gigantic integrated nozzle for expanding the scramjet's exhaust. There were small wings to trim the aircraft and provide control. Its overall configuration was ideal for efficient high-speed flight but gave poor lift at low speeds and in particular for taking off.

Much new technology was required, including a lightweight, composite-material hydrogen tank and advanced computer programs for modeling the airflow around the aircraft and through the engines. The hydrogen fuel would require to be carried in the form of a slush (liquid and ice mixture) to limit the volume of the propellant tanks and therefore the size, and most importantly weight, of the aircraft's structure (a kilogram of hydrogen ice has a lower volume than a kilogram of liquid hydrogen). Heat-resisting carbon materials would be needed for the aerodynamic surfaces that would endure temperatures over 1,700 degrees Celsius (3,000 degrees Fahrenheit) during hypersonic ascent and atmosphere re-entry, and titanium aluminide panels for most of the fuselage. A major hurdle was the development of the scramjet in which liquid hydrogen would be injected into the combustion chamber and be ignited by the hot compressed air rushing in at hypersonic speed. The exhaust would primarily consist of water vapor and be environmentally friendly but the decision to use hydrogen as fuel was mainly driven by the need for high performance and high efficiency.

The NASP design incorporated the clever idea of using the atmospheric heating of the vehicle to increase the thrust of its scramjet engines: by circulating hydrogen fuel through the plane's skin to warm it up prior to injection into the engine the energy generated by atmospheric drag was effectively added to the thrust of the scramjet. At the same time the cryogenic hydrogen flow would cool the aircraft. It was initially believed that this revolutionary scramjet propulsion and temperature control system would make it possible for NASP to reach Mach 25 in the high atmosphere; enough for it to achieve orbit without the need for additional rocket engines. However, as the development progressed it became clear that the maximum speed attainable would be about Mach 17, as at higher speeds the weight of the active thermal control system would exceed that which would be required to add conventional rocket engines and propellant. Hence for NASP to be used as a single-stage-to-orbit spaceplane it would need additional rocket propulsion. In fact, rocket propulsion would be required also to get NASP up to Mach 3 for the scramjets to take over. And rocket propulsion would be required to perform the deorbit maneuver at the end of an orbital mission. To keep the vehicle sufficiently light, it was planned to use a new type of rocket engine that, apart from onboard fuel, would be fed with air liquefied during atmospheric flight by a combination of ram-compression and cooling by liquid hydrogen; an airbreathing rocket engine known as LACE ('Liquefied Air Cycle Engine'). Updated requirements stated that the X-30 must carry a crew of two and that, although it was an experimental vehicle, it must also be able to deliver a small payload into orbit. Hence rather than a demonstrator, the X-30 was to be a semi-operational vehicle. It was supposed to be able to fly a mission every 72 hours (compared to a maximum of one per month for the Space Shuttle) and require only 100 workers for its operations. The ensuing fully operational launch vehicle derivative was expected to cut the cost per kg of payload by a factor of ten.

Meanwhile the estimated cost of the full development into an operational vehicle had increased far beyond the originally projected $3.3 billion for a relatively modest technology demonstrator. By the early 1990s the projected cost for the demonstrator was $17 billion and the fully operational launch vehicle would require another $10 to $20 billion (in 2011 dollars those numbers would be close to 27 and 16 to 32 billion, respectively). It was also expected that another two decades of development would be required to master all the relevant technological issues prior to building a working prototype. Moreover, the argument that NASP would enable airplane-like operations that would result in low costs per flight and rapid turn-around times sounded awfully similar to those predicted but never achieved for the Space Shuttle. The use of liquid hydrogen alone would mean a complete departure from conventional airport storage and distribution facilities, essentially ruling out the use of normal airfields because it would be prohibitively expensive to equip a sufficiently large network of them with the required production, storage and handling facilities. The severe cost increase and schedule stretch, uncertain operational benefits, and the necessity to comply with too many (civil space, commercial airline and military) requirements eroded support for NASP below the critical level. In addition, the collapse of the Soviet Union severely reduced the push in the US for ambitious technical and military programs. Inevitably, it was terminated in 1993.

NASP was initially superseded by the Hypersonic Systems Technology Program (HySTP) in which NASA and the Department of Defense continued with technology development on a less ambitious scale, but when the Air Force withdrew in 1995 the development of a US spaceplane pretty much expired.

During the winter of 1993-1994 the Air Force's Phillips Laboratory conducted a six-week study of an interesting alternative called 'Black Horse' which called for a single-stage rocket propelled spaceplane using hydrogen peroxide and kerosene. This non-cryogenic propellant combination is better suited than cryogenic fluids such as liquid hydrogen and liquid oxygen for rapid-reaction launches for military purposes, and furthermore has a much higher density leading to smaller tanks and therefore an overall smaller, lighter aircraft. But it lacks the specific impulse required for a single-stage spaceplane. To compensate for this lack of performance (and made practicable by the use of storable fluids) once the Black Horse was airborne it would rendezvous with a large subsonic tanker aircraft (a converted airliner) for aerial refueling prior to continuing into orbit. The spaceplane would effectively start its flight not from the ground at zero speed but from 13 km (43,000 feet) and Mach 0.85 (a more advanced system could involve a newly developed supersonic tanker able to do the refueling at a much higher speed). As a pure rocket plane the structure weight constraints for this Aerial Propellant Transfer (APT) vehicle where nevertheless severe. And although aerial refueling is standard procedure for agile military jet fighters, doing it with an 84,000 kg (185,000 pound) rocket plane would be a different matter entirely. In early 1994 Martin Marietta investigated a near-term suborbital X-plane with turbofans and liquid oxygen/kerosene rocket engines to demonstrate the APT concept. Flying up to about half orbital speed, Mach 12, this aircraft could fire an expendable upper stage to insert a satellite into low orbit, or use a Sänger-type atmospheric bounce trajectory to fly for long distances or drop bombs. Being about half the size of the Black Horse design it was called 'Black Colt'. Several people who worked on these studies went on to establish the Pioneer Rocketplane company in 1996 to develop a commercial satellite launcher based on ATP called 'Pathfinder'. However, they quickly found out that a switch from hydrogen peroxide to liquid oxygen as oxidizer was required to turn Black Colt into an effective launch vehicle. In the late-1990s, just as the size and complexity of the vehicle increased, the intended market of launching constellations of communications satellites evaporated due to developments in less costly terrestrial mobile phone networks. Without a strong demand, the Pathfinder could no longer be pursued.

The British HOTOL (short for 'Horizontal Take-Off and Landing') was a concept for an unmanned, single-stage-to-orbit spaceplane able to use ordinary runways. It would require the development of a novel propulsion system that combined turbojet, ramjet and rocket engine elements. As with the Sänger-II and X-30/NASP, it would use hydrogen as fuel and draw upon atmospheric oxygen as much as possible in order to save the weight and volume of oxidizer that the vehicle would otherwise be required to carry.

In 1982 rocket engineer Alan Bond came up with the idea for a new type of engine combining airbreathing and rocket propulsion that he thought would enable a spaceplane to consist of only one single vehicle (like NASP). Around the same time Robert Parkinson of British Aerospace (BAe) was conducting a study of a reusable launch vehicle. Bond's engine design, the rights for which had by then been bought by Rolls Royce, was combined with BAe's launch vehicle concept and in 1985 the BAe/Rolls Royce HOTOL project was born. It became an official national program in 1986 when the government decided to fund a 24 month proof-of-concept study led by Parkinson and John Scott of Rolls Royce.

The design that emerged looked somewhat like a torpedo with small delta wings at the rear, a single moveable fin up near the nose, and air intakes at the aft-bottom fuselage. It would be powered by four of the Rolls Royce RB545 Swallow engines, which in the atmosphere would use liquid hydrogen to pre-cool the hot air entering the engine and thereby make unusually high compression possible. The air initially entering the engine would have a temperature of about 1,000 degrees Celsius (1,800 degrees Fahrenheit) because it would arrive at high speed and then be slowed down almost to a standstill, resulting in almost all kinetic energy being converted into heat. Subsequent compression in the engine for efficient combustion and thrust generation would increase the air temperature even further, so starting off with the hot air would quickly lead to unacceptably high temperatures in the engine. However, using pre-cooling the compression would start with air which had been chilled to minus 130 degrees Celsius (minus 200 degrees Fahrenheit), yielding less extreme temperatures after high compression. In addition, cooling the air would avoid the need for heavy, high-temperature materials in the compressor section of the engine. Some of the now relatively warm hydrogen coming out of the air pre-cooler system would be used to drive the engine's hydrogen turbopump while the rest would be burned together with the cooled air in the turbojet section of the engine at relatively low speeds and in the ramjet section at supersonic speeds. At an altitude of 26 km

The British HOTOL concept [BAe].

(85,000 feet) and flying at Mach 5 the engines would be switched to pure rocket mode to burn hydrogen with onboard liquid oxygen instead of air. Unlike NASP, there would not be any scramjet. The return flight would be completely unpowered.

From the very start, HOTOL was to be able to fly into space and back completely automatically, without a crew and with limited contact with ground control stations. Any astronaut passengers would merely be payload in a special container that would fit into the cargo bay located in the middle of the fuselage.

The development of HOTOL soon came up against a typical spaceplane issue that must have also plagued NASP and Sänger-II: the large movement of the center of gravity as the flight progressed. Most of the propellant was housed in the forward and center fuselage whilst the heavy and fixed weight of the propulsion system was all necessarily placed at the aft end of the vehicle. This meant that the plane's center of gravity moved significantly aft as propellant was consumed. At the same time the large range of Mach numbers meant the aerodynamic center of air pressure moved significantly during the flight: it being aft while accelerating during the ascent, and forward while slowing down upon return. Another complicating factor was that the plane would take off with virtually full tanks but would later approach the airport and land with empty tanks and very little payload, so the center of gravity would thus be completely different at the same relatively low velocities at the start and the end of a mission. Balancing the centers of gravity and pressure became very tricky for the HOTOL lay-out, forcing the designers to modify the fuselage and to locate the wings far aft in order that their lift could counteract the weight of the engines and keep the plane stable at all speeds and propellant loads. But the new configuration that solved the balancing problems required a reduction of the payload. This was compensated by abandoning the conventional aircraft take-off in favor of a rocket propelled trolley (resulting in a similar take-off approach to that Eugen Sänger had in mind for his Silbervogel). This was a serious departure from the originally envisaged aircraft-like operations capability and significantly lowered the commercial attractiveness of the project. And when the design team nevertheless found that they had to resort to all manner of untried, experimental materials and structure technologies to maintain a decent payload at a reasonable flight price this caused the government to withdraw its support in 1988, and shortly thereafter Rolls Royce also pulled out. In an effort to save the project in 1990 BAe approached the Soviets to study launching a HOTOL-derived vehicle off the back of their Antonov An-225 (as planned for their MAKS concept). This so-called 'Interim HOTOL' would also abandon the complex RB545 engines and instead use conventional rocket engines, possibly of Russian pedigree. Thus HOTOL would become a pure rocket plane launched by a subsonic carrier first stage that would release it at an altitude of 9 km (30,000 feet) and a starting velocity of about Mach 0.7. Antonov also studied the possibility of fitting the An-225 with two additional jet engines (turning it into an An-325) to carry a larger version of the vehicle. But neither the UK government nor the European Space Agency expressed any serious interest and the disintegrating USSR could ill afford to participate, so in 1992 the project was canceled.

Whereas Sänger-II was to be a jet-engined hypersonic carrier aircraft for a rocket propelled upper stage, and NASP was initially conceived to be an airbreathing-only hypersonic spaceplane, the original HOTOL was to fly itself into space using rocket motors and would therefore have been a real rocket plane in that respect (albeit with additional jet propulsion). On the other hand it was to fly unmanned, making it more like a missile than a plane capable of being controlled by a pilot at will. For these spaceplanes the distinction between a jet aircraft, a pre-programmed launcher and a rocket plane becomes rather ambiguous.

The Soviets also had their own aerospaceplane projects. From the mid-1970s to the late 1980s the Myasishchev Experimental Design Bureau worked on the MG-19, a rather megalomaniac concept involving a triangular lifting body with a 500,000 kg take-off weight and a 40,000 kg payload to low orbit. It would use turbojets and then scramjets to get to Mach 16, and subsequently rocket propulsion with hydrogen fuel superheated by a nuclear reactor to achieve orbit. The complexity of the vehicle, the dangers involved in flying a nuclear reactor in a hypersonic aircraft, and the priorities of the Buran space shuttle project meant that by the 1990s this project was gone. The idea of using nuclear power in a spaceplane was not new (since the aforementioned Martin Astroplane with its nuclear magnetohydrodynamic engines had already been proposed in the US in 1961) but it is surprising that Myasishchev still considered it to be a sane solution as late as the 1980s.

The USSR's specific response to NASP was the Tupolev Tu-2000, a long-range heavy bomber and single-stage-to-orbit vehicle whose turbojet/scramjet/rocket engines used liquid hydrogen and liquid oxygen. Development of this spaceplane started in 1986, aiming for an initial experimental two-person design called the Tu-

2000A that would be capable of reaching Mach 6. After the collapse of the Soviet Union work continued in Russia with two tests of an experimental, sub-scale scramjet at subsonic and supersonic speeds up to Mach 6 using S-200 tactical missiles in 1991 and 1992. However, in 1992 the project was suspended owing to a lack of funds. From the few concept images available it appears that the Tu-2000 would have looked very similar to NASP.

In Japan some spaceplane development was ongoing during the late 1980s and early 1990s under the designation 'Japanese Single Stage To Orbit' (JSSTO). Four LACE propulsion units were to drive the vehicle to Mach 5, then six scramjets would accelerate it to Mach 12. Further acceleration to orbital speed would also be achieved using the LACE engines, but now being above the 'sensible' atmosphere the engines would be fed air that was liquefied and stored in tanks earlier in the ascent, a process called ACE ('Air Collection Engine'). Mitsubishi Heavy Industries tested a sub-scale scramjet in a hypersonic wind tunnel in 1994 but the work does not seem to have progressed much further than that.

In a research effort named STS (Space Transportation System) 2000, Aerospatiale in France in the late 1980s and early 1990s investigated a single-stage ramjet/rocket spaceplane that looked similar to the Concorde, as well as a Sänger-II-like concept in which a ramjet/rocket aircraft carried a rocket propelled second stage spaceplane that would be separated at Mach 6. Another French company, Dassault Aviation, worked on a Mach 7 scramjet aircraft that would air-launch an expendable Ariane 5 second stage carrying a Hermes-derived shuttle. This was called STAR-H, for 'Système de Transport spatial Aérobie Réutilisable – Horizontal' (i.e. airbreathing reusable space transportation system for horizontal take-off and landing). These French concepts were not particularly new, all resembling (aforementioned) ideas from the mid-1960s such as the European Space Transporter, the Dassault Aerospace Transporter and the various two-staged spaceplanes studied by the British Aircraft Corporation.

Of all the spaceplane projects of the 1980s, NASP, a single-stage-to-orbit vehicle with (at least initially) only airbreathing propulsion, appears to have been the most ambitious and most complex. Nevertheless, despite the large amount of technology development that was carried out, not much of this is in the public record; the project was so advanced that even today many of its details remain secret. The Sänger-II was theoretically the least complex by using two stages instead of the more constraining single-stage-to-orbit approach and incorporating airbreathing and rocket propulsion in two separate vehicles. Sänger-II's first stage would not fly fast enough to require exotic thermal protection materials and scramjets; conventional titanium panels (with additional carbon layers at the hot-spots) and relatively simple ramjets would suffice. During re-entry, the vehicle, being mostly large and now empty tanks, would have a significantly lower density than the Space Shuttle Orbiter (hence a larger surface area in comparison to its weight) and so would encounter relatively benign temperatures. In contrast to NASP it would have been able to put substantial payloads into orbit and operate as a real launch vehicle, potentially beating everybody else to the market. However in the end all of the spaceplane concepts proved to be too far ahead of their time: scramjet and combined propulsion technology was not yet sufficiently developed and the computer programs for simulating airflow and combustion at high speeds and temperatures were not mature enough. An article on NASP in the magazine *Flight International* in October 1987 had already included an ominous warning: "The odd thing is that the excitement is based almost entirely on theoretical research and small-scale laboratory work. Nobody has run a Mach 25 scramjet continuously for more than a few seconds, and no powered atmospheric vehicle has attained anything like the speeds envisioned for NASP." Nevertheless, in March 1992 *Popular Mechanics* enthused "Space Race 2000 is on", anticipating one of the international spaceplane contenders developing a real vehicle by the turn of the century. But by the mid-1990s all the spaceplane projects had more or less gone: the ideal, airliner-like spaceplane as described in the Introduction of this book was (and still is) a long way off.

Part of the issue is that whereas for a conventional aircraft the design can be split into elements like fuselage, wings and engines, and each can (to some extent) follow an independent development track, a spaceplane requires a fully integrated approach in which any small change of one item can have dramatic implications for the rest of the design. For HOTOL for instance, it turned out that for each additional kilogram of inert weight that the vehicle design gained (say, an additional piece of electronics or thermal insulation) some 25 kg of additional propellant would be needed to restore the vehicle's performance; and additional tank and structure mass would be required to house this extra propellant, which itself further increased the overall weight. It was a vicious circle. Another parameter with a dangerously strong growth factor was the specific impulse of the rocket propulsion, where a change of just 1% would impose a 4% change in the gross lift-off weight. Therefore a slight increase in rocket engine performance would pay off but over-optimism concerning the engine's performance could make the vehicle considerably heavier, larger and more expensive. Accepting a slight reduction in speed to compensate for weight growth is possible in the design of normal aircraft, but a no-go for a spaceplane that needs to achieve a fixed minimum velocity if it is to enter orbit. As only 3% of HOTOL's take-off weight would consist of payload, it would not require many design changes to consume all of this weight allocation for useful cargo and thus render the vehicle completely useless. The Concorde had a 5% payload but its design was considerably less sensitive to changes because it flew much slower than HOTOL and had more scope to trade-off between payload, speed and range; flying slightly slower than originally envisaged does not a ruin an aircraft design, but for a spaceplane it means the difference between achieving orbit or not.

The very high sensitivity of its performance to the input assumptions makes the design of a single-stage-to-orbit vehicle particularly difficult, and obliges engineers to pay very close attention to the fine detail. It also requires that a wide range of new technologies be advanced to maturity before the spaceplane design can be finalized, because any small discrepancy in expected weight or performance can have major consequences for the entire vehicle design. All this, plus the need to make the vehicle easy to maintain and operate, makes clear just how daunting the task of developing a spaceplane is.

MODERN CONCEPTS

Although the large spaceplane programs were canceled, some related developments did survive. Work on hypersonic scramjet propulsion in the US was continued in NASA's Hyper-X program, and resulted in two test flights of the small X-43A. This unmanned experimental vehicle was launched from the nose of a Pegasus rocket that was itself dropped from a converted airliner, and set new records for an airbreathing vehicle by achieving Mach 9.6 and a scramjet burn of 10 seconds on its latest flight in November 2004. Scramjet development tests are continuing with the similar X-51, which flew for the first time in May 2010 and reached a speed of 'only' Mach 5 but a much longer powered-flight time of 200 seconds. The second flight, in June 2011, was unsuccessful due to a failure of the scramjet engine. Some people insist that the US military already has a highly secret orbital spaceplane in operation (such as the 'Blackstar' reported by *Aviation Week*) but the evidence is unconvincing.

Shortly after HOTOL foundered in 1988, members of the engine design team led by Alan Bond set up a new company (Reaction Engines Ltd) to continue to develop the HOTOL concept, focusing initially on an improved version of the RB545 engine called the SABRE ('Synergistic Air Breathing Engine') and in particular the crucial precooler section. The spaceplane concept currently being worked on is the Skylon presented in the Introduction of this book as a "perfect spaceplane" (it is named after a futuristic art structure included in the 1951 Festival of Britain, which the fuselage strongly resembles). The designers reckon they have fixed the flaws in the HOTOL design, in particular the stability problem due to the heavy engines in the aft part of spaceplane. The Skylon solution is to locate the engines in the middle of the vehicle, housing them in nacelles at the tips of the delta wings in the same way envisaged for the Keldysh Bomber in 1947. This prevents the center of gravity from moving aft as the propellant tanks are depleted. Moreover, since the engines are not fully integrated with the fuselage they can be tested separately from the remainder of the vehicle. The engine nacelles have a peculiar banana-shape because their air intakes have to point directly into the airflow, whereas the spaceplane's wings and body must fly at an angle to create lift. Each engine will give a maximum thrust of 1,350,000 Newton in airbreathing mode, and 1,800,000 Newton in rocket mode.

According to the company their SABRE propulsion would make Skylon very safe and reliable, and enable it to take off without the rocket trolley that would have been necessary for HOTOL. But this meant Skylon would need a sturdy undercarriage as well as strong brakes to stop itself before the end of the runway if a problem were to occur just before take-off. It was decided to cool the brakes by water, which would boil away and dissipate the heat caused by the braking friction. The cooling water would be jettisoned following a successful take-off, thus reducing the weight of the undercarriage by several metric tons. At landing Skylon would be empty and hence fairly light, so the brakes would not need water cooling in order to be able to stop the plane without catching fire. Due to its aerodynamic characteristics upon re-entry, the vehicle would slow down at higher altitudes than the Space Shuttle Orbiter, keeping the skin of the vehicle significantly cooler, hence requiring only a durable reinforced ceramic for most of its skin. The turbulent airflow around the wings during re-entry

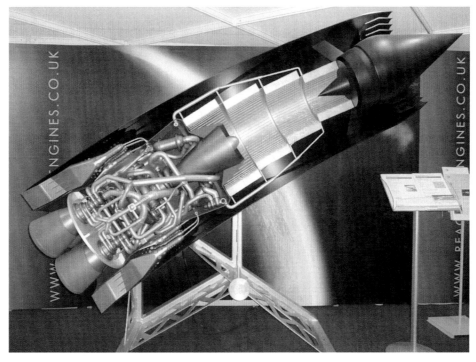

Model of the SABRE engine [Reaction Engines Limited].

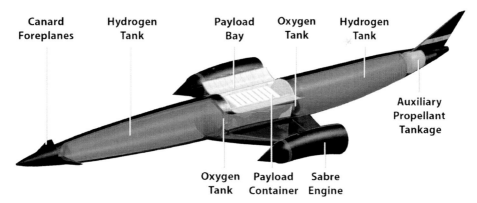

Cutaway of the Skylon concept [Adrian Mann & Reaction Engines Limited].

would, however, necessitate active cooling of some parts of the vehicle. Skylon is expected to be able to put 12,000 kg of payload into low orbit. Its take-off noise is expected to be acceptable for taking off from regular airports in populated areas but the runways would have to be extended to 5.6 km (3.5 miles) in length, of which the first 4 km (2.5 miles) would require to be stronger than usual to cope with the heavily laden Skylon rolling at high speed.

An independent review by the European Space Agency, which is also funding part of the technology development for Skylon, concluded in 2011 that the overall design "does not demonstrate any areas of implausibility". Reaction Engines is confident that Skylon will soon reach a technical maturity sufficient to convince investors that it is a valid commercial opportunity which warrants funding to full development. The project's cost estimates indicate that if a fleet of 90 vehicles were produced it would be possible to buy a Skylon for about $650 million, which is roughly comparable to a large jet airliner. Early customers would pay $30 to $40 million per flight but with more aircraft flying and an increasing total number of flights, the price should fall to around $10 million per launch. In comparison, a current Ariane 5 expendable rocket costs around $150 million per launch and puts less mass into a low orbit.

At the time of writing, Reaction Engines is doing tests on an experimental version of the precooler and plans to build a sub-scale version of the SABRE to demonstrate (on the ground) the complete engine's airbreathing and rocket modes as well as the transition between these. Tests of the nacelle in which the SABRE is to be housed are to be performed using a Nacelle Test Vehicle (NTV). This is planned to be launched from the ground and use rocket engines to get up to Mach 5, at which speed internal ramjet combustion systems will simulate the operation of the air-breathing engine. The remainder of the nacelle test article will be as close as possible to the real thing, including the control systems and internal flow ducts. The NTV is also to get some data on shock interactions between the nacelle and the Skylon's wing. According to Bond, "we could have a Skylon plane leaving Heathrow airport sometime during this century".

In 1991 Bristol Spaceplanes, another small aerospace company in the UK, began working on rocket plane concepts. Their 'Spacecab' design involves a Concorde-like carrier aircraft that uses four ordinary turbojet engines to take off and accelerate to Mach 2 and two rocket engines to reach Mach 4. At that speed a small, delta-winged rocket propelled orbiter carrying two pilots and either six passengers or cargo would separate and climb into orbit. The company insists this is a very conservative design that does not require any new technology to be developed. A next-generation vehicle called 'Spacebus' would fit the carrier aircraft with newly developed turbo-ramjets to achieve Mach 4 and rocket engines for Mach 6. Its enlarged orbiter would have room for fifty passengers. David Ashford, the company's managing director, has published his ideas in two popular science books: *Your Spaceflight Manual – How You Could be a Tourist in Space Within Twenty Years* (written with space tourism promoter Patrick Collins) and *Spaceflight Revolution*.

Private companies like Reaction Engines and Bristol Spaceplanes keep working on spaceplane technology, and the development of hypersonic and scramjet engines for military applications is strongly supported in the US (specifically in the Falcon program). Non-military government space agencies such as NASA and the European Space Agency have not completely given up on reusable launch vehicles either. The Future European Space Transportation Investigation Programme (FESTIP) study by ESA in 1994-1998 for instance, investigated many basic reusable launcher concepts, several of which were spaceplanes. In 2005 ESA also ran a small internal study (in which I participated as System Engineer) for a small rocket plane called Socrates. This was intended to fly at speeds up to Mach 12 and be operated for about 30 flights. It was specifically to investigate spaceplane operations and maintenance, such as how long it takes to replace rocket engines, how an onboard health monitoring system could help speed up maintenance, and how long thermal protection materials could last. A reusable rocket plane's ability to make many test flights at relatively low cost should result in a higher reliability in comparison to modern expendable launchers which are usually deemed operational after only a single qualification flight but typically have a failure rate of 2 to 3% (meaning that two or three payloads per hundred are lost, in turn pushing up insurance costs).

India, a country that is making great strides in spacecraft and launcher technology, is investigating a concept known as AVATAR (for 'Aerobic Vehicle for hypersonic Aerospace TrAnspoRtation'). This vehicle would take off using airbreathing turbo-ramjet engines and full tanks of fuel but none of the liquid oxygen that it would later need for its rocket propelled flight phase. Instead, during atmospheric flight separate ram air intakes collect air that is subsequently liquefied using liquid hydrogen-cooled heat exchangers; similar to the ACE principle of the JSSTO spaceplane concept. But unlike JSSTO, AVATAR involves another step in which the liquid oxygen in the air is mechanically extracted and stored in the previously empty oxidizer tanks so that they will be full by the time AVATAR requires to switch over to rocket propulsion. Although this is an extremely ambitious project, India is developing its capabilities at a rapid pace.

BUSINESS CASE

In spite of the multiplicity of studies and technology developments, government and industry funding for spaceplanes and other types of reusable launch vehicles remains modest at best: space agencies and the launch vehicle companies always seem to opt for a conventional, expendable rocket as their next generation launcher; spaceplanes are perpetually the next-next generation, with the result that for the last 50 years their full development initiation has always been 20 years in the future. This isn't due to a lack of concepts, because in addition to the spaceplane proposals described in this chapter there are literally hundreds of ideas and designs at a variety of levels of maturity and realism. David Ashford of Bristol Spaceplanes reckons that the lack of progress on orbital spaceplanes can be attributed to an entrenched mind-set of the world's space agencies and the vested interests of launcher industries, leading them to continue to pursue improvements and cost-reductions for their expendable launch vehicles rather than to replace these with something better. But it seems to me that really the same issues that killed the high-profile spaceplane concepts of the 1980s are still the root of the problem: uncertain economic benefits, very high development costs and great technical and financial risk; a lethal mix for any project that is not driven by a strong military or political agenda such as the Manhattan atomic bomb development of the 1940s and the Apollo lunar program of the 1960s respectively.

Reusable launchers are more expensive to develop than expendable ones, since in addition to the difficulty of developing something that can go into orbit a spaceplane must also be designed to come back, which involves re-entry into the atmosphere, a descent phase and a soft landing. Furthermore, using such vehicles requires a large infrastructure involving a long runway, facilities for vehicle and engine maintenance, (cryogenic) propellant production factories and storage tanks, and logistical systems to manage the distribution of spare parts. Large aircraft like the Airbus 380 and large launchers like Ariane 5 typically cost $10 billion to develop and it is hard to imagine how a spaceplane that combines the functions of both these vehicles is going to cost less. In fact, it is easier to see how it would cost significantly more. For instance, the reusable Venture Star SSTO abandoned in 2001 due to the expensive problems with its X-33 precursor was expected to cost close to $35 billion to make operational, and its price would certainly have increased if the additional developments to overcome the X-33 problems were carried out.

As regards the recurring costs (i.e. the costs which are imposed for every flight), a reusable launcher requires inspection and maintenance prior to each mission and, as experience with the Space Shuttle showed, this will be more complicated and hence more expensive than for a conventional aircraft. Furthermore, it is currently expected that due to the high strains involved in launch and re-entry in combination with the need to keep the vehicle's structure extremely light, spaceplanes will be able to make at most several hundred flights before they must be scrapped. Replacement rates and costs will therefore be higher than for airliners, where a single aircraft may undertake over 10,000 flights before having to be withdrawn from service. The operations costs for reusable systems therefore run the risk of turning out to be actually greater than for expendable rockets. The high development, infrastructure and maintenance costs mean that operating reusable launchers can only translate into attractive launch prices if they perform many flights per year. It is just like with commercial airlines, which keep their planes in the air for as many hours as possible in order to keep their costs down. A rapid turn-around is required to limit the size and hence the buy-cost of the vehicle fleet. Yet to be able to make many flights there must be a large number of customers who require many more payloads to be launched than is currently the case: today there are about 70 launches per year, although some carry multiple satellites; in contrast, on any normal day there are close to 30,000 airliners in the skies above the USA alone. But the launch market will only significantly increase in size (with space tourists and new satellite applications which are currently prohibitively expensive) if launch prices fall by a factor of 10 or so, which in turn requires efficient reusable systems with low maintenance overheads. It is a difficult Catch-22 situation: launches should become cheaper when the market is sufficiently large, but the market cannot dramatically increase until launch prices drop significantly. How the launch market will grow as a function of launch price reductions is debated heavily, and seems to be driven more by opinions than by hard statistical data.

In addition, current spaceplane designs are only capable of reaching low orbits so expendable rocket stages would still be required to boost satellites into higher orbits. And of course these 'kick stages' eat up payload volume and weight. Most current spaceplane concepts would not be able to place today's telecommunication satellites into geosynchronous orbit, this being the most profitable part of the non-government satellite launch market. Spaceplanes therefore need a large new market in low orbit, something that space tourism could provide if the flights were sufficiently affordable and safe; failing that, they will have to rely upon the increased use of small satellites intended to work in low orbit. Whether new markets would be sufficient to justify the development of a reusable launch vehicle is the $10 billion-plus question that is very difficult to answer right now. Even in the mature and well understood airline market, aircraft companies are generally betting the farm when engaging in the development of large new aircraft like the Airbus 380 and the Boeing 787 Dreamliner. You can imagine what the risks will be in trying to enter a relatively new, poorly understood market like that of future space launches with projects having costs on such a scale. In addition, spaceplanes face competition from smart low-cost expendable launchers, especially at low flight rates. For instance, Reaction Engines estimates that at a flight rate of 70 missions per year, a single flight of their Skylon spaceplane would cost in the order of $30 to 40 million. Therefore in terms of cost per kilogram into orbit this means Skylon might be beaten by the SpaceX company's Falcon Heavy expendable launcher whose development budget was much lower than that for Skylon.

An additional problem is the very long time to bring a complicated vehicle like a spaceplane into operational service: the Airbus 380 took about 13 years to develop, the Ariane 5 rocket about 12 years, and the F-22 fighter aircraft around 20 years. Any aircraft which incorporates as many new technologies as an airbreathing spaceplane will take at least two decades to advance from conceptual design to fully operational system, and that is not counting the additional time to develop and fly any sub-scale pathfinder test vehicles.

It is therefore not difficult to appreciate why spaceplane projects find it very hard to attract private investors; they are generally not interested in high-risk ventures that might deliver some unknowable return on investment after several decades. To limit the risks it seems to make sense to first build one or more smaller, less complex, less expensive demonstrators before committing to the development of a fully operational spaceplane. This philosophy appears to have been adopted in the US, where several hypersonic and scramjet test vehicles are currently being developed and flown. But while the investments are still substantial, such prototypes typically do not have any commercial use. Governments often finance at least the development and prototype phase; indeed this is how most expendable launch vehicle developments started, and is how the Concorde came into being. Government organizations could certainly play an important role in the development of the basic technology, such high-temperature materials and scramjets, just as in the early days of aviation the basic airfoil shapes were developed by NACA (the forerunner of NASA) and subsequently employed in almost all aircraft, even today. Apart from purely economic reasons the development of new strategic technology, the generation of high-quality jobs, the guaranteeing of national and especially military access to space, and indeed national prestige, can all serve as stimuli for governments to invest in high-risk technological projects such as spaceplanes. But new expendable launchers can also satisfy many of these desires at potentially lower costs and risks.

If it is difficult to close the business case for orbital spaceplanes, is it possible that a suborbital vehicle may make more sense? A smart concept that was investigated as part of the FESTIP study of the mid-1990s was that of a 'Suborbital Hopper' that involves a reusable vehicle which releases its payload into space at a speed just short of that required for orbit. A small rocket stage then gives the cargo the final kick to enter orbit. Such a launcher saves huge amounts of propellant by not having to boost its own weight into orbit, and is less constrained by the need to keep structure weight to the absolute minimum. In addition, such a vehicle may find profitable markets in rapid point-to-point transportation of people and time-sensitive cargo all across the world, space tourism, and undertaking short-duration microgravity and high-altitude experiments that need more time than can be provided by sounding rockets (point-to-point transportation may account for only a small fraction of the commercial aviation industry but because that industry is enormous it might still be far bigger than any short-term space launch market). In a military role it could act as a strategic bomber, uncatchable spy plane, rapid-reaction satellite launcher and rapid intervention vehicle capable of delivering special forces anywhere in the world within 2 hours. This last application is currently being studied by the US Marines under the name SUSTAIN, for 'Small Unit Space Transport And INsertion'. In fact, most work currently done in the field of hypersonic flight and propulsion is primarily for military purposes and is not intended to make space available to civilian travel and commerce. A suborbital spaceplane able to (almost) fly around the globe once would be something in between the short-range suborbital rocket planes described in the next chapter and a fully orbital spaceplane: a lower-cost, lower-risk project paving the way for a truly orbital spaceplane both in terms of technology and market development.

The 'Astroliner' suborbital rocket plane launch system proposed by Kelly Space & Technology in the US in the 1990s was a similar concept, with the addition of a Boeing 747 serving as a first stage. The jetliner would tow the rocket plane to an altitude of 6 km (20,000 feet) and Mach 0.8. The Astroliner would separate and shoot up to 110 km (360,000 feet) in order to release an expendable upper stage through a nose door and place several metric tons of payload into a low orbit. The rocket plane itself would continue its suborbital trajectory, re-enter the atmosphere and land on a conventional runway. The Astroliner would have jet engines for tow-flight assist and powered final descent and landing, and three Russian RD-120 liquid kerosene/liquid oxygen rocket engines for the zoom into space. During 1997 and 1998 the company conducted tests of the tow-launch concept at Edwards using a modified F-106 Delta Dart jet fighter towed behind a large C-141 Starlifter transport aircraft. Apart from this, the project does not seem to have progressed much although the concept is still advertised on the company's website.

Most current launchers are not exactly environmentally friendly because they burn large amounts of kerosene and rubber-like solid propellants on every flight. However, since the worldwide launch rate is very low their impact when compared to airplanes or cars is fairly negligible. Several modern rockets use liquid oxygen and hydrogen as propellants for at least some of their rocket stages, the combustion of which results in nothing more than water vapor. But what if spaceplanes are launching into orbit on a regular basis? The good news is that owing to the need for high performance, these vehicles will very probably also use hydrogen as fuel and burn it with oxygen drawn from the air during airbreathing flight phases and then with liquid oxygen for rocket propulsion. They would not emit any carbon dioxide or toxic gases. However, even water vapor may not be completely harmless when emitted at massive rates: at high altitudes it may linger for a long time, and it is not yet clear what the environmental impact would be. The water condensation trails left in the sky by high-flying jets have, for instance, already been shown to have a measurable effect on the amount of sunlight which reaches the ground. Moreover, liquid hydrogen is difficult to produce; it currently requires around 15 kilowatt-hours of energy per kilogram, so the source of the energy for making the fuel becomes very important. But that is not a particular spaceflight problem, it is part of the overall clean-energy issue.

Compensating for pollution by spaceplanes might be an increase in environment-monitoring satellites as a result of a fall in launch prices, data from which may well increase our understanding of weather and climate and result in the proper measures being taken to protect our world. In addition, astronauts generally return from space deeply impressed with the notion of how small the Earth really is and how thin the atmosphere appears from orbit. Flying more people into space may greatly increase awareness of the fragility of our planet. Finally, the heavy usage of hydrogen fuel by spaceplanes may boost the world's hydrogen industry. Spaceplanes could very well become the first large-scale commercial users of liquid hydrogen, reducing hydrogen prices and stimulating the development of efficient production, transportation and storage technologies. The economy and practicality of clean, hydrogen-powered cars could be improved by this. At the very least, spaceplane operators could incorporate energy-efficient systems and renewable energy sources into their ground operations; a new industry has the advantage that it can adopt sustainability and environmental awareness right from the start. Of course, the fact that spaceplanes would be reusable should save much energy and materials for the production of vehicles in comparison to expendable rockets, the valuable structures and other equipment of which are lost when they burn up in the atmosphere or crash into the sea.

In spite of all the potential benefits, developing and possibly flying a spaceplane or any type of reusable launch vehicle is still an (economic and technical) adventure rather than an everyday routine. And, as stated in the quote that opened this chapter, that is inhibiting success.

9

Joyriding a rocket plane

"Ah, but a man's reach should exceed his grasp, or what's a heaven for?" –
Robert Browning

On 4 October 2004 the (unofficial) airplane altitude record of 107.8 km (353,700 feet)
established by the X-15 in 1963 was finally broken. Not by a new, large-budget
government rocket aircraft but by the privately developed SpaceShipOne rocket
plane which pilot Brian Binnie flew to an altitude of 112.0 km (367,400 feet). The
project was entirely funded by a private sponsor and the vehicle was developed and
flown by a small commercial aircraft company.

This revolutionary development in rocket planes and spaceflight had its roots in
the X prize, a $10 million reward announced in 1996 for the first private enterprise to
develop and launch a suborbital vehicle. The competition's rules dictated it had to be
capable of carrying three people to the 'edge of space', and this, in accordance with
International Aeronautical Federation regulations, was defined as an altitude of 100
km (62 miles). An X prize vehicle would give its passengers a thrilling ride, enabling
them to view the curvature of the Earth and enjoy several minutes of weightlessness
in the same way as experienced by X-15 pilots. To prove its reusability, the X prize
organization required the same vehicle to make a second flight within two weeks of
the first launch with at most 10% of its dry weight being replaced. In addition to the
pilot, it had to be capable of carrying two passengers, but the flights required to win
the prize could be made by the pilot only.

The purpose of the prize (in 2004 renamed the 'Ansari X Prize' following a multi-
million dollar donation from entrepreneurs Anousheh Ansari and Amir Ansari) was
to encourage the development of suborbital space tourism and thus kick-start a non-
governmental human spaceflight industry. It was modeled after the aviation prizes of
the early twentieth century which tremendously boosted aviation, such as the Orteig
Prize for crossing the Atlantic that was won by Charles Lindbergh and the Schneider
Trophy that encouraged the development of extremely fast seaplanes (the heritage of
which was evident in several fighter planes of the Second World War, most notably
the Spitfire). Twenty-six teams from around the world declared their participation in
the competition, with some intending to employ relatively simple rockets (one even a
modern derivative of the A4/V2 design) launched from the ground or slung beneath

stratospheric balloons, others choosing rocket powered spaceplanes, and others rather exotic concepts such as pulse-jet driven flying saucers. One group even imagined a do-it-yourself suborbital rocket plane which you would be able to assemble in your own garage and launch from the nearest airfield.

The big surprise, however, was Scaled Composites, the Californian company of famous aircraft designer Burt Rutan, which initially shied away from publicity but in April 2003 revealed a project that was far ahead of its competitors. Not only did the company have a good plan but also real hardware: a fully operational twin-engined turbojet high-altitude carrier plane called the White Knight, a mobile mission control center, a mobile propulsion test facility, and a prototype of the Space-ShipOne air-launched, three-seat rocket plane. The company was by then already known for its innovative small aircraft designs, among them the Voyager aircraft that in 1986 flew around the Earth in just over 9 days without refueling or landing (73% of its weight at take-off consisted of fuel, leading to design constraints somewhat similar to those faced by spaceplane designers).

SpaceShipOne is primarily built using composite materials, a signature of Scaled Composites' designs, as indicated by the name of the company. The fuselage is bullet shaped, similar in appearance to the X-1. Its stubby wings have a slightly swept-back

SpaceShipOne in a glide flight [Scaled Composites, LLC].

SpaceShipOne carried under the White Knight aircraft [Scaled Composites, LLC].

leading edge, a straight trailing edge, and a vertical fin bearing a single horizontal stabilizer at each wingtip. The total length is 8.5 meters (28 feet), the wingspan is 8.2 meters (27 feet), and the total take-off weight is 2,900 kg (6,380 pounds). Its flight profile resembles that of the X-15 by involving an air-drop, boosted ascent, ballistic trajectory into space, re-entry and glide back to the ground. But it is intrinsically a much simpler aircraft, designed not for cutting-edge research flights but purely as a precursor for commercial tourism flights, and it benefits from an additional 40 years' of developments in aerodynamics, materials and avionics (as well as the considerable experience of the X-15 program). Although its maximum speed is Mach 3 rather than the X-15's Mach 6.7 and the mission does not call for extreme speed, it does call for extreme altitude. And whereas the X-15 could not survive a steep descent into the atmosphere and so had to fly a 40 degree ascent and descent trajectory over a horizontal distance of some 500 km (300 miles), SpaceShipOne flies up and down almost vertically so that its entire flight occurs within 40 km (25 miles) of its base. This greatly simplifies its operations by not requiring a large network of ground stations, chase planes and emergency landing sites.

SpaceShipOne is propelled by a single, revolutionary rocket motor which is a mix of a solid rocket booster and a liquid propellant motor. This SpaceDev SD010 hybrid motor uses a solid rubber-like HTPB (hydroxyl-terminated polybutadiene) grain as fuel, but in combination with liquid nitrous oxide (also known as laughing gas). The main benefit over a solid propellant booster is that this hybrid engine can be throttled and shut down at any moment by varying the amount of liquid oxidizer that enters the combustion chamber. Without the liquid oxidizer, it is totally safe from explosion during transport and handling. Furthermore these propellants have a higher specific impulse. The hybrid is also simpler than a liquid propellant rocket engine by having only one valve and redundant igniters. In contrast to the complex

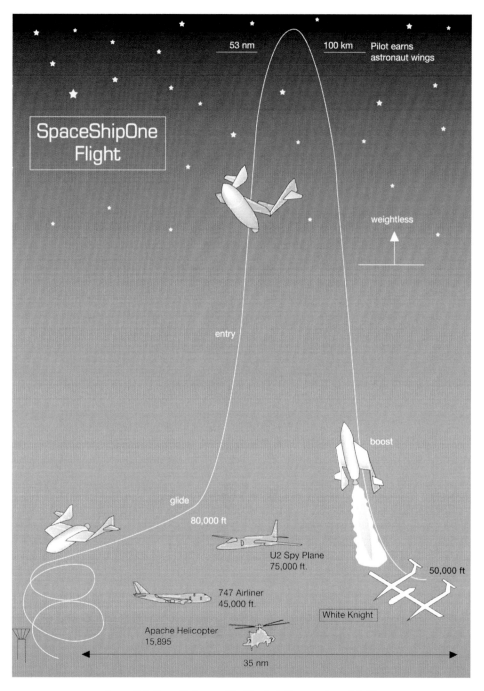

SpaceShipOne flight profile [Scaled Composites, LLC].

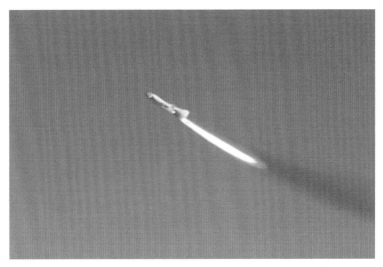

SpaceShipOne shoots up under rocket power [Scaled Composites, LLC].

XLR99 engine of the X-15 the SD010 uses only non-toxic, easy-to-handle propellants and it has never failed to start. It has a maximum thrust of 75,000 Newton, a specific impulse of 250 seconds, and a maximum total burn duration of 87 seconds.

Prior to re-entering the atmosphere the plane's two tail booms and the rear half of the wings fold upward on a hinge that runs the length of the wing. This 'feathered' position gives the aircraft a high-drag that allows a safe, stable "carefree, hands-off" penetration of the atmosphere which greatly reduces aerodynamic and aerothermal loads. For this innovative solution Rutan was inspired by a badminton shuttlecock, which always orients itself correctly with the direction of flight. The cockpit has a spacecraft-like environmental control system and features many windows to provide a good view for the pilot and passengers (although no passengers were carried). The aircraft has three flight control systems: a direct manual control for subsonic speeds, an electric control system for supersonic speeds (where muscle power alone is unable to handle the aerodynamic forces), and a reaction control system for high altitudes. The thrusters emit non-toxic cold gas (there is no combustion involved). State-of-the-art instrumentation provides the pilot with the precise guidance information he needs to manually fly SpaceShipOne during the critical boost and re-entry phases. Flight test data is sent to a mission control center during each flight, where it is recorded for careful post-flight analyses.

The only SpaceShipOne aircraft was registered as N328KF, with N the prefix for US-registered aircraft and 328KF chosen by Scaled Composites to stand for 328 K (for kilo, meaning thousand) feet, corresponding to the 100 km altitude goal (registry number N100KM was already taken).

The White Knight plane, SpaceShipOne's carrier, is itself an innovative aircraft. It too is made mostly out of composite materials. It has two afterburning turbojets, thin wings that have a total span of 25 meters (82 feet) and two tail booms. Most of the cockpit, instrumentation and other internal equipment are identical to those installed on SpaceShipOne, enabling it to flight-qualify much of the equipment intended for SpaceShipOne, thereby sharing the development costs for the two aircraft. The White Knight could be used as a trainer aircraft for SpaceShipOne pilots. The high thrust from its turbojets with afterburners in combination with the low weight, as well as it enormous speed brakes for rapid deceleration meant that rocket plane pilot trainees could use the White Knight to rehearse SpaceShipOne's boost flight, approach and landing very realistically.

On 21 June 2004 the White Knight took the diminutive SpaceShipOne with 62-year-old pilot Mike Melvill to an altitude of 14 km (46,000 feet). The spaceplane was dropped into a gliding flight, then fired its rocket motor for 76 seconds. Shortly after ignition of the rocket motor, wind shear suddenly made the aircraft roll 90 degrees to the left. Melvill attempted to correct it and unexpectedly rolled 90 degrees to the right. He then managed to level the plane again and proceed with the steep but still somewhat unstable powered boost to a maximum speed of Mach 2.9. During the rocket burn Melville reported a loud bang that was later realized to have been caused by the overheating and subsequent crumpling of a new aerodynamic fairing that had been fitted around the rocket nozzle. Fortunately the fairing's collapse did not affect the flight. After burn-out of the engine the plane continued unpowered to an altitude in excess of 100 km (62 miles). This coasting phase and the following free-fall back to Earth lasted about 3.5 minutes, during which time Melvill opened a bag of M&Ms and watched them float weightlessly around the cockpit. At the highest point of the trajectory the vehicle's speed was almost zero. Then it began to fall, accelerating to a maximum speed of Mach 2.9 (the same as its maximum speed going up, as potential energy converted back to kinetic energy). During the fall, the two tail booms and rear parts of the wings were put in a vertical position to achieve the high-drag configuration that facilitated a safe, stable penetration of the atmosphere. The thickening air then decelerated the vehicle, and subjected Melvill to a tolerable 5 G deceleration. The re-entry air temperatures remained less than 600 degrees Celsius (1,100 degrees Fahrenheit) owing to the large area of the underside of the aircraft and the relatively modest velocity. There was no need for heat shields or tiles because the hot re-entry phase was brief and the air at high altitude too tenuous to transfer a lot of heat; the skin of the aircraft remained much cooler than the surrounding air (the X-2 flew its 'heat barrier' research flights at similar speeds but at much lower altitudes, while the X-15 and orbital vehicles returning from space endure much higher temperatures as a result of their faster entry speeds). In fact, SpaceShipOne's structure hardly contains any metal parts. At 17 km (57,000 feet) the wings and tail were repositioned and the aircraft reverted to a conventional glider for its descent to the runway in the Mojave Desert in California.

"It was a mind blowing experience, it really was; absolutely an awesome thing," Melvill said after landing. With this flight he became the first private civilian to fly an aircraft into space, as well as the first person to leave the atmosphere in a non-government sponsored vehicle. (All rocket aircraft except the early, pre-war rocket-boosted gliders were developed under government contracts for military or research purposes.) Measured by the number of world newspapers that carried the story above the fold, the flight was the second largest news event of the year, being topped only by the capture of Saddam Hussein in Iraq.

Work on Scaled Composites' suborbital spaceplane concept began right after the X Prize announcement in 1996 and the full development program was initiated in April 2001, hidden from the public and the competitors by the inhospitable Mojave Desert. To finance the project the company got a $30 million grant from Paul Allen, Microsoft cofounder and third-wealthiest person in America. Since the X Prize was $10 million, Allen could not expect to get a return on his investment any time soon but he was in it for the sense of adventure rather than for the money. The overall plan was to mature the concept, then sell improved vehicles to a space tourism company. "Spaceflight is not only for governments to do," Allen said. "Clearly, there's an enormous pent-up hunger to fly into space and not just dream about it."

SpaceShipOne made its first captive flight on 20 May 2003 and shortly thereafter Rutan announced the project to the public. After a second captive flight there were seven successful glide drop tests before pilot Brian Binnie made the first powered flight on 17 December of the same year (deliberately marking the 100th anniversary of the first ever powered aircraft flight by the Wright brothers). A short burn of the rocket motor pushed the aircraft to Mach 1.2 and an altitude of 21 km (68,000 feet). The left main gear collapsed due to a roll oscillation upon landing but the damage was minor and Binnie was uninjured. After another glide test flight there was a series of progressively faster and higher flights, culminating in the one in June 2004 that put Mike Melvill into space. During the test program SpaceShipOne also became the first privately funded aircraft to exceed Mach 2. All of the flights took place from the Mojave Airport Civilian Flight Test Center, the runway close to Scaled Composites' premises. The four pilots that flew SpaceShipOne came from a variety of aerospace backgrounds: Mike Melvill was a test pilot, Brian Binnie a former Navy pilot, and both Doug Shane and Peter Siebold were company engineers. They all trained to fly SpaceShipOne using a flight simulator (like the X-15 pilots) as well as by flying the White Knight and other aircraft produced by Scaled Composites.

After Melvill's space flight, everything was deemed ready to try for the X Prize by making two such flights within a fortnight. On 29 September 2004 Melvill shot up to an altitude of 103 km (338,000 feet), which was slightly less than planned due to a serious roll instability during the rocket-boost phase, but was still above the 100 km requirement. It was quickly followed on 4 October (specifically chosen to mark the 47th anniversary of the launch of Sputnik) by Brian Binnie's fully successful flight to the record altitude of 112.014 km (367,500 feet) that won the X Prize for Scaled Composites and also made SpaceShipOne the first privately funded aircraft to exceed Mach 3: when the motor cut off at over 61 km altitude (200,000 feet) the maximum speed was Mach 3.09, an equivalent velocity of 3,490 km per hour (2,170 miles per hour). Melvill and Binnie, the two pilots who flew above the 100 km (330,000 feet) mark were issued the first commercial 'astronaut wings' by the US Federal Aviation Administration.

No further flights were made, as the prize had been won and the concept and the technology proven. For commercial space tourism flights, Rutan wanted to develop a larger rocket plane that could seat more passengers and incorporate more

SpaceShipOne in the National Air and Space Museum [Photo by Eric Long, National Air and Space Museum, NASM WEB 10516-2005, Smithsonian Institution].

redundant systems and aerodynamic stability for increased safety. In addition, he did not wish to risk damaging the unique and now historic SpaceShipOne. Since 2005 the small rocket plane has hung on display in the main atrium of the National Air and Space Museum in Washington D.C., between the Wright Flyer, the Spirit of St. Louis and the Bell X-1, and near the first X-15. As a tribute to SpaceShipOne's achievement, in 2006 a small piece of its carbon fiber material was cut off and launched on the New Horizons probe heading for Pluto. An attached inscription reads: "To commemorate its historic role in the advancement of spaceflight, this piece of SpaceShipOne is being flown on another historic spacecraft: New Horizons. New Horizons is Earth's first mission to Pluto, the farthest known planet in our solar system. SpaceShipOne was Earth's first privately funded manned spacecraft. SpaceShipOne flew from the United States of America in 2004."

A fiberglass replica of SpaceShipOne created using the same molds used to make the original can be found in the AirVenture Museum in Oshkosh. Another full-scale replica is on display in the William Thomas Terminal at Meadows Field Airport in Bakersfield, while a third is in the Mojave Spaceport's Legacy Park, and a fourth is hanging above the stairs in the main entrance of Building 43 of Google's Googleplex campus (Google cofounder Larry Page was a trustee on the X Prize board) and a card taped to the nozzle implores, "Attention Googlers: Please do NOT launch. Thanks." SpaceShipOne also became a popular model rocket, with Estes Industries currently offering several SpaceShipOne models that you can launch from your own back yard repeatedly by replacing the little solid propellant rocket motor.

SpaceShipTwo and White Knight Two [Scaled Composites, LLC].

Rutan's company has now teamed up with the Virgin Group, famous for its airline and its entertainment and communications companies, as well as its charismatic and adventurous head, Sir Richard Branson. Under the name 'The Spaceship Company', the Virgin Group and Scaled Composites have set up a joint venture to develop the SpaceShipTwo and White Knight Two aircraft which will be operated by a company called Virgin Galactic. At the time of writing, the 'spaceline' plans to operate a fleet of five SpaceShipTwo vehicles starting no earlier than 2012. They have been taking bookings at $200,000 per passenger for the early flights, and by late 2011 had over 450 paid customers. It is expected that ticket prices will drop significantly as flight operations mature, increasing the size of the space tourist market.

SpaceShipTwo, based on the same principle, concept and shape as SpaceShipOne is roughly twice the size in order to house two pilots and six passengers. It will be propelled by a larger hybrid rocket motor named 'RocketMotorTwo' delivering over 230,000 Newton of thrust. Development of the new rocket plane was delayed when in 2007 an explosion occurred during an oxidizer flow test that was being conducted at the Mojave Air & Space Port. Three staff were killed and another three severely injured; rocket engines are still potentially dangerous devices that have to be handled with great care. White Knight Two is an innovative twin-hull aircraft that carries the SpaceShipTwo rocket plane between its fuselages. It is also designed to operate as a zero-G parabolic-flight aircraft for SpaceShipTwo passenger training or micro-gravity science flights, and as a high-altitude research plane. It could potentially launch other rockets than SpaceShipTwo, such as small sounding rockets with instruments for scientific research.

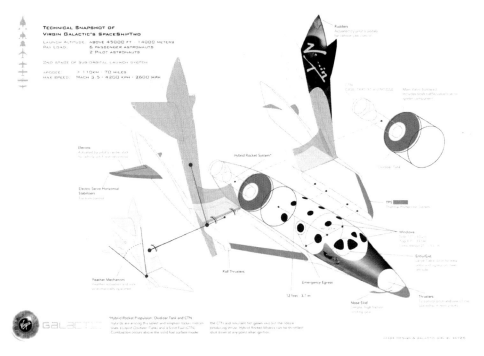

SpaceShipTwo technical diagram [Virgin Galactic].

Unlike previous rocket plan projects, environmental impact is now an important issue in aviation. With respect to carbon dioxide (CO_2) emissions the hybrid engine is not exactly 'green' but according to Virgin Galactic, "CO_2 emissions per passenger on a spaceflight will be equivalent to approximately 60% of a per-passenger return commercial London/New York flight." This is about 500 kg of carbon dioxide per passenger per flight. So even if SpaceShipTwo flights eventually number 1,000 per year the resulting carbon dioxide emissions would be in the order of one-thousandth of what a major airline typically expels into the atmosphere during a year. Virgin Galactic nevertheless accepts that the environmental impact of their operations could have serious implications for the image and success of their business, and the larger Virgin empire is committed to being as environmentally friendly as is practical. The company therefore plans to run its spaceport(s) with as much renewable energy as possible, which may even make them a net energy producer and potentially "carbon negative" by preventing more emissions of carbon dioxide than its vehicles produce. White Knight Two's jet engines will initially burn kerosene but are also capable of running on butanol, a biofuel that can be made from algae.

The first SpaceShipTwo, christened VSS (Virgin Space Ship) 'Enterprise' (after the legendary *Star Trek* starship) made its first glide flight on 10 October 2010, being launched by the first White Knight Two aircraft VMS (Virgin Mother Ship) 'Eve', and it performed its first 'feathered' flight on 4 May 2011. To date, a total of 16 glide flights have been made, and round 100 test flights are expected before the first passengers will be carried. The first commercial flight is expected no earlier than 2012. The company will initially operate from Spaceport America, a brand new $210 million airport for suborbital vehicles located in New Mexico. There are also plans for a sister spaceport in northern Sweden. Singapore and the United Arab Emirates have both also shown interest in establishing suborbital flight facilities.

A SpaceShipTwo flight will be an incredible adventure offering the possibility, albeit brief, to experience what astronauts (and X-15 pilots) feel and see, without the heavy workload. You will be dropped from the carrier aircraft at an altitude of 15 km (50,000 feet) and then go supersonic within 8 seconds. After 70 seconds of powered flight, during which you attain a maximum speed of just over Mach 3 (equivalent to about 3,500 km per hour, or 2,100 miles per hour) the rocket plane will coast to a peak altitude of 110 km (360,000 feet). The virtually drag-free parabolic trajectory will last for 3.5 minutes, during which you will be able to float about in the relatively spacious cabin and admire the view of Earth below and the curvature of the horizon through the large windows.

Other companies are also working on suborbital rocket planes for space tourism, with microgravity science and high-altitude experiments (as on the X-15) forming a secondary market. XCOR Aerospace, which is based on the same Mojave airfield as Scaled Composites, is developing its 'Lynx' rocket plane (superseding its earlier and similar 'Xerus' design). Unlike SpaceShipTwo this double-delta-winged vehicle will take off from a runway on its own power and hence will not require a carrier aircraft. This simplifies the development and operations (one rather than two planes) but it means the rocket aircraft has to carry all the propellant for the entire flight itself. The Lynx Mark-I prototype aircraft is considerably smaller than SpaceShipTwo and will only be able to reach an altitude of 60 km (200,000 feet) carrying a pilot and a single paying passenger. A more advanced Mark-II production version is to be able to reach the milestone of 100 km (330,000 feet). The passenger will have to remain strapped in his seat, as the cockpit is too small for weightless acrobatics. On the other hand, the initial ticket price announced by the company is about half that of a SpaceShipTwo flight. XCOR appears to be well advanced in the general development of liquid propellant rocket engines, and has reported that its 13,000 Newton XR-5K18 liquid oxygen and kerosene rocket engine (four of which will be needed to power the Lynx) is almost ready for flight. But propulsion is only one part of a rocket plane, and although the company has done extensive wind tunnel testing using a scale model of the Lynx, its announcement that it expects to start the test flight campaign of its Mark-I prototype in 2012 appears rather optimistic.

XCOR modified an existing canard configuration (i.e. tailless) 'Long EZ' sports aircraft to demonstrate its rocket engine capabilities by installing two 1,800 Newton restartable, pressure-fed, regeneratively cooled rocket engines which burn isopropyl alcohol and liquid oxygen. This 'EZ-Rocket', which is a modest-performance rocket plane in its own right, has made a total of 26 flights including a number of air show demonstrations. In December 2005 the EZ-Rocket set the world record for 'Distance without Landing' for a ground-launched rocket powered aircraft with a flight from Mojave to California City, a distance of 16 km (9.94 miles). "That was the shortest long-distance record flight ever!" pilot Dick Rutan exclaimed. XCOR also built and

Artistic impression of the Lynx rocket plane [XCOR Aerospace].

flew the 'X-Racer', a sleek rocket aircraft based on the airframe of the 'Velocity SE' canard sports plane. This was a prototype for aircraft to compete in rocket plane races organized by the Rocket Racing League, an organization that seeks to promote rocket aircraft development by flying competitions. The X-racer is equipped with an XR-4K14 restartable, pump-fed rocket engine that burns liquid oxygen and kerosene with a thrust of 6,600 Newton. It made its first flight on 25 October 2007. The test program has now been completed after a total of 40 flights and demonstrations. The X-Racer holds claim to several (unofficial) records including the most flights made in a single day by a manned rocket powered aircraft, and the fastest turn-around for a manned rocket powered vehicle.

Armadillo Aerospace, the small aerospace company of computer game developer John Carmack, who made his fortune by developing popular games such as Doom and Quake, has also made a rocket engine for the Rocket Racing League. It equipped the Rocket Racing League's current Mark-II and Mark-III Rocket Racers, which are also based on the Velocity airframe (in this case the Velocity XL FG version) with a home-grown rocket engine that is fed with liquid oxygen and ethanol and develops a maximum thrust of 11,000 Newton. Seven successful test flights were made by the Mark-II aircraft during August 2008 and both machines are currently used for flight demonstrations. The Rocket Racing League hopes to generate sufficient interest for a number of teams to build or purchase similar rocket aircraft in order to participate in rocket propelled air races. In the meantime, you can download a video game that puts you in the cockpit of a Rocket Racer.

EZ-Rocket [XCOR Aerospace].

In March 2002 the Space Adventures company that organizes 'flight participant' missions to the International Space Station, unveiled a mockup of the 'Cosmopolis XXI' (C-21) lifting body-type suborbital rocket plane at Zhukovskiy Air Base near Moscow. This was to be developed by the Russian Myasishchev Design Bureau, be launched from the design bureau's existing M-55X 'Geofizika' high altitude aircraft, and be able to carry a pilot and two passengers into space at $98,000 per ticket with the first flight in 2004. The carrier aircraft with the C-21 attached would first slowly climb to an altitude of 17 km (56,000 feet) and then gather speed in order to make a vertical climb to 20 km (66,000 feet) to release the C-21. The C-21 would then ignite its expendable solid propellant rocket motor. When this motor burned out it would separate and fall away, leaving the C-21 to follow a ballistic arc to a peak altitude of 100 km (330,000 feet). The rocket plane would glide back to the airport and make a parachute-assisted touchdown. But Space Adventures has abandoned its plans to use the C-21 and instead contracted Armadillo Aerospace to develop a vertical launched, vertically landing suborbital rocket capsule to implement its planned suborbital flight services.

The giant European space company EADS Astrium announced in 2007 that it was to develop a suborbital rocket plane for space tourism. This single-stage, straight-winged plane would take a pilot and four passengers to the edge of space and offer a great view through many large windows and a roomy cabin for weightless antics. It would take off from a normal airport and climb to an altitude of 12 km (39,000 feet) with jet engines, then ignite a Romeo liquid oxygen-methane rocket

The Mark-III Rocket Racer in flight [Rocket Racing League].

engine to reach 60 km (200,000 feet) in just 80 seconds with enough velocity to continue unpowered to its 100 km (330,000 feet) apogee. As the plane fell back the pilot would use small thrusters to control its attitude for re-entry into the atmosphere prior to restarting the jet engines to return to the airport. Jet engines use 10 to 20 times less propellant than rocket motors of the same thrust over the same time and are much more efficient for the first and final phases of a flight (SpaceShipTwo's carrier aircraft uses jet engines for the same reason) but when they are not providing thrust at high altitudes they are dead weight. SpaceShipTwo effectively leaves them behind once it separates from its carrier. Jet engines are also handy in case of a failure of the rocket engine, as well as for ferry flights between airfields. Astrium expected to require around 1 billion euro to develop their system (much more than SpaceShipTwo is estimated to cost), flights to begin in 2012, and tickets to cost up to 200,000 euro. "The development of a new vehicle able to operate in altitudes between aircraft (20 km) and below satellites (200 km) could well be a precursor for rapid transport point-to-point vehicles, or quick access to space," the company said.

Artistic impression of the take-off of the EADS Astrium suborbital rocket plane [EADS Astrium & Marc Newson Ltd].

Famous designer Marc Newson was to take care of the aesthetics of the design, and the images in the brochure published by Astrium sure are beautiful. As Astrium builds the Ariane 5 launcher and its mother company EADS develops and produces the famous Airbus airliners as well as the Eurofighter military jet, the company seems ideally suited to pursuing a suborbital rocket plane project: it has all the necessary knowledge, experts and facilities in-house, and could incorporate a lot of existing EADS aircraft and spacecraft equipment such as cockpit instrumentation, undercarriage and control thrusters.

After their 2007 announcement, however, Astrium remained awfully quiet about their rocket plane, making it appear to have been merely a publicity stunt rather than a real project. But early in 2011 the company announced that it was indeed working on the concept and that after having placed work on hold for several years due to the global economic downturn it was planning to spend a further 10 million euro on it in 2011; a considerable sum but not much in comparison with the 1 billion euro that it had predicted for full development. "We continue to mature the concept, maintaining the minimum team in order that when we find the relevant partnership we are ready and have progressed sufficiently," Astrium CEO François Auque told reporters in January 2011. Once it has secured the required financial and industrial partners, the company expects to be able to put the rocket plane into service within five years.

In 2004 another big European aeronautics company, Dassault Aviation in France, announced its own suborbital rocket plane design called VSH. This was based on an earlier design for an automated air-launched reusable hypersonic vehicle known as VEHRA ('Véhicule Hypersonique Réutilisable Aéroporté') but was intended to be manned and therefore VSH stood for 'VEHRA Suborbital Habité'. The delta-winged rocket plane would be carried into the air by a commercial aircraft, be released at an altitude of 7.6 km (25,000 feet) and a speed of Mach 0.7, and ignite a liquid oxygen-kerosene rocket engine to climb to the milestone altitude of 100 km (330,000 feet). Design work is progressing in the context of the K-1000 project that Dassault is self-financing with several industrial partners in Switzerland.

Bristol Spaceplanes, mentioned earlier for its Spacecab and Spacebus projects, is working on a rocket plane called 'Ascender'. This is a delta-winged aircraft with two jet engines and a single rocket motor similar in concept to that of EADS Astrium but only able to seat a pilot and a single passenger. Ascender's rocket engine, a prototype of which is to fly on a sounding rocket, will use hydrogen peroxide and kerosene as propellants. As such it resembles the Spectre rocket engines developed in the 1950s to power the SR.53 and SR.177. Ascender is also to pave the way for the company's orbital spaceplane concepts (discussed above). However progress is slow because the company is waiting for a serious investor so that it can afford to appoint a full-time team of engineers.

Virgin Galactic would seem to be the most advanced company in terms of making and flying suborbital rocket planes, but if the space tourism market really takes off there ought to be room for several aircraft manufacturers and operators to compete. This would hopefully lower ticket prices further, resulting in ever more people being able to afford a flight to the edge of space.

The next step foreseen by Burt Rutan is an orbital rocket plane for space tourism, but that poses a tremendous challenge because although the 100 km (330,000 feet) altitude reached by SpaceShipTwo will be sufficiently above the atmosphere to circle the Earth a couple of times, the speed of the vehicle falls far short of that required to

Artistic impression of the Ascender rocket plane [Bristol Spaceplanes].

enter orbit. To achieve orbit at that height, a vehicle must have a horizontal speed of 7.8 km per second (4.8 miles per second); i.e. 28,000 km per hour (17,500 miles per hour). SpaceShipTwo reaches Mach 3 at engine burn-out in a steep climb but at the top of its parabolic arc its speed is virtually zero (as all its energy has been converted into altitude). Compared to SpaceShipTwo's maximum speed of 0.9 km per second (0.6 miles per second) an orbital rocket plane needs to go over 8 times faster; and as kinetic energy increases with the square of the speed that means a propulsion system capable of delivering almost 70 times as much energy! This is why satellite launchers and orbital spaceplane concepts are so much larger than suborbital rocket planes such as SpaceShipTwo and Lynx; even though they all reach space, in terms of energy and thus propellant volume the difference is huge. Weight constraints are also much more demanding for an orbital spaceplane. Whereas a suborbital rocket plane's dry weight can be approximately 40% of the vehicle's overall weight including propellant, the energy needed to go into orbit demands that a plane's empty weight be no more than 10% of its take-off weight (for both types of vehicle these percentages diminish if multiple stages and/or airbreathing propulsion are employed but the large difference remains). This also has consequences for safety: where normal aircraft structures are usually designed to be able to withstand 1.5 times the highest load expected to occur during the plane's lifetime (and even 2 times for the undercarriage) this margin will be extremely difficult to meet for reusable orbital spaceplanes. Even for expendable launchers, which are less constrained regarding empty weight, this factor is typically only 1.2, except for crewed launchers where it is 1.4 according to NASA standards.

In short, the difficulty in achieving orbit is not so much to get up to high altitude, it is rather to attain the necessary high velocity with a structure weight that provides a reasonable amount of reliability and safety. Factoring in the much more extreme re-entry temperatures that will require heat shields, and that 'feathering' cannot be used for hypersonic re-entry, clearly indicates that an orbital SpaceShipThree will not be merely an upgrade of SpaceShipTwo but a completely new, much larger, and more complicated spaceplane that will be vastly more expensive to develop and operate.

10

Man versus robot

"Man is the best computer we can put aboard a spacecraft... and the only one that can be mass produced with unskilled labor." – Wernher von Braun

Will a future hypersonic plane have pilots on board or be fully automatic? Nowadays aerodynamics and rocket propulsion are fairly well understood and can be accurately modeled and simulated. As a result, the areas where direct pilot intervention may be needed due to unexpected behavior of an aircraft/spaceplane are rapidly decreasing. This is especially true for vehicles that have a fixed, pre-determined trajectory such as missiles and launch vehicles. So-called Unmanned Aerial Vehicles have become important operational military assets, and these aircraft are steered from the ground or fly their missions completely autonomously as aerial robots. It is therefore likely that future space planes will be flown by a computer under human supervision from the ground rather than directly by a human pilot, particularly as hypersonic vehicles tend to be aerodynamically unstable and therefore require sophisticated avionics for efficient and safe control. For instance, Skylon is to fly automatically; any astronauts to be transported into orbit will be housed inside its payload bay.

Especially on a satellite launch vehicle with relatively little margin for errors and malfunctions, operating without pilots results in a simpler and thus cheaper design; a crew requires a comfortable cabin with regulated pressure and temperature, requires to have an escape capability if the spaceplane is less reliable than a regular aircraft (which rocket vehicles invariably are) and requires higher safety margins to be built into the design. Not having any people on board potentially makes the vehicle less expensive and saves weight and space that can be used for more payload. Moreover, catastrophic failures have less grave consequences; compare the dramatic aftermaths of the losses of the Space Shuttles Challenger and Columbia with those of the many but almost forgotten failures of unmanned expendable launchers.

However, on many occasions having an exceptionally skilled pilot on board saved the X-15 and earlier rocket aircraft. So any fully automatic flight control system on a versatile hypersonic aircraft intended for various types of missions must be smart and capable of reacting very rapidly to unexpected situations and emergencies. That may be difficult, as programming a computer for unforeseen events is near to impossible whilst the human brain excels at improvisation (although developments in so-called neural networks may result in self-learning computers that can quickly react to new situations). And what if a spaceplane carries astronauts onboard? Even if they are not flying the vehicle themselves, the aircraft will still need to incorporate the additional equipment and various reliability enhancing redundancies that an unmanned vehicle can do without. Would it be acceptable for them to ride into space in a fully robotic hypersonic launcher? Or would a pilot with a manual override capability be required, as for astronauts launched on current expendable rockets like the Soyuz and even the Space Shuttle, if only for psychological reasons? The impact on the design would be limited if one of the transported crew could fly the vehicle in an emergency, in order that no additional seat need be assigned to a pilot.

Talking of people on board spaceplanes and rocket planes in general, what about vehicle safety? The early rocket propelled aircraft like the Me 163 were extremely hazardous. Four pilots died and two were severely injured during the X-1, X-2 and X-15 programs and there were also many less serious accidents. Of the 16 individual airframes involved, 10 were completely or largely destroyed in accidents: not a very good safety record given that the X-planes only made a total of some 415 flights, a total that can be readily accumulated by a single airliner in 6 months of operations. Does this mean that rocket planes are inherently dangerous and hence ought never to be used for suborbital space tourism and/or mass transportation into orbit? Surely we have learned much about high-speed, high-altitude flight since those days, and rocket propulsion has also greatly matured. Suborbital flight in particular, benefits not only from the experience gained from the experimental rocket planes but also from high-performance jet aircraft in general.

Furthermore, whilst the high losses among pilots flying rocket planes may appear high today, they were not particularly exceptional compared to the accident rate in experimental aviation and the general testing of prototype aircraft. In the late 1940s and the 1950s test pilot loss rates in the US were in the order of one per week. And crashes of military jets in operational service occurred frequently. Nowadays crashes and aircraft explosions are very rare, even for new types, so there is no real reason to expect suborbital rocket aircraft like SpaceShipTwo to suffer from anything like the loss rates of early jets. However, a suborbital launch is certainly more hazardous than a regular airline flight, and orbital spaceflight even more so. In part this is due to the extreme speeds, altitudes and temperatures involved, in combination with the need to keep the vehicle as light as possible, and in part due to the still experimental nature of human spaceflight. At the time of writing, the number of crewed space missions is less than 290, well below the number of planes in the air on a typical day. There have been even fewer suborbital rocket plane flights into FAI-certified space. Indeed, only two X-15 flights and the recent three missions of SpaceShipOne ascended above the milestone altitude of 100 km (62 miles), and another eleven X-15 flights exceeded 80 km (50 miles). In today's world of health and safety regulations, the relatively low trustworthiness of rocket vehicles is certainly a business risk. People have come to expect that even radically new aircraft will not kill anyone, and that suborbital space tourists riding rocket planes should not feel that they are putting their lives on the line. On the other hand, perhaps it is the risk that provides the sense of adventure.

11

Conclusions

"Nothing ever built arose to touch the skies unless some man dreamed that it should, some man believed that it could, and some man willed that it must." – Charles Kettering

The 'golden age' of the rocket plane, whether it is defined in terms of the number of aircraft, speed of progress or number of flights, kicked off with the He-176 in 1939, essentially at the same time as the jet age, and arguably ended with the final flight of the X-15 in 1968. Successful rocket aircraft projects of that period were based upon three vital ingredients: a good aircraft design, a good rocket engine, and a great pilot. If any part of this fundamental triangle was lacking, the outcome was often disaster: aircraft pitched over due to Mach tuck, engines blew up, and pilots overshot landing fields and crashed their expensive aircraft. The extreme speeds that rocket aircraft achieved and the new aerodynamic territories they ventured into meant things could go wrong very fast and very unexpectedly. Pilots who let their powerful, sleek planes get ahead of them often did not make it back. And extensive flight experience did not mean that pilots were safe from making mistakes. For each new, experimental rocket aircraft every pilot was essentially inexperienced. The same applied to the designers, but at least they rarely lost their lives due to a fault in their aircraft or engine.

However, even while they were in the limelight, airplanes with rocket propulsion were rapidly rendered obsolete by improved turbojet engines. The Me 163B was the only pure rocket fighter that ever entered military service, while the only operational mixed-power fighters were the Mirage IIIC, -E, and -S, and for most of the time even these flew without their optional rocket packs. Altogether the rocket propelled fighter plane does not have a very impressive track record when taking into account all the development effort on experimental aircraft and prototypes.

Rocket planes were soon realized only to be really useful as research aircraft to fly at extremely high speeds and altitudes. The X-15 set incredible records for aircraft speed and altitude but the data it collected at the extremes of its flight envelope was so far beyond what was required for operationally useful manned aircraft that there was no need to make a successor to push the boundaries even further. Orbital rocket planes, the logical next step, proved to be too complex, too costly, and ultimately not really needed. Rocket aircraft development therefore stopped at the end of its infancy and at the peak of its success, and so never matured into really operationally useful series-produced planes. Instead, new military planes relied on advanced jet engines and spacecraft kept using vertical take-off launchers that were usually expendable or at best included a reusable shuttle that was able to glide back from orbit.

By the mid-1980s it seemed that a second golden age was about to begin, with a number of ambitious spaceplanes and hypersonic airliners such as the Sänger-II and HOTOL following up on the experimental rocket planes of the 1950s and 1960s. But these new vehicles would not be pure rocket planes, as they were to rely on airbreathing propulsion for the first part of their flight. Indeed, NASP was initially expected to do without rocket motors. In that respect, they were more similar to the various mixed-power interceptors of the 1960s.

But the revival proved to be a false start. While routine hypersonic flight into orbit appeared to many people to be imminent, the unforgiving numbers in the engineers' weight budgets and the managers' cost estimates said otherwise. In part the optimism appears to have been inspired by the ease of imagining a spaceplane flying into orbit as a natural extension of high-speed and high-altitude aviation: an X-15, just flying a bit faster. Looked at like this, the intrinsic difficulties seemed smaller than they really are. Spaceplanes are inherently large because of the enormous volumes of propellant required. It is possible to make a small, relatively low cost aircraft, but not a small, cheap orbital spaceplane (indeed, people build simple aircraft in their garage but it is very unlikely that one day your neighbor will roll a hypersonic satellite launcher out of his shed).

Dr. Richard Hallion, a former Chief Historian of the US Air Force recently said of the apparent lack of progress in hypersonic flight (and thus the spaceplanes discussed in this book): "The hope of hypersonics thus became inextricably caught up in what might be termed a hypersonic hype. This led, over time, to a cycle of fits and starts that has largely worked to discredit the potential of the field and taint it with an image of waste and futility. Typically, a program has begun with great fanfare and promise, increased in complexity, and when realistic performance, schedule, and cost estimates are derived, its appeal quickly fades."

In addition to the canceled X-20 Air Force project, NASA has a long history of abandoned hypersonic projects, including the X-30, X-33, X-34 and X-38. The space agency seems essentially to have given up on spaceplanes, shuttle-type space gliders, and indeed reusable launchers in general for the near future. The Russians had their single Buran flight but never progressed beyond paper studies for real spaceplane concepts. At present, neither NASA nor the Russian Space Agency, nor indeed any other space agency, is willing to risk burning its hands on another shuttle, let alone a spaceplane project.

A modest resumption of interest in the rocket plane was kicked off by the success of SpaceShipOne. Hopefully other suborbital rocket propelled aircraft will soon fly. However, it seems that brief suborbital flights represent the last niche in aeronautics for the pure rocket powered plane to play a useful role: any future hypersonic aircraft or orbital spaceplane will primarily rely on advanced forms of jet propulsion, perhaps in combination with rocket power if really necessary. In fact, even the early

X-planes like the X-1, X-2 and X-15, as well as the D-558-2 Skyrocket, were launched from large turbojet aircraft that can be regarded as airbreathing first stages.

The development of hypersonic launch vehicles will be expensive but by using a one-step-at-a-time approach, also known as 'crawl-walk-run', it may be technically feasible as well as affordable with or without government funding. The logic is clear: start with a suborbital aircraft such as SpaceShipTwo, advance to a suborbital hopper that can launch payloads into low orbit at the apogee of its ballistic flight into space, and finally make an orbital spaceplane. Each of these steps could be a commercially viable project in its own right, earning the money needed to fund the next step on the road to a fully reusable launch vehicle. In this regard, NASA's Commercial Orbital Transportation Services program (which encourages private companies to introduce crew and cargo transportation vehicles to service the International Space Station) is of interest since one of the participants, SpaceDev, is developing the aforementioned Dream Chaser mini-shuttle.

At the same time, the US military's desire for a long-range hypersonic missile (or even an attack aircraft) able to reach any place on Earth in no-time and fly too fast to be shot down is generating a lot of spaceplane technology. Perhaps in the foreseeable future the quest for the first operational hypersonic aircraft able to routinely fly into orbit and back will finally be concluded. Meanwhile the old quip in the US military remains valid that "hypersonics is the future of airpower and always will be".

A proper airplane should have pilots on board, but the more that time passes the lower the chance that future spaceplanes will be directly piloted by anyone present. For launch vehicles flying only cargo it seems certain no crew will be required, and pilots may not be needed even for transporting astronauts. The Sänger-II and NASP spaceplanes would have had cockpits and flight crews but HOTOL was specifically intended to fly without, as indeed is its Skylon successor. The technology of orbital spaceplanes will be just as exciting as that of the X-15, but without pilots the concept loses a lot of its glamour and sense of adventure.

So what are the chances of there being a fully reusable, crewed rocket plane (with or without airbreathing engines) like the Euro 5 discussed in the Preface of this book blasting through the air anytime soon? Unfortunately, I think it looks like it will take quite while. The only operational rocket aircraft in the near future will be suborbital. When orbital spaceplanes eventually come around (hopefully) it is likely they will be fully automatic vehicles rather than resembling a hypersonic fighter aircraft piloted by gallant astronauts.

Still, the required technology and the possibilities that are on offer are extremely exciting: if spaceplanes really can dramatically reduce the cost of putting things into orbit then they will at long last open up space for large-scale commerce, production, moonbases, space solar energy satellites, space hotels and other marvelous ideas that are currently wishful artistic impressions and science fiction.

I just cannot accept that the expensive, wasteful expendable rocket as we know it today is the best that we can do and therefore represents the final answer in access to space. Spaceplanes truly represent the last great aeronautical frontier.

Appendix:

Aircraft maximum velocity and altitude evolution

The illustrations show the maximum velocity and the maximum altitude that aircraft have achieved over the years, and as such they encapsulate much of the story told in this book.

Since 1939 the (unofficial) maximum velocity records have all been set by rocket propelled aircraft, with the trend being steeply exponential then concluding with the X-15 in 1967. Around the same time that the X-15 program ended, the maximum velocities attained by turbojet and ramjet aircraft also reached their limits. It will be possible to fly faster using airbreathing propulsion but it will require scramjets (work on experimental versions of which continues to this day). It is also interesting to note that velocities that were initially achieved by mixed-propulsion interceptors using jet

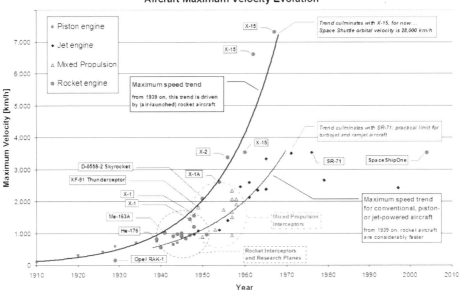

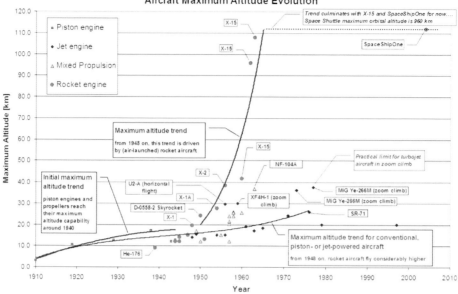

Aircraft Maximum Altitude Evolution

as well as rocket engines were soon surpassed by jet-power-only aircraft (rendering mixed propulsion obsolete by about the end of the 1950s).

The (unofficial) maximum altitude records have been exclusively the province of rocket aircraft since 1948, with the exponential trend once again culminating with the X-15. Turbojet/ramjet aircraft cannot fly at altitudes above 30 km (100,000 feet) for extended times and are only able to surpass this during short zoom climbs. Sustained airbreathing flight at higher altitudes will require scramjets.

Given the exponential growth of the maximum velocity and altitude achieved by aircraft over time, it is understandable that many people expected these trend lines to continue into the 1970s and beyond with aircraft reaching orbital altitudes as well as orbital velocities within a decade or two. Of course the Space Shuttle actually did so in 1981 but it was a vertical take-off, rocket-launched space glider rather than a true rocket plane. Real spaceplanes possessing rocket engines, sophisticated airbreathing engines or combinations of the two, have yet to progress beyond the drawing board.

SpaceShipOne managed to exceed the highest altitude achieved by the X-15 but got nowhere near that aircraft's record velocity; it travels about as fast as the fastest airbreathing aircraft. But SpaceShipOne was the first aircraft in four decades to reach the edge of space.

Bibliography

BOOKS AND REPORTS

Ashford, David and Patrick Collins, *Your Spaceflight Manual – How You Could be a Tourist in Space Within Twenty Years*, Simon & Schuster, UK, 1990

Ashford, David, *Spaceflight Revolution*, Imperial College Press, UK, 2002

Brown, Eric, *Wings on My Sleeve*, Phoenix, London, UK, 2007

Brown, Eric, *Wings of the Luftwaffe; Flying the Captured German Aircraft of World War II*, Hikoki Publications, Manchester, UK, 2010

Bowman, John Normal, *The Handbook of Rockets and Guided Missiles*, Perastadion Press, USA, 1957

Buttler, Tony, *American Secret Projects; Fighters & Interceptors 1945-1978*, Midland Publishing, Hinckley, UK, 2007

Dressel, Joachim, *Natter Bachem Ba 349 and other German rocket fighter projects*, Schiffer Publishing, Atglen, USA, 1994

Dyer, Edwin M., *Japanese Secret Projects; Experimental Aircraft of the IJA and IJN 1939-1945*, Midland Publishing, Hersham, UK, 2009

Gooden, Brett, *Projekt Natter, Last of the Wonder Weapons*, Ian Allen Publishing, Hersham, UK, 2006

Gordon, Yefim, *Soviet Rocket Fighters*, Midland Publishing, Hinckley, UK, 2006

Green, William, *Rocket Fighter*, Ballantine Books, New York, USA, 1971

Griffin, David J., *Hawker Hunter 1951 to 2007*, Lulu.com, 2006

Hagerty, Jack and Jon C. Rogers, *Spaceship Handbook; Rocket and Spacecraft Designs of the 20th Century: Fictional, Factual and Fantasy*, ARA Press, Livermore, USA, 2001

Hendrickx, Bart and Bert Vis, *Energiya-Buran, The Soviet Space Shuttle*, Springer/Praxis, Chichester, UK, 2007

Herwig, Dieter and Heinz Rode, *Luftwaffe Secret Projects: Ground Attack & Special Purpose Aircraft*, Midland Publishing, Hinckley, UK, 2003

Hogg, Ian V., *German Weapons of the Second World War*, Greenhill Books, London, UK, 2002

Kempel, Robert W., Weneth D. Painter and Milton O. Thompson, *Developing and*

Flight Testing the HL-10 Lifting Body: A Precursor to the Space Shuttle, NASA Reference Publication 1332, USA, 1994

Koelle, D.E. and H. Kuczera, Sänger, *An Advanced Launcher System for Europe, IAF-87-207*, 38th Congress of the International Astronautical Federation, UK, October 1987

Koelle, D.E., *Handbook of Cost Engineering for Space Transportation Systems, including Transcost 7.3*, TransCostSystems, Ottobrun, Germany, 2010

Larson, W.J., L.K. Pranke, J. Connoly, R. Giffen et al., *Human Spaceflight Mission Analysis and Design*, McGraw-Hill, USA, 1999

Ley, Willy, *Rockets, Missiles and Men in Space*, Signet Books, New York, USA, 1968

Libis, Scott and Tony Landis, *Lockheed NF-104A Aerospace Trainer*, Air Force Legends Number 204, Steve Ginter, California, USA, 1999

McElyea, T., *A Vision of Future Space Transportation; A Visual Guide to Future Spacecraft Concepts*, Apogee Books, Burlington, Canada, 2003

Millard, Douglas, *The Black Arrow Rocket, a history of a satellite launch vehicle and its engines*, Science Museum, London, 2001

Miller, Jay, *The X-Planes, X-1 to X-45*, Third Edition, Midland Publishing, Hinckley, UK, 2001

Miller, Ron, *The Dream Machines, a pictorial history of the spaceship in art, science and literature*, Krieger Publishing Company, Malabar, Florida, USA, 1993

Morgan, Hugh and John Weal, *German Jet Aces of World War 2*, Osprey Publishing, UK, 1998

Myhra, David, *Messerschmitt Me 263*, Schiffer Military History, USA, 1999

Myhra, David, *Sänger – Germany's Orbital Rocket Bomber in WWII*, Schiffer Military History, USA, 2002

Pace, Steve, *Republic XF-91 Thunderceptor Rocket Fighter (Air Force Legends Number 210)*, Ginter Books, USA, 2000

Ransom, Stephen and Hans-Hermann Cammann, *Jagdgeschwader 400, Germany's Elite Rocket Fighters*, Osprey Publishing, UK, 2010

Raymer, Daniel P., *Aircraft Design: A Conceptual Approach*, Third Edition, AIAA Education Series, Reston, USA, 1999

Ry, Marek, *German Air Project 1935-1945* vol.1, Mushroom Model Publications, Redbourn, UK, 2003

Reed, R. Dale, *Wingless Flight, the Lifting Body Story*, NASA History Series, Washington, USA, 1997, http://ntrs.nasa.gov/archive/nasa/casi.ntrs.nasa.gov/19980169231_1998082126.pdf

Reitsch, Hanna, *The Sky My Kingdom, Memoirs of the Famous German WWII Test-Pilot*, Greenhill Books, London, UK, 1997

Rose, Bill, *Secret Projects – Military Space Technology*, Midland Publishing, Hersham, UK, 2008

Sänger, Eugen and Irene Bredt, *A Rocket Drive for Long Range Bombers*, Technical Information Branch, Bauer Navy Department, 1952, www.astronautix.com/data/saenger.pdf

Sänger, Eugen, *Space Flight; Countdown for the Future*, McGraw-Hill, USA, 1965

Smith, Melvyn, *Space Shuttle; An Illustrated History of US winged spacecraft: X-15 to Orbiter*, Haynes Publishing Group, Sparkford, UK, 1985

Thompson, Milton O., *At the Edge of Space; the X-15 Flight Program*, Smithsonian Institution, USA, 1992

Warsitz, Lutz, *The First Jet Pilot, the story of German test pilot Erich Warsitz*, Pen and Sword Books, South Yorkshire, UK, 2008

Winchester, Jim, *The World's Worst Aircraft*, Grange Books, Rochester, UK, 2005

Winchester, Jim, *Concept Aircraft; Prototypes, X-Planes and Experimental Aircraft*, Grange Books, Rochester, UK, 2005

Yeager, Chuck and Leo Janos, *Yeager, and Autobiography*, Bantam Books, USA, July 1985

Yeager, Chuck, Bob Cardenas, Bob Hoover and Jack Russell, *The Quest for Mach One: A First-Person Account of Breaking the Sound Barrier*, Penguin Studio, USA, October 1997

Zaloga, Steven J., Hugh Johnson and Chris Taylor, *German V-Weapon Sites 1943-45*, Osprey Publishing, UK, 2007

Zaloga, Steven J. and Ian Palmer, *Kamikaze, Japanese Special Attack Weapons 1944-45*, Osprey Publishing, UK, 2011

Ziegler, Mano, *Rocket fighter: the story of the Me 163* (translation of *Raketenjager Me 163*), Macdonald, London, UK, 1963

ARTICLES, PAPERS AND PRESENTATIONS

Anonymous, Will Rocket Planes Reach the Stars?, *Popular Mechanics*, USA, November 1928, http://books.google.nl/books?id = wd4DAAAAMBAJ&print sec = frontcover&hl = en&source = gbs_v2_summary_r&cad = 0#v = onepage&q = &f = false

Anonymous, First Rocket Glider Launched Successfully in Actual Tests, *Modern Mechanics*, USA, September 1931, http://blog.modernmechanix.com/2010/01/19/first-rocket-glider-launched-successfully-in-actual-tests

Anonymous, Rocket Turbine Will Drive Sensational High Speed Plane, *Popular Science*, USA, December 1931

Anonymous, Vertical Take-Off Rocket at Woomera, *The Canberra Times*, Australia, 25 May 1953, http://trove.nla.gov.au/ndp/del/article/2882679

Anonymous, Armstrong Siddeley Snarler, *Flight*, UK, 6 August 1954

Anonymous, Super Sprite, The First British Production Type A.T.O. Rocket Motor, *Flight*, UK, 5 August 1955

Anonymous, Armstrong Siddeley Screamer, *Flight*, UK, 27 July 1956

Anonymous, Aerospaceplane: NASA's flame rekindled, *Flight International*, UK, 10 October 1987, www.flightglobal.com/pdfarchive/view/1987/1987 percent20- per cent202001.html

Anonymous, Japan Joins Scramjet Club, *Popular Mechanics*, USA, October 1994

Anonymous, SpaceShipOne Joins the Icons of Flight on Display at Smithsonian's National Air and Space Museum, Smithsonian's National Air and Space

Museum press release, Washington D.C., USA, 5 October 2005, www.nasm. si.edu/events/pressroom/releaseDetail.cfm?releaseID = 138

Anonymous, Cooperation and competition in the worldwide space transport sector; an overview of current trends, ESA/PB-LAU(2010)5, European Space Agency Launcher Programme Board, Paris, March 2010

Anonymous, Firm rockets into space tourism; The European aerospace giant EADS is going into the space tourism business, BBC News, UK, June 13, 2007

http://news.bbc.co.uk/2/hi/science/nature/6749873.stm

Adolphus, David Traver, Opel RAK 2: enough to blow up a whole neighborhood

Hemmings Auto Blogs, USA, 10 May 2008, http://blog.hemmings.com/index.php/ 2008/05/10/opel-rak-2-enough-to-blow-up-a-whole-neighborhood

Amos, Jonathan, Space tourism jet work continues, BBC News, UK, 12 January 2011

Amos, Jonathan, www.bbc.co.uk/news/science-environment-12176754

Amos, Jonathan, Skylon spaceplane approaches decision time, BBC News, UK, 21 September 2010, www.bbc.co.uk/blogs/thereporters/jonathanamos/2010/09/it-is-one-of-those.shtml

Amos, Jonathan, Skylon: Ending 40 years of hurt, BBC News, UK, 24 May 2011, www.bbc.co.uk/news/science-environment-13520948

Anderson, John D., Research in Supersonic Flight and the Breaking of the Sound Barrier, Chapter 3 of *SP-4219 From Engineering Science to Big Science*, NASA History Series, Washington D.C., US, 1998 , http://history.nasa.gov/SP-4219/ Chapter3.html

Arndt, Rob, Fi-166 High-Altitude Rocket Fighter, Predecessor to the Bachem 349, http://discaircraft.greyfalcon.us/Fi.htm

Ashford, David, Space Tourism – The Key to Low Cost Access to Space, ESA Explospace, Sardinia, Italy, October 1998, www.spacefuture.com/archive/space_ tourism_the_key_to_low_cost_access_to_space.shtml

Ashford, David, An aviation approach to space transportation, *The Aeronautical Journal of the Royal Aeronautical Society*, Volume 113, UK, August 2009

Ashford, David, The Great Space Scandal, *Journal of the Institution of Engineering Designers*, UK, November 2009, www.bristolspaceplanes.com/library/BSP_ The_Great_Space_Scandal.pdf

Atwood, Tom, Komet Me 163; Chief test pilot Rudy Opitz tells it like it was, *Flight Journal*, August 2007

Barnes, T.D. and R.E. Friedrichs, Northrop's Roach Dry Lake Site, http:// area51specialprojects.com/roach_lake/northrop.htm

Baugher, Joe, FJ-4F, www.astronautix.com/craft/fj4f.htm

Baugher, Joe, Northrop XP-79B, www.astronautix.com/craft/xp79.htm

Baugher, Joe, Republic XF-91 Thunderceptor, 21 November 1999, http:// www.joebaugher.com/usaf_fighters/p91.html

Bedard, Andre, Composite Solid Propellants, http://www.astronautix.com/articles/ comlants.htm

Bell, Jeffrey F., Rocket Plane Roulette, Spacedaily.com, 7 March 2007, www.space daily.com/reports/Rocket_Plane_Roulette_999.html

Birkenstock, Wolfgang, Breaking the Sound Barrier, *Flug Revue* 11, 1999, www.flug-revue.rotor.com/frheft/FRH9911/FR9911f.htm

Bogar et al., Hypersonic Airplane Space Tether Orbital Launch System, Boeing, St. Louis, USA, 7 January 2000, www.spaceelevator.com/docs/355Bogar.pdf

Bombeau, Bernard, The early years of the delta adventure, *Interavia Business & Technology*, July, 2001, http://findarticles.com/p/articles/mi_hb3126/is_655_56/ai_n28851436/?tag = content;col1

Braas, Nico, SNCASO SO-9000 Trident and SO-9050 Trident II, Let Let Let Warplanes website, 28 June 2008, www.letletlet-warplanes.com/2008/06/28/sncaso-so-9000-trident-and-so-9050-trident-ii

Braas, Nico, Saunders Roe SR.53 and 177, Let Let Let Warplanes website, 7 June 2008, www.letletlet-warplanes.com/2008/07/06/saunders-roe-sr53-and-177

Braun, Wernher von, The Spaceplane That Can Put YOU in Orbit, *Popular Science*, USA, July 1970

Crossfield, Scott et al., X-15 Pilot's Panel, 8 June 1989, USA, www.hq.nasa.gov/pao/History/x15conf/pilots.html

Dana, Bill, A history of the X-15, Charles A. Lindbergh Lecture at the National Air and Space Museum, Smithsonian Institution, Washington D.C., USA, 1988 www.nasa.gov/centers/dryden/history/Speeches/x-15_speech/x15-1spch.html

Danne, Harold A., Across the Atlantic in a Rocket Plane, The American Interplanetary Society Bulletin, June-July 1931, http://epizodsspace.no-ip.org/bibl/inostr-yazyki/bais/1931/bais_no_010.pdf

Day, Dwayne A., The X-15 and Hypersonics, U.S. Centennial of Flight Commission website, 2003, www.centennialofflight.gov/essay/Evolution_of_Technology/X-15/Tech28.htm

Day, Dwayne A., Six blind men in a zoo: Aviation Week's mythical Blackstar, *The Space Review*, 13 March 2006, www.thespacereview.com/article/576/1

Day, Dwayne A. and Robert G. Kennedy III, Soviet Star Wars, *Smithsonian Air & Space*, USA, 1 January 2010, www.airspacemag.com/space-exploration/Soviet-Star-Wars.html

Day, Dwayne A., Fire in the sky: the Air Launched Sortie Vehicle of the early 1980s, *The Space Review*, 22 February 2010, www.thespacereview.com/article/1569/1

Day, Dwayne A., Prophets of science fiction, *The Space Review*, 20 June 2011, www.thespacereview.com/article/1868/1

Dinerman, Taylor, Is the RLV industry emerging from hibernation?, *The Space Review*, 26 October 2009, www.thespacereview.com/article/1497/1

Dinkin, Sam, X-15 and today's spaceplanes, *The Space Review*, 9 August 2004, www.thespacereview.com/article/204/1

Dyason, Anton, Blackburn Buccaneer S.Mk.50 SAAF, IPMS SA Media Group, 5 March 2000, http://newsite.ipmssa.za.org/content/view/111/28/1/0

Easterbrook, Gregg, Beam Me Out Of This Death Trap, Scotty, *The Washington Monthly*, USA, April 1980

Fawkes, Steven, Space tourism and carbon dioxide emissions, *The Space Review*, 19 February 2007, www.thespacereview.com/article/813/1

Foust, Jeff, Is there a business case for RLVs?, *The Space Review*, 2 September 2003, www.thespacereview.com/article/44/1

Foust, Jeff, First steps towards point-to-point spaceflight, *The Space Review*, 23 February 2009, www.thespacereview.com/article/1311/1

Furniss, Tim, Sänger Aerospaceplane Gains Momentum, *Flight International*, vol. 136, UK, 12 August 1989

Garrison, Peter, The Real X-Men; Life came at you fast when you flew the X-15, *Air & Space Magazine*, USA, 1 November 2007, www.airspacemag.com/military-aviation/x-men.html

Gernsback, Hugo, Berlin to New York in less than One Hour!, *Everyday Science and Mechanics*, USA, November 1931, http://blog.modernmechanix.com/2007/05/02/berlin-to-new-york-in-less-than-one-hour

Godwin, Matthew, Interview with David Ashford (transcript), ESA oral history of Europe in space, Bristol, UK, 22 October 2007, www.eui.eu/HAEU/OralHistory/pdf/INT628.pdf

Goebel, Greg, Crusader in Action, Air Vectors website, www.vectorsite.net/avcrus_2.html

Goebel, Greg, The Mikoyan MiG-19, Air Vectors website, www.vectorsite.net/avmig15_3.html

Goebel, Greg, The North American A-5/RA-5 Vigilante, www.vectorsite.net/ava5.html

Goebel, Greg, The Zero-Length Launch Fighter, Air Vectors website, www.vectorsite.net/avzel.html

Griswold, Wesley S., Mile-a-Second Rocket Plane Will Fly 100 Miles High, *Popular Science*, June 1958

Grosdemange, H. and G. Schaeffer, The SEPR 844 Reusable liquid Rocket Engine for Mirage Combat Aircraft, AIAA 90-1835, 26th AIAA/ASME/SAE/ASEE Joint Propulsion Conference, Orlando, USA, July 1990

Hallion, Richard P., Hypersonic Power Projection, Mitchell Institute Press, USA, 2010, www.afa.org/mitchell/reports/MP6_Hypersonics_0610.pdf

Harlow, John (editor) et al., UK Manned Rocket Aircraft, *Space Chronicle*, Vol. 55, Suppl.2, British Interplanetary Society, London, UK, 2002

Harpole, Tom, White Elephant; How the Soviet Buran space shuttle helped the United States win the cold war, *Air & Space Magazine*, USA, January 2003, www.airspacemag.com/space-exploration/cit-harpole.html

Hedman, Eric, The wit and wisdom of Burt Rutan, *The Space Review*, 22 August 2011, www.thespacereview.com/article/1910/1

Hempsell, Mark and Roger Longstaff, Skylon Users' Manual, Reaction Engines Limited, UK, September 2009, www.reactionengines.co.uk/downloads/SKYLON_User_Manual_rev1-1.pdf

Hoyt, Robert P., Design and Simulation of Tether Facilities for the HASTOL Architecture, 36th AIAA/ASME/SAE/ASEE Joint Propulsion Conference, Huntsville, USA, July 2000

Humphrey, Hal and Joe Baugher, North American FJ-4 Fury, 4 January 2008, www.joebaugher.com/usaf_fighters/p86_24.html

Jarvis, Calvin R. and Wilton P. Lock, Operational Experience with the X-15 Reaction Control and Reaction Augmentation Systems, NASA Technical Note D-2864, Flight Research Center Edwards, California, USA, June 1965, http://www.nasa.gov/centers/dryden/pdf/87709main_H-364.pdf

Kamm, Richard W., Early Jet Aircraft Mechanic, Greater Saint Louis Air & Space Museum, USA, 2006, www.airandspacemuseum.org/EarlyJetAircraftKramm.htm

Kempel, Robert W. and Richard E. Day, A Shadow over the Horizon, the Bell X-2, *American Aviation Historical Society Journal*, USA, Spring 2003, www.bellx-2.com/sightings/horizon/article.html

Kerstein, Aleksander and Drago Matko, Eugen Sänger: Eminent Space Pioneer

Acta Astronautica, Volume 61, Issues 11-12, December 2007, www.sciencedirect.com/science/article/pii/S0094576507001336

Kightly, James, Messerschmitt Me 163 Komet Pilot, *Aeroplane*, London, UK, September 2010

Klein, Israel, Rocket Auto May Lead to One Hour Trips, New York to Berlin, *The Evening News*, San Jose, California, USA, 20 June 1928, http://news.google.com/newspapers?nid = 1977&dat = 19280620&id = mDAiAAAAIBAJ&sjid = EqQFAAAAIBAJ&pg = 1039,7301005

Lasser, David, The Conquest of Space, Penguin Press, New York, 1931 (reprinted by Apogee Books in 2002)

Lumsden, Marshall, Ed Maloney's Mission; the man behind, beside, and all over, the Planes of Fame Air Museum, *Smithsonian Air & Space*, USA, 1 March 2008, www.airspacemag.com/history-of-flight/planes_of_fame.html

Maksel, Rebecca, The Mysterious Second Seat, *Smithsonian Air & Space*, USA, September 2010

Morris, William W., Yeager vs. Crossfield dueling test pilots: the quest for Mach 2, *Airpower*, USA, November 2003

Perry, Robert L., The Antecedents of the X-1, AIAA Paper No. 65-453, USA, AIAA Second Annual Meeting, San Francisco, USA, 1965, www.rand.org/pubs/papers/2009/P3154.pdf

Pope, Gregory T., Space Race 2000, *Popular Mechanics*, USA, March 1992

Pike, John, X-30 National Aerospace Plane (NASP), Federation of American Scientists website, June 1997, www.fas.org/irp/mystery/nasp.htm

Rickard, J., Hawker P.1072, Military History Encyclopedia on the Web, June 2010 www.historyofwar.org/articles/weapons_hawker_P1072.html

Rothmund, Christophe and John Harlow, A History of European Liquid-Propellant Rocket Engines for Aircraft, IAA-99-2901, 35th AIAA/ASME/SAE/ASEE Joint Propulsion Conference, Los Angeles, USA, June 1999

Rothmund, Christophe, Reusable Man-rated Rocket Engines; The French Experience 1944-1996, IAC-04-IAA-6.15.3.02, 55th International Astronautical Congress, Vancouver, Canada, 2004

Sänger, Hartmut E. and Alexandre D. Szames, From the Silverbird to Interstellar Voyages, IAC-03-IAA.2.4.a.07, 54th International Astronautical Congress, Bremen, Germany, 2003

Santovincenzo, A., L. Innocenti and M.O. van Pelt, The Socrates Vehicle – ESTEC

CDF Design, AIAA 2005-3341, AIAA/CIRA 13th International Space Planes and Hypersonics Systems and Technologies Conference, Capua, Italy, 2005

Sarigul-Klijn, Marti and Nesrin Sarigul-Klijn, Flight Mechanics of Manned Suborbital Reusable Launch Vehicles with Recommendations for Launch and Recovery, Mechanical and Aeronautical Engineering Department, University of California, USA, 2003, http://mae.ucdavis.edu/faculty/sarigul/AIAA_2003_0909_revised_Sep03.pdf

Scott, Jeff, Turboramjet, Aerospaceweb "ask a rocket scientist", USA, 1 July 2001, www.aerospaceweb.org/question/propulsion/q0175.shtml

Scott, Jeff, HOTOL and Skylon, Aerospaceweb "ask a rocket scientist", USA, 14 November 2004, www.aerospaceweb.org/question/spacecraft/q0202.shtml

Sheppard, Ian, Towards hypersonic flight, *Flight International*, 11 November 1997, www.flightglobal.com/articles/1997/11/26/30037/towards-hypersonic-flight.html

Snead, Mike, Assessing the practicality of scramjet-powered, single-stage aero-spaceplanes, *The Space Review*, 31 March 2008, http://www.thespacereview.com/article/1092/1

Strickland, John K., Current strategies towards air-breathing space launch vehicles, *The Space Review*, 1 August 2011, www.thespacereview.com/article/1894/1

Varvill, Richard and Alan Bond, A Comparison of Propulsion Concepts for SSTO Reusable Launchers, *Journal of the British Interplanetary Society*, Vol. 56, UK, 2003, www.reactionengines.co.uk/downloads/JBIS_v56_108-117.pdf

Wade, Mark, Astroplane, www.astronautix.com/lvs/astplane.htm

Wade, Mark, Keldysh Bomber, www.astronautix.com/lvs/kelomber.htm

Wade, Mark, LR54, www.astronautix.com/engines/lr54.htm

Wade, Mark, MiG 105-11, www.astronautix.com/craft/mig10511.htm

Wade, Mark, Mustard, www.astronautix.com/lvs/mustard.htm

Wade, Mark, Opel, www.astronautix.com/lvs/opel.htm

Wade, Mark, Saenger I, www.astronautix.com/lvs/saengeri.htm

Wade, Mark, Saenger II, www.astronautix.com/lvs/saengerii.htm

Wade, Mark, Tu-2000, www.astronautix.com/craft/Tu2000.htm

Wade, Mark, X-15B, www.astronautix.com/craft/x15b.htm

Wade, Mark, XCALR-2000A-1, www.astronautix.com/engines/xca000a1.htm

Wainfan, Barnaby, Komet Me 163, A Fighter Ahead of Its Time, *Flight Journal*, August 2007

Westman, Juhani, Global Bounce (article on Sänger's Silbervogel), April 2008, www.pp.htv.fi/jwestman/space/sang-e.html

Wilson, G.P., HOTOL, Aerospaceplane for Europe, IAF-87-208, 38th Congress of the International Astronautical Federation, UK, October 1987

Winter, Frank H. and James, S. George, Highlights of 50 Years of Aerojet, a Pioneering American Rocket Company, 1942-1992, *Acta Astronautica*, Vol. 35, No. 9-11, Elsevier Science Ltd., 1995

Winter, Frank H. and Robert van der Linden, Out of the Past, *Aerospace America*, USA, May 2005, www.aiaa.org/Aerospace/images/articleimages/pdf/oopmay05.pdf

Yamazaki, Akio, Tail of the Tiger; Japan's "Shusui" Interceptor, *Air Enthusiast*, UK, No. 115, January/February 2006

Yoon, Joe, Jet Engine Types, Aerospaceweb "ask a rocket scientist", USA, 1 July 2001, www.aerospaceweb.org/question/propulsion/q0033.shtml

Zettl, Heinz H. and Ernst-Peter Berresheim, 80th Anniversary of Fritz von Opel's Record-Setting RAK 2 Ride, *Opel News*, Germany, 7 May 2008, http://archives.media.gm.com/archive/documents/domain_82/docId_45531_pr.html#more

Zubrin, Robert M. and Mitchell Burnside Clapp, Black Horse: One Stop to Orbit, *Analog Magazine*, USA, June 1995

WEBSITES

Armadillo Aerospace website: www.armadilloaerospace.com, accessed August 2011

Bristol Spaceplanes website: www.bristolspaceplanes.com, accessed August 2011

Les avions-fusée Russes: http://xplanes.free.fr/florov/afr-1.html, accessed September 2010

Ba-349 Project Natter replica website: http://ba-349.com/de/projekt.html, accessed October 2010

Bayerische Flugzeug Historiker "Deutsche Raketenflugzeuge bis 1945" website: www.bayerische-flugzeug-historiker-ev.de/bayflughist/projekte/2009_jets/bfh_dtsch_rakflz.html, accessed March 2010

British rocket interceptors website by Nicolas Hill: www.spaceuk.org/sr53/sr53.htm, accessed November 2010

Canberra Association webpage on WK163: www.bywat.co.uk/wk163.html, accessed November 2010

Canberra WT207 crash webpage by Alan Clark: www.peakdistrictaircrashes.co.uk/pages/peakdistrict/peakdistrictwt207.htm, accessed November 2010

Dassault Aviation webpage on its VHS/K-1000 suborbital vehicle project: www.dassault-aviation.com/en/aviation/press/in-the-air/in-the-air-2010/suborbital-aviation-on-the-very-edge-of-space.html?L = 1&cHash = 8328870e9f, accessed August 2011

European Space Agency's "Tribute to the Space Shuttle": www.esa.int/SPECIALS/shuttle, accessed July 2011

George Lucas' website on his full-scale airplane models: www.luftwaffewest.com, accessed September 2010

Gottlob Espenlaub website: www.drachenarchiv.de/Seiten/e_espenlaub.html, accessed March 2010

HIJMS Submarine I-29 Tabular Record of Movement: www.combinedfleet.com/I-29.htm, accessed June 2010

Hurricat page of the Fleet Air Arm Archive website: www.fleetairarmarchive.net/aircraft/seahurricane.htm, accessed August 2010

Imperial Japanese Navy Page, by Bob Hackett and Sander Kingsepp, HIJMS Submarine RO-501 Tabular Record of Movement: www.combinedfleet.com/RO-501.htm, accessed June 2010

Introduction to Future Launch Vehicle Planes [1963-2001] by Marcus Lindroos www.pmview.com/spaceodysseytwo/spacelvs, updated June 2001, accessed August 2010

Lockheed F-104 G Zell website by Wolfgang Bredow: www.bredow-web.de/Luftwaffenmuseum/Kampfjets/Starfighter/starfighter.html, accessed September 2010

Luft 46 website: www.luft46.com, accessed January 2010

Kamakura Ohka Monument website by Bill Gordon: http://wgordon.web.wesleyan.edu/kamikaze/monuments/kamakura/index.htm, accessed June 2010

Kelly Space & Technology Astroliner website: www.kellyspace.com/launchvehicle, accessed November 2011

Me 163 website by Rob de Bie: www.xs4all.nl/~robdebie/me163.htm, accessed April 2010

Mikoyan/Gurevich Ye-50 website: http://www.aviastar.org/air/russia/mig-e50.php, accessed September 2010

Myhra's Aviation History, by Dr. David Myhra: http://www.german-aircrafts.com, accessed April 2011

NASA Dryden Flight Research Center Fact Sheets: www.nasa.gov/centers/dryden/news/FactSheets/index.html, accessed July 2011

NASA Dryden Flight Research Center X-1 Flight Summary: http://www.nasa.gov/centers/dryden/history/HistoricAircraft/X-1/fltsummary.html,accessed December 2010

National Museum of the United States Air Force Me 163B webpage: www.nationalmuseum.af.mil/factsheets/factsheet.asp?id = 508, accessed April 2010

NF-104A website by its primary test pilot USAF Lt. Col. Robert W. Smith: www.nf104.com, accessed February 2011

Nova Online "Faster Than Sound" website: www.pbs.org/wgbh/nova/barrier/men.html, accessed December 2010

OKB MIG Design Bureau Unofficial Reference Website, MiG 19SU page: http://wp.scn.ru/mig_okb/planes-19-sm50, accessed October 2010

OKB MiG Design Bureau Unofficial Reference Website, I-270 page: http://wp.scn.ru/mig_okb/planes-exp-i270g, accessed September 2010

OKB MiG Design Bureau Unofficial Reference Website, Ye-50 page: http://wp.scn.ru/mig_okb/planes-mig21-e50, accessed September 2010

Opel.tv website, RAK-2 rocket car movie: www.opel.tv/index.php?channel = opelclassic&seite = 0&mo = 23-Mai-1928_de, accessed January 2010

Opel.tv website, RAK-3 rocket rail car movie: www.opel.tv/index.php?channel = opelclassic&seite = 0&mo = 23-Jun-1928_de, accessed January 2010

Opel.tv website, RAK-1 rocket airplane movie: www.opel.tv/index.php?channel = opelclassic&seite = 0&mo = RAK_Protokoll_QT, accessed January 2010

Reaction Engines Ltd. Website: http://www.reactionengines.co.uk, accessed May 2011

Rocket interceptor website "Les intercepteurs à moteur-fusée": http://jpcolliat.free.fr/trident/trident-1.htm, accessed October 2010

Rocket Racing League website: www.rocketracingleague.com, accessed August 2011

Russian Aviation Museum website pages on the RP-318 by Alexandre Savine: http://

ram-home.com/ram-old/rp-318.html and http://ram-home.com/ram-old/rp-318-1.html, accessed July 2010

Scaled Composites website: www.scaled.com, accessed August 2011

Star-Raker website by Ryan Crierie: www.alternatewars.com/SpaceRace/Space Race.htm, accessed November 2011

Sukhoi Su-7 website: http://www.aviastar.org/air/russia/su-7.php, accessed September 2010

Virgin Galactic website: www.virgingalactic.com, accessed August 2011

Walter rockets website by Shamus Reddin: http://www.walterwerke.co.uk, accessed August 2010

Erich Warsitz website: www.firstjetpilot.com, accessed March 2010

The X-Hunters website, on aircraft crashes near Edwards Air Force Base: www.thexhunters.com, accessed March 2010

X-2 website: http://bellx-2.com, accessed January 2011

X-15 interactive NASA website: www.nasa.gov/externalflash/x15_interactive, accessed March 2011

X-15 ejection seat webpage by Kevin Coyne: www.ejectionsite.com/x15seat.htm, accessed March 2011

X-15 website by Paul Raveling: www.sierrafoot.org/x-15/x-15.html, accessed June 2011

XCOR Aerospace website: www.xcor.com accessed August 2011

XF-91 pictures website: http://1000aircraftphotos.com/Contributions/PippinBill/5110.htm, accessed October 2010

XF-91 fact sheet of the National Museum of the US Air Force: http://www.nationalmuseum.af.mil/factsheets/factsheet.asp?id = 584, accessed October 2010

Yakovlev Yak 27 website: http://www.aviastar.org/air/russia/yak-27.php, accessed September 2010

ZELMAL: http://www.globalsecurity.org/military/systems/aircraft/f-84d-zelmal.htm, accessed September 2010

DOCUMENTARIES AND MOVIES

Planes that Never Flew, Alba Communications Limited/DD Video, 2003

Dean, Elvis, *Rocketmen*, Eureka Media/Aerocinema.com, 2009, http://aerocinema.com/component/content/article/41-features-rocketmen.html

Dean, Elvis, *The Komet & the Natter*, Eureka Media/Aerocinema.com, 2009, http://aerocinema.com/component/content/article/25-features-komet-and-the-natter.html

Duffy, James, *Wings of Fire, the Messerschmitt Me-163*, Rocket.aero, 2006, www.rocket.aero/me163.html

Duffy, James, *Mach 2, D-558 and X-2*, Rocket.aero, 2007, www.rocket.aero/mach2.html

The History Channel, *Secret Allied Aircraft of WWII*, A&E Television Networks/Alba Home Vision, 2005

The History Channel, *Secret Japanese Aircraft of WWII*, A&E Television Networks/
Alba Home Vision, 2005

The History Channel, *Secret Luftwaffe Aircraft of WWII*, A&E Television
Networks/Alba Home Vision, 2005

The History Channel, *Secret Russian Aircraft of WWII*, A&E Television Networks/
Alba Home Vision, 2005

The History Channel, *Kamikaze, episode of Dogfights, Season Two*, A&E Television
Networks, 2008

Technik Museum Speyer, Buran, *History and Transportation of the Russian Space
Shuttle OK-GLI to the Technik Museum Speyer*, Elser Film, 2008

Index

Printed by Publishers' Graphics LLC